IUTAM Symposium on Size Effects on Material and Structural Behavior at Micron- and Nano-Scales

SOLID MECHANICS AND ITS APPLICATIONS
Volume 142

*Series Editor:* G.M.L. GLADWELL
*Department of Civil Engineering*
*University of Waterloo*
*Waterloo, Ontario, Canada N2L 3GI*

*Aims and Scope of the Series*

The fundamental questions arising in mechanics are: *Why?*, *How?*, and *How much?* The aim of this series is to provide lucid accounts written by authoritative researchers giving vision and insight in answering these questions on the subject of mechanics as it relates to solids.

The scope of the series covers the entire spectrum of solid mechanics. Thus it includes the foundation of mechanics; variational formulations; computational mechanics; statics, kinematics and dynamics of rigid and elastic bodies: vibrations of solids and structures; dynamical systems and chaos; the theories of elasticity, plasticity and viscoelasticity; composite materials; rods, beams, shells and membranes; structural control and stability; soils, rocks and geomechanics; fracture; tribology; experimental mechanics; biomechanics and machine design.

The median level of presentation is the first year graduate student. Some texts are monographs defining the current state of the field; others are accessible to final year undergraduates; but essentially the emphasis is on readability and clarity.

*For a list of related mechanics titles, see final pages.*

# IUTAM Symposium on Size Effects on Material and Structural Behavior at Micron- and Nano-Scales

Proceedings of the IUTAM Symposium held in Hong Kong, China, 31 May–4 June, 2004

Edited by

Q. P. SUN

*Department of Mechanical Engineering,*
*The Hong Kong University of Science and Technology,*
*Hong Kong, China*

and

P. TONG

*Bioengineering Department,*
*University of California at San Diego*
*La Jolla, California, U.S.A.*

Springer

A C.I.P. Catalogue record for this book is available from the Library of Congress.

ISBN-10 1-4020-4945-5 (HB)
ISBN-13 978-1-4020-4945-3 (HB)
ISBN-10 1-4020-4946-3 (e-book)
ISBN-13 978-1-4020-4946-0 (e-book)

---

Published by Springer,
P.O. Box 17, 3300 AA Dordrecht, The Netherlands.

*www.springer.com*

*Printed on acid-free paper*



Printed in the Netherlands.

# Contents

(Participants and organizers of the IUTAM Symposium on *Size Effects on Material and Structural Behavior at Micron- and Nano-Scales* at the Conference Hall of the University Center in the campus of Hong Kong University of Science and Technology on 1 June 2004).

# Preface

The Symposium on "Size effects on material and structural behavior at micron- and nano-scales" was held from 31 May – 4 June, 2004 at the Hong Kong University of Science and Technology, Hong Kong. The aim of this symposium was to bring engineers, researchers and scientists from material science, biology, physics and mechanics together to discuss different aspects of and the latest advances in this active multi-discipline field. Sponsored by the International Union of Theoretical and Applied Mechanics (IUTAM), this 4-day symposium followed the tradition of IUTAM held in the single session format of IUTAM symposia. The scientific presentations and discussions focused on the following topics:

1. Behaviors of materials and structures at micron- and nanometer-scales;
2. Physical bases of size effects;
3. Adaptive and multi-functional behaviors of materials at small scales;
4. Size effects in fracture and phase transformation of solids;
5. Multi-scale modeling and simulation;
6. Microstructure and deformation with moving interfaces;
7. Size effects in material instability and its propagation.

Thirty five oral presentations were made by distinguished scholars. Theoretical, experimental and computational aspects of the subject were discussed and addressed in the symposium. Two round table discussions were held on the future research directions. The open and friendly environment provided excellent opportunities for stimulating discussion and intensive exchange of ideas among all participants. Also, local research graduate students participated in the meeting and involved in the organization of the Symposium.

A total of sixty registered participants (some of them are PhD students and post-doctorates) attended the technical sessions of the Symposium. Thirty five invited speakers came from eight countries: Australia (1), China mainland (10), France (3), USA (7), China Hong Kong (10), UK (1), Sweden (1), Germany (1), Poland (1), and Taiwan (1).

This volume collects twenty five written contributions to the Symposium from the invited speakers. Time and effort spent by these authors in participating in the meeting and preparing their manuscripts for this book is greatly appreciated. As in every successful scientific conference, financial support is an essential part of the symposium. Thanks are due to the IUTAM Bureau, the Research Grants Council of Hong Kong SAR, the National Natural Science Foundation of China, the US Army Research Office-Far East, the Hong Kong University of Science and Technology, and the Hong Kong Society of Theoretical and Applied Mechanics for sponsoring the meeting and providing partial funding to bring the international participants to Hong Kong.

Finally, the editors would like to thank all the speakers and contributors to this symposium for their invaluable contributions. The editors also wish to thank the

Organizing Committee for their dedicated work. Special thanks also are due to Kluwer Academic Publishers, for their effort and cooperation to produce this attractive Symposium Proceedings.

Qingping SUN, Hong Kong, China
*and*
Pin TONG, USA

October 2005

## International Scientific Committee of the Symposium

P. Tong (Chairmen, Hong Kong, China),
N. Fleck (UK),
H. Gao (Germany),
J. Hutchinson (USA),
K.C. Hwang (Beijing, China),
W. Nix (USA),
A. Zaoui (France),
W. Zhong (Dalian, China),
L.B. Freund (IUTAM Representative).

## Local Organizing Committee

P. Tong (USA)
Q.P. Sun (Co-Chairman, Hong Kong, China)
Ellie Ho (Secretary, HKUST)
P. Feng (HKUST)
Y.J. He (HKUST)
K.L. Ng (HKUST)

## Sponsors

- IUTAM (The International Union of Theoretical and Applied Mechanics)
- RGC (Research Grant Council of Hong Kong SAR)
- US National Science Foundation (NSF)
- Kluwer Academic Publishers
- Hong Kong University of Science and Technology
- HKSTAM (The Hong Kong Society of Theoretical and Applied Mechanics)

## List of Participants

| | |
|---|---|
| Agde Tjernlund J. | KTH Solid Mechanics, Royal Institute of Technology, Osquars backe 1, 100 44 Stockholm, Sweden |
| Ball J. | University of Oxford, UK |
| Chen Z. | Department of Civil and Environmental Engineering, University of Missouri-Columbia, Columbia, MO 65211-2200, USA |
| Cherkaoui M. | LPMM-CNRS University of Metz, Ile du Saulcy, 57045, Metz, Cedex1 France |
| Dai H.H. | Department of Mathematics, City University of Hong Kong, Kowloon, Hong Kong |
| Fan J.-H. | Alfred University, USA |
| Feng X.-Q | Department of Engineering Mechanics, Tsinghua University, Beijing 100084, China |
| Guo W.-L. | Institute of Nano Science, Nanjing University of Aeronautics and Astronautics, Nanjing, 210016, P. R. China |
| Han C.-S. | Max Planck Institute for Metals Research, Heisenbergstr. 3, D-70569 Stuttgart, Germany |
| He Y.J. | Department of Mechanical Engineering, The Hong Kong University of Science and Technology, Hong Kong |
| Hu G.-K. | Department of Applied Mechanics, Beijing Institute of Technology 100081, Beijing, China |
| Huang R. | Research Center for Mechanics of Solids, Structure & Materials, Department of Aerospace Engineering and Engineering Mechanics, The University of Texas, Austin, TX 78712, USA |
| Huo Y. | Department of Mechanics and Engineering Sciences Fudan University, Shanghai, 200433, China |
| Hwang K.C. | FML, Department of Engineering Mechanics, Tsinghua University, Beijing 100084, China |
| Lee S. | National Tsing Hua University, Taiwan |

| | |
|---|---|
| Li J.-Y. | University of Nebraska-Lincoln, USA |
| Lu C.-S. | Centre for Advanced Materials Technology (CAMT), School of Aerospace, Mechanical and Mechatronic Engineering J07, The University of Sydney, NSW 2006, Australia |
| Ngan A.H.W. | Department of Mechanical Engineering, The University of Hong Kong, Pokfulam Road, Hong Kong, P.R. China |
| Petryk H. | Institute of Fundamental Technological Research, Polish Academy of Sciences, Swietokrzyska 21, 00-049 Warsaw, Poland |
| Qian L.M. | Department of Mechanical Engineering, The Hong Kong University of Science and Technology, Hong Kong |
| Rajendran A.M. | U. S. Army Research Office, RTP, NC 27709-2211, USA |
| Sun Q.P | Department of Mechanical Engineering, The Hong Kong University of Science and Technology, Hong Kong, China |
| Tong P. | Bioengineering Department, University of California at San Diego, La Jolla, California, USA |
| Wang B. | School of Physics and Engineering, Sun Yat-sen University, Guangzhou, China |
| Wang M.Y. | Department of Automation & Computer-Aided Engineering, The Chinese University of Hong Kong, Shatin, NT, Hong Kong |
| Wei Y.-G. | LNM, Institute of Mechanics, Chinese Academy of Sciences, Beijing 100080, China |
| Woo C.H. | Department of Mechanical Engineering, The Hong Kong Polytechnic University, Hung Hom, Kowloon, Hong Kong |
| Xiang Y. | The Hong Kong University of Science and Technology, Hong Kong |
| Yang W. | Department of Engineering Mechanics, Tsinghua University, Beijing 100084, CHINA |
| Yu T.X. | Department of Mechanical Engineering, The Hong Kong University of Science and Technology, Hong Kong |
| Zaoui A. | Laboratorire de Mecanique des Solides, Ecole Polytechnique, CNRS 91128 Palaiseau cedex, France |

| | |
|---|---|
| Zhang T.-Y. | Department of Mechanical Engineering, The Hong Kong University of Science and Technology, Hong Kong |
| Zhang X.X. | The Hong Kong University of Science and Technology, Hong Kong |
| Zhao H. | Laboratoire de Mécanique et Technologie, ENS-Cachan/CNRS/ University Paris 6, 61, Avenue du président Wilson, 94235 Cachan cedex, France |

# List of Contributors


| | |
|---|---|
| Abdennadher S. | Laboratoire de Méani que et Technologie, ENS-Cachan/CNRS/ University Paris 6, 61, Avenue du président Wilson, 94235 Cachan cedex, France |
| Agde Tjernlund J. | KTH Solid Mechanics, Royal Institute of Technology, Osquars backe 1, 100 44 Stockholm, Sweden |
| Bai Y.-L. | State Key Laboratory of Nonlinear Mechanics (LNM), Institute of Mechanics, Academia Sinica, Beijing 100080, P.R. China |
| Capolungo L. | LPMM-CNRS University of Metz, Ile du Saulcy, 57045, Metz, Cedex1 France |
| Chen Z. | Department of Civil and Environmental Engineering, University of Missouri-Columbia, Columbia, MO 65211-2200, USA |
| Cherkaoui M. | LPMM-CNRS University of Metz, Ile du Saulcy, 57045, Metz, Cedex1 France |
| Dai H. H. | Department of Mathematics, City University of Hong Kong, Kowloon, Hong Kong |
| Dingreville R. | George W. Woodruff School of Mechanical Engineering, Atlanta Georgia 30332-0405 USA |
| Feng G. | Department of Materials Science and Engineering, Stanford University, Stanford, CA 24305, USA |
| Feng X.-Q. | Department of Engineering Mechanics, Tsinghua University, Beijing 100084, China |
| Gamstedt E. K. | KTH Solid Mechanics, Royal Institute of Technology, Osquars backe 1, 100 44 Stockholm, Sweden |
| Gao H. | Max Planck Institute for Metals Research, Heisenbergstr. 3, D-70569 Stuttgart, Germany |
| Gudmundson P. | KTH Solid Mechanics, Royal Institute of Technology, Osquars backe 1, 100 44 Stockholm, Sweden |
| Guo W.-L. | Institute of Nano Science, Nanjing University of Aeronautics and Astronautics, Nanjing, 210016, P.R. China |

Guo Y.-F. Institute of Nano Science, Nanjing University of Aeronautics and Astronautics, Nanjing, 210016, P.R. China

Han C.-S. Max Planck Institute for Metals Research, Heisenbergstr. 3, D-70569 Stuttgart, Germany

He Y.J. Department of Mechanical Engineering, The Hong Kong University of Science and Technology, Hong Kong

Herve E. Laboratorire de Mecanique des Solides, Ecole Polytechnique, CNRS 91128 Palaiseau cedex, France

Hong W. Department of Engineering Mechanics, Tsinghua University, Beijing 100084, CHINA

Hu G.-K. Department of Applied Mechanics, Beijing Institute of Technology 100081, Beijing, China

Huang R. Research Center for Mechanics of Solids, Structure & Materials, Department of Aerospace Engineering and Engineering Mechanics, The University of Texas, Austin, TX 78712, USA

Huang Y. Department of Mechanical and Industrial Engineering, University of Illinois, Urbana, IL 61801, USA

Huo Y. Department of Mechanics and Engineering Sciences Fudan University, Shanghai, 200433, China

Hwang K.C. FML, Department of Engineering Mechanics, Tsinghua University, Beijing 100084, China

Im S.H. Research Center for Mechanics of Solids, Structure & Materials, Department of Aerospace Engineering and Engineering Mechanics, The University of Texas, Austin, TX 78712, USA

Karagiozova D. Institute of Mechanics, Bulgarian Academy of Sciences, Acad. G. Bonchev Street, Block 4, Sofia 1113, Bulgaria

Keung L.H. Department of Mechanical Engineering, the Hong Kong University of Science and Technology, Clear Water Bay, Hong Kong, People's Republic of China

Lam D.C.C. Department of Mechanical Engineering, the Hong Kong University of Science and Technology, Clear Water Bay, Hong Kong, People's Republic of China

| | |
|---|---|
| Li H. | Department of Mechanical Engineering, The University of Hong Kong, Pokfulam Road, Hong Kong, P.R. China |
| Liu C. | Materials Science and Technology Division, Los Alamos National Laboratory, Los Alamos, NM 87545 |
| Lu C.-S. | Centre for Advanced Materials Technology (CAMT), School of Aerospace, Mechanical and Mechatronic Engineering J07, The University of Sydney, NSW 2006, Australia |
| Ma X.-L. | Department of Engineering Mechanics, Tsinghua University, Beijing 100084, CHINA |
| Maciejewski G. | Institute of Fundamental Technological Research, Polish Academy of Sciences, Swietokrzyska 21, 00-049 Warsaw, Poland |
| Mai Y.-W. | Centre for Advanced Materials Technology (CAMT), School of Aerospace, Mechanical and Mechatronic Engineering J07, The University of Sydney, NSW 2006, Australia |
| Marcadon V. | Laboratorire de Mecanique des Solides, Ecole Polytechnique, CNRS 91128 Palaiseau cedex, France |
| Müller I. | Technische Univ. Berlin, FB6, Sekr HF2, Thermodynamik, Strasse des 17. Juni, 135, D-10623 Berlin, Germany |
| Ngan A.H.W. | Department of Mechanical Engineering, The University of Hong Kong, Pokfulam Road, Hong Kong, P.R. China |
| Nix W. D. | Department of Materials Sciences and Engineering, Standford University, Stanford, CA 94305, USA |
| Petryk H. | Institute of Fundamental Technological Research, Polish Academy of Sciences, Swietokrzyska 21, 00-049 Warsaw, Poland |
| Qian L.M. | Department of Mechanical Engineering, The Hong Kong University of Science and Technology, Hong Kong |
| Qu J. | George W. Woodruff School of Mechanical Engineering, Atlanta Georgia 30332-0405 USA |
| Qu S. | Department of Mechanical and Industrial Engineering, University of Illinois, Urbana, IL 61801 |

| | |
|---|---|
| Shen L.-M. | Department of Civil and Environmental Engineering, University of Missouri-Columbia, Columbia, MO 65211-2200, USA |
| Stupkiewicz S. | Institute of Fundamental Technological Research, Polish Academy of Sciences, Swietokrzyska 21, 00-049 Warsaw, Poland |
| Sun Q.P. | Department of Mechanical Engineering, The Hong Kong University of Science and Technology, Hong Kong, China |
| Tang B. | Department of Mechanical Engineering, The University of Hong Kong, Pokfulam Road, Hong Kong, P.R. China |
| Tong P. | Bioengineering Department, University of California at San Diego, La Jolla, USA |
| Wang B. | School of Physics and Engineering, Sun Yat-sen University, Guangzhou, China |
| Wang H.T. | Department of Engineering Mechanics, Tsinghua University, Beijing 100084, CHINA |
| Wang M. Y. | Department of Automation & Computer-Aided Engineering, The Chinese University of Hong Kong, Shatin, NT, Hong Kong |
| Wei Y.-G. | LNM, Institute of Mechanics, Chinese Academy of Sciences, Beijing 100080, China |
| Wo P.C. | Department of Mechanical Engineering, The University of Hong Kong, Pokfulam Road, Hong Kong, P.R. China |
| Woo C.H. | Department of Mechanical Engineering, The Hong Kong Polytechnic University, Hung Hom, Kowloon, Hong Kong |
| Xiao X.-D. | Department of Physics and Institute of Nano Science and Technology, The Hong Kong University of Science and Technology, Hong Kong |
| Xu G.-S. | Department of Mechanical Engineering, University of California at Riverside, CA 92521 |
| Yang W. | Department of Engineering Mechanics, Tsinghua University, Beijing 100084, China |
| Yu T.X. | Department of Mechanical Engineering, The Hong Kong University of Science and Technology, Hong Kong |

Zaoui A. — Laboratorire de Mecanique des Solides, Ecole Polytechnique, CNRS 91128 Palaiseau cedex, France

Zhao H. — Laboratoire de Méanique et T echnologie, ENS-Cachan/CNRS/ University Paris 6, 61, Avenue du président Wilson, 94235 Cachan cedex, France

Zheng Y. — School of Physics and Engineering, Sun Yat-sen University, Guangzhou, China

Zhou S. — Department of Automation & Computer-Aided Engineering, The Chinese University of Hong Kong, Shatin, NT, Hong Kong

## Program of the Symposium:

### MONDAY MAY 31, 2004

09:00-09:30 **OPENING CEREMONY:** *P. Tong and Q.P. Sun*

Welcome address from the University Administration

09:30-10:00 Deformation and Diffusion in Nano-grained Metals
*Wei Yang, XinLing Ma, HongTao Wang, Wei Hong, Tsinghua University, China*

10:00-10:30 **BREAK**

**Chair:** *K.C. Hwang*

10:30-11:00 Multiscale Analysis of Strength Enhancement of Cellular Materials under Impact Loading

*H. Zhao, S. Abdenadher, I. Nasri, Laboratoire de Mécanique et Technologie, ENS-Cachan/CNRS/University, France*

11:00-11:30 Micromechanics of Micropolar Composites
*Gengkai Hu, Beijing Institute of Technology, China*

11:30-12:00 An MD Investigaton of the Size Effect on Multiscale Simulation of Thin Film Delamination

*Zhen Chen, Luming Shen, University of Missouri-Columbia, USA*

12:00-14:00 **LUNCH**

**Chair:** J.H. Fan

| | |
|---|---|
| 14:00-14:30 | Fracture Analysis in the Conventional Theory of Mechanism-based Strain Gradient (CMSG) Plasticity<br>*K.C. Hwang, Tsinghua University, China*<br>*S. Qu, Y. Huang, H. Jiang, University of Illinois, Urbana*<br>*C. Liu, Materials Science and Technology Division, Los Alamos National Laboratory, Los Alamos*<br>*P.D. Wu, Alcan International Limited, Kingston Research and Development Center, Canada* |
| 14:30-15:00 | Fractal and Scaling Phenomena on Fracture at Micro-scales<br>*Chunsheng LU, Yiu-Wing MAI, The University of Sydney, Australia* |
| 15:00-15:30 | **BREAK** |

**Chair:** Y.W. Mai

| | |
|---|---|
| 15:30-16:00 | A Multi-Scale Modeling Scheme for Sub-Micron Size Effects on Cyclic Plasticity<br>*Jinghong Fan, Zhihui Gao, Xiangguo Zeng, Chongqing University, China*<br>*Jinghong Fan, Alfred University, USA* |
| 16:00-16:30 | Level Set Simulations of the Formation of Dislocation Networks and Junctions<br>*Yang Xiang, Hong Kong University of Science and Technology, HK*<br>*David J. Srolovitz, Weinan E, Princeton University, USA*<br>*Li-Tien Cheng, University of California, San Diego, USA* |
| 16:30-17:00 | Issues and Challenges in Multiscale Modeling of Bio/Nano Systems<br>*A.M. Rajendran, U.S. Army Research Office, RTP, NC 27709-2211, USA* |
| 18:00-19:00 | **DINNER** |

**TUESDAY JUNE 1, 2004**

**Chair:** *H. Petryk*

| | |
|---|---|
| 09:00-09:30 | Nucleation of Phases and Local Minimizers of Energy<br>*John Ball, University of Oxford, UK* |

09:30-10:00 Solid-solid Phase Transformations: Non-existence of One-dimensional Stress Problems, Model Equation and Uniqueness Conditions

*Hui-Hui Dai, City University of Hong Kong, Hong Kong*

10:00-10:30 **BREAK**

**Chair**: *H.H. Dai*

10:30-11:00 Morphological Instability and Kinetics of an Elastic Film on a Viscoelastic Substrate

*R. Huang, The University of Texas, USA*

11:00-11:30 Magnetization Reversal and Hysteresis in Nanocrystalline Ferromagnets

*Jiangyu Li, Heliang Qu, University of Nebraska-Lincoln, USA*

11:30-12:00 Effects of the Structural Topology and Connection Size on the Strength of Circular Honeycombs under In-plane Compression

*T.X. Yu, D. Karagiozova, The Hong Kong University of Science & Technology, HK*
*D. Karagiozova, Bulgarian Academy of Sciences, Bulgaria*

12:00-14:00 **LUNCH**

**Chair:** *J.Y. Li*

14:00-14:30 Size Effects of Nanoindentation Creep

*H. Li, A.H.W. Ngan, The University of Hong Kong, HK*

14:30-15:00 Multi-scale Characterizations for Ductile Thin Film Delamination

*Yueguang Wei, Chinese Academy of Sciences, China*
*Guanshui Xu, University of California at Riverside, USA*

15:00-15:30 **BREAK**

**Chair:** *A.H.W. Ngan*

15:30-16:00 An Experimental Investigation of the Relationship between Molecular Structure and Length Scales in Inelastic Deformation of an Amorphous Thermoplastic

*Jessica Agde Tjernlund, E. Kristofer Gamstedt, Zhi-Hui Xu, Royal Institute of Technology (KTH), Sweden*

16:00-16:30 Impact of Phase Transition on the Nano-indentation Hardness and Microwear Behavior of NiTi Shape Memory Alloy

*Linmao Qian, Xudong Xiao, Qingping Sun, Tongxi Yu, The Hong Kong University of Science and Technology, HK*

16:30-17:30 **Round Table Discussion On Future Research Directions** (Chair: *K.C. Hwang and P. Tong*)

18:00-20:00 **Banquet**

**WEDNESDAY JUNE 2, 2004**

**Chair:** *W.L. Guo*

09:00-09:30 Particle Size Effects in NANOcomposites

*A. Zaoui, V. Marcadon, E. Hervé, Ecole Polytechnique, France*

09:30-10:00 Mechanism-based Strain Gradient Crystal Plasticity

*Chung-Souk Han, Huajian Gao, Max Planck Institute for Metals Research, Germany*
*Chung-Souk Han, William D. Nix, Stanford University, USA*
*Yonggang Huang, University of Illinois, USA*

10:00-10:30 **BREAK**

**Chair:** *X.Q. Feng*

10:30-11:00 Size Effect in Magnetic Materials

*X.X. Zhang, Hong Kong University of Science and Technology, HK*

11:00-11:30 On Modeling Deformation Instability and Pattern Formation during Phase Transition in NiTi Microtubing

*Y. J. He, Hong Kong University of Science and Technology, HK*

11:30-12:00 Plastic Deformation of Rough Surfaces in the Nanoindentation Test

*Tong-Yi Zhang, Hong Kong University of Science and Technology, HK*

12:00-14:00 **LUNCH**

**Chair:** *T.Y. Zhang*

14:00-14.30 Multiscale Mechanics of Carbon Nanotubes and Their Composites

*Xi-Qiao Feng, Tsinghua University, China*

14:30-15:00 A Phase-Field Method for Optimization of Solid Structures: Cahn-Hilliard Model

*Michael Yu Wang Shiwei Zhou, The Chinese University of Hong Kong, HK*

15:00-15:30 **BREAK**

**Chair:** *B. Wang*

15:30-16:00 Modeling of Austenite/martensite Laminates with Interfacial Energy Effect

*H. Petryk, S. Stupkiewicz, G. Maciejewski, Polish Academy of Sciences, Poland*

16:00-16:30 Size Effect in Nano/molecular Electromechanical Systems: Function over Strength

*Wanlin Guo, Nanjing University of Aeronautics and Astronautics, China*

16:30-17:30 **Round Table Discussion On Future Research Directions** (Chair: *Z. Zaoui and H. Petryk*)

18:00-19:00 **DINNER**

**THURSDAY JUNE 3, 2004**

**Chair:** *M. Cherkaoui*

09:00-09:30 Size Effects of Phase Transition in Thin Films

*C.H. Woo, B. Wang, Z. Man, Hong Kong Polytechnic University, HK*
*Hanchen Huang, Rensselaer Polytechnic Institute, USA*

09:30-10:00 Tunable Ferroelectric Phase Transition

*Biao Wang, C.H. Woo, The Hong Kong Polytechnic University, HK*

10:00-10:30 **BREAK**

**Chair:** *S.Q. Shi*

10:30-11:00 Mechanical Strengths of Low-K Dielectric Thin Films

*Sanboh Lee, T.C. Liu, National Tsing Hua University, Taiwan*
*B.-T. Chen, S. M. Jang, Taiwan Semiconductor Manufacturing Company, Advanced Module Technology division, Taiwan*

11:00-11:30 Energy Penalties in Materials with Microstructures of Phase Transitions

*Yongzhong Huo, Fudan University, China*

| | |
|---|---|
| 11:30-12:00 | Homogenization Method for Strength and Inelastic Behavior of Nanocrystalline Materials<br>*Laurent. Capolungo, Jianmin Qu, University of Metz, Ile du Saulcy, France*<br>*Mohammed. Cherkaoui, George W. Woodruff, School of Mechanical Engineering, USA* |
| 12:00-14:00 | **LUNCH** |

**Chair:** *Y.Z. Huo*

| | |
|---|---|
| 14:00-14:30 | Effects of Surface Energy on the Elastic Behavior of Nano-particles<br>*Remi Dingreville, Jianmin Qu, Georgia Institute of Technology, USA*<br>*Mohammed Cherkaoui, University of Metz, France* |
| 14:30-15:00 | Size Effect in Formation of Junctions and Defects in Carbon Nanotubes<br>*San-Qiang Shi, F.Y. Meng, L.G. Zhou, The Hong Kong Polytechnic University, HK*<br>*F.Y. Meng, R. Yang, Institute for Metal Research, Chinese Academy of Sciences, Shenyang, China* |
| 15:00-15:30 | **BREAK** |

**Chair:** *Q.P. Sun*

| | |
|---|---|
| 15:30-16:00 | Size effect of Hair Sized Structures — Numerical Results<br>*P. Tong, University of California at San Diego, USA* |

**CLOSING**

# TUNABLE FERROELECTRIC PHASE TRANSITION

BIAO WANG[1], C.H. WOO[2] and YUE ZHENG[1]
[1]*School of Physics and Engineering,*
*Sun Yat-sen University, Guangzhou, China*
[2]*Department of Mechanical Engineering,*
*The Hong Kong Polytechnic University,*
*Hung Hom, Kowloon, Hong Kong*

## Abstract

Ferroelectricity is a collective phenomenon, the characteristics of which depend on the combined effects of many factors, such as the ambient temperature, boundary conditions, sample dimensions, and misfit epitaxial stresses, etc. It is well known that ferroelectric thin films generally exhibit characteristics that are different from their bulk counterparts. By considering ferroelectric thin films on a compliant substrate, we found that the characteristics of the phase transition, such as, the Curie temperature, the order of the phase transition, etc., can be tuned simply by changing the relative thickness of the film and substrate. Our system is described using the time-dependent Ginzburg-Landau equation. The relaxation effect of the free surface is considered using the "extrapolation length" parameter. Based on such a model, we can predict the change of the Curie temperature versus the film and substrate thickness, and determine the critical thickness of the film, under which the ferroelectric transition will be totally suppressed. For some first-order ferroelectric materials, there exists a critical thickness of the substrate, above which the transition becomes second-order due to the constraint of the substrate.

## 1. Introduction

It is well-known that different ferroelectric materials have different phase transition characteristics, such as different phase transition temperature, different order of phase transition, etc. Recently it has been revealed that the ferroelectric phase transition also depends on the so-called 'extrinsic factors', such as the size and the boundary conditions to which the ferroelectric materials are subjected. Taking ferroelectric thin films as an example, one can find that for most ferroelectric materials, the Curie temperature will decrease with decreasing the thickness of thin films due to the constraint of the surfaces, and it may reduce to absolute zero degree when the thickness of the film reaches some critical value, thus the ferroelectric phase transition was suppressed totally due to the size



*Q. P. Sun and P. Tong (eds.), IUTAM Symposium on Size Effects on Material and Structural Behavior at Micron- and Nano-Scales,* 1–12.

effect.[1-3] Generally speaking, one cannot get rid of the misfit stress in the film fabrication process, and such stresses in the thin film play an important role in determining the Curie temperature[4,5]. By reducing the thickness of the substrate, the misfit stresses can be relaxed to some extend, and the corresponding Curie temperature can be changed in some range. In such way, the phase transition temperature can be tuned simply by changing the relative thickness of the film and substrate. For $PbTiO_3$ film on $SrTiO_3$ substrate, we find that the Curie temperature can be tuned in the range of $0°K$ to $1000\,°K$.

Accompanying the paraelectric to ferroelectric phase transition, the transformation strains will be induced. By constraining the strain, the substrate will restrict the ferroelectric phase transition of the film. Although many ferroelectric materials exhibit the first-order phase transition, some of them may behavior as the second-order transition materials due to the constraint of the substrate. To reduce the thickness of the substrate, the constraint will be reduced. In such way, one can obtain the first-order transition thin film or the second-order transition thin film from the same material simply by changing the relative thickness of the film and substrate. For $PbTiO_3$ film on $SrTiO_3$ substrate system, the critical thickness of the substrate, above which the transition becomes second-order due to the constraint of the substrate, was determined.

## 2. General Formulation

Generally speaking, the ferroelectric phase transition is caused by a change of the crystallographic structures associated with transformation strain. If the polarization P is along one direction, the transverse transformation strain can be expressed as $\varepsilon^T = QP^2$, where $Q$ is the electrostrictive coefficient. Due to the constraint of the substrate and the inhomogeneous distribution of the polarizations in the film, characteristics of the phase transition may be affected, such as the Curie temperature, or even the order of the transitions, etc. The elastic strain energy induced by the transformation strain can be calculated using the following integration,

$$\begin{aligned} U_s &= \frac{1}{2}\int_0^{H+h}\frac{\sigma_{xx}^2+\sigma_{yy}^2}{G}dz = \frac{1}{G}\int_0^h \sigma^2(z)dz + \frac{1}{G_0}\int_h^{H+h}\sigma^2(z)dz \\ &= GQ^2\int_0^h P^4 dz + g_1[\int_0^h P^2 dz]^2 + g_2[\int_0^h P^2(z-a)dz]^2 + g_3\int_0^h P^2 dz\int_0^h P^2(z-a)dz \end{aligned} \quad (1)$$

$$g_1 = Q^2[\frac{Gh+HG^2/G_0}{(H+h)^2} - \frac{2G}{H+h}],$$

where $$g_2 = Q^2\{-\frac{2G}{I} + \frac{G}{3I^2}[a^3+(h-a)^3] + \frac{G^2}{3G_0 I^2}[(H+h-a)^3-(h-a)^3]\}, \quad (2)$$

$$g_3 = \frac{Q^2}{(H+h)I}\{G[-a^2+(h-a)^2] + G^2/G_0[(H+h-a)^2-(h-a)^2]\}.$$

where $G = (C_{11}+C_{12}-2C_{12}^2/C_{11})$, $I=(H+h)^3/12$ is the area moment of inertia, and the neutral axis can be determine as follows:

$$a=\frac{\int_0^{H+h}G(z)zdz}{\int_0^{H+h}G(z)dz}=\frac{1}{2}\bullet\frac{Gh^2+G_0H^2+2G_0Hh}{Gh+G_0H} \tag{3}$$

Consider a ferroelectric thin film and substrate system (Fig.1). Both the film and the substrate are treated as cubic elastic body with elastic Modulus $C_{11}, C_{12}, C_{11}^0, C_{12}^0$ and thickness $h, H$ , respectively.

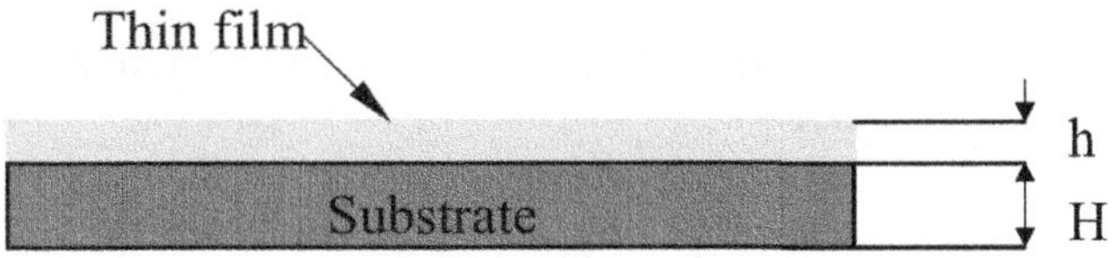


*Figure 1.* Schematic of ferroelectric thin film and substrate system.

During fabrication, the misfit between the film and the substrate creates a misfit stress $\sigma^0=\sigma_{xx}=\sigma_{yy}$ in the film. If the misfit strain is $\varepsilon^0=(a_s-a_f)/a_f$ , where $a_s$, $a_f$ are the lattice constants of the substrate and the film, respectively, the misfit stress in the film can be derived as,

$$\sigma^0=G\varepsilon^0[\frac{h}{H+h}+\frac{h^2-2ah}{2I}(z-a)-1] \tag{4}$$

For thick substrates, first two terms in equation (4) vanish as $H\to\infty$ and only the last term remains.

The misfit stresses here play the role of applied loading, and the interaction energy between them and the transformation strains is given as,

$$U_e=-\int_0^h(\sigma_{xx}\varepsilon^T+\sigma_{yy}\varepsilon^T)dz=-2Q\int_0^h\sigma^0P^2dz \tag{5}$$

According to the Landau phase transition theory, the "intrinsic free energy" related with the ferroelectric thin film could be written in the form as,

$$U_p=\int_0^h[\frac{A}{2}(T-T_{c0})P^2+\frac{B}{4}P^4+\frac{C}{6}P^6+\frac{D}{2}(\frac{\partial P}{\partial z})^2-\frac{1}{2}E_dP]dz+\iint_S\frac{D\delta^{-1}}{2}P^2dxdy \tag{6}$$

where $A, B, C$ and $D$ are expansion coefficients of the corresponding bulk material. $T_{c0}$ is the Curie temperature of the bulk crystal, $S$ represent the upper and lower unit surface planes of the film, $\delta$ is the extrapolation length that measures the strength of the surface effect on the polarization6, and $E_d$ is the depolarizing field, which depends on the electric boundary condition, and can be expressed in the following form as[1],

$$E_d=-\frac{1}{\alpha}(P-\frac{\phi}{h}\int_0^hPdz) \tag{7}$$

where $\alpha$ is the dielectric constant of the film, and $\phi$ takes on the value of 0 or 1, respectively, when (1) the ferroelectric film is between two metallic electrodes in short-circuit conditions; (2) the ferroelectric film is between two dielectric substrates.

The evolution equation corresponding to the energy equations (1), (5) and (6) can be established through the following variational derivative,

$$
\begin{aligned}
-K\frac{\partial P}{\partial t} &= \frac{\delta(U_s+U_e+U_p)}{\delta P} \\
&= (\Lambda_1 z+\Lambda_2)P+\Lambda_3 P^3+CP^5-D\frac{d^2P}{dz^2}-R
\end{aligned}
\tag{8}
$$

where $K$ is the kinetic coefficient related to the domain wall mobility, and

$$\Lambda_1 = 4g_2\Delta_2+2g_3\Delta_1-\frac{2QG\varepsilon^0}{I}(h^2-2ah),$$

$$\Lambda_2 = A(T-T_{c0})+\frac{1}{\alpha}-4QG\varepsilon^0\left(\frac{h}{H+h}-\frac{ah^2-2a^2h}{2I}-1\right)+(4g_1-2ag_3)\Delta_1+(2g_3-4ag_2)\Delta_2,$$

$$\Lambda_3 = B+4GQ^2,$$

$$R=\frac{\phi}{\alpha h}\int_0^h P dz \tag{9}$$

with

$$\Delta_1=\int_0^h P^2 dz,\quad \Delta_2=\int_0^h P^2(z-a)dz \tag{10}$$

and the boundary conditions are

$$\frac{\partial P}{\partial z}=\frac{P}{\delta},\ \text{for } z=0;\quad \frac{\partial P}{\partial z}=-\frac{P}{\delta},\ \text{for } z=h \tag{11}$$

In the following, through determining the bifurcation point of equation (8), one can obtain the phase transition temperature. Then by using a perturbation approach, the effect of film dimensions and surface conditions on the system behavior of ferroelectric thin films near the paraelectric-ferroelectric phase transition can be investigated. Based these results, one can find that the phase transition characteristics of ferroelectric films, such as the Curie temperature, the order of transition, etc. can be tuned only through changing the 'extrinsic conditions' such as the thickness of the film and substrate.

## 3. Curie Temperature of Ferroelectric this Film on A Compliant Substrate

In this part, we only focus on the Curie temperature, or the supercooling temperature for the first-order transition. Therefore it is enough to consider only the linearized Landau-Ginzburg equation. We have carried out extensive calculation, and found that the effect of the last three terms of the elastic energy related with the phase transition in equation (1) on the Curie temperature can be neglected, and the main contributions of the elastic energy are on the higher order terms of the Landau free energy expansion.

Therefore the simplified linearized evolution equation corresponding to the energy equations (1), (5) and (6) can be established through the following variational derivative,

$$\begin{aligned} -K\frac{\partial P}{\partial t} &= \frac{\delta(U_s+U_e+U_p)}{\delta P} \\ &= (\Lambda_1 z+\Lambda_2)P - D\,\frac{d^2P}{dz^2} - R \end{aligned} \tag{12}$$

where $K$ is the kinetic coefficient related to the domain wall mobility, and

$$\begin{aligned} \Lambda_1 &= -\frac{2QG\varepsilon^0}{I}(h^2-2ah), \\ \Lambda_2 &= A(T-T_{c0}) + \frac{1}{\alpha} - 4QG\varepsilon^0\left(\frac{h}{H+h} - \frac{ah^2-2a^2h}{2I} - 1\right), \\ R &= \frac{\phi}{\alpha h}\int_0^h P dz, \end{aligned} \tag{13}$$

and the boundary conditions are

$$\frac{\partial P}{\partial z} = \frac{P}{\delta}, \text{ for } z=0; \qquad \frac{\partial P}{\partial z} = -\frac{P}{\delta}, \text{ for } z=h \tag{14}$$

It is obvious that $P=0$ is a trivial solution of equation (12), representing the stationary non-ferroelectric state. If the temperature is reduced below some critical value, the trivial zero solution will bifurcate yielding $P \neq 0$ solutions, corresponding to the transformation of the paraelectric state to the ferroelectric state. Therefore, the Curie temperature can be established through the bifurcation point. From the bifurcation theory of the partial differential equation, the bifurcation point can be determined from the eigenvalue $\lambda$ of the linearized equation:

$$(\Lambda_1 z+\Lambda_2+\lambda)P - D\,\frac{d^2P}{dz^2} - R = 0 \tag{15}$$

where the eigenvalue means $\lambda$ values for which non-zero solutions of equations (15) and boundary condition (14) exist. The bifurcation point, and the Curie temperature is the point at which $\lambda = 0$, where the stable $P=0$ solution starts to become unstable. Since $R$ is a constant, we can derive the solution for the homogenous equation first, and then determine $R$. The equation is simplified with the changed variable

$$\begin{aligned} \eta &= \left(\frac{D}{\Lambda_1}\right)^{2/3}\frac{1}{D}(\Lambda_1 z+\Lambda_2+\lambda) \\ \frac{d^2P}{d\eta^2} &- \eta P = -R \end{aligned} \tag{16}$$

The general solution of equation (16) is a linear combination of the two Airy functions $Ai(\eta), Bi(\eta)$, therefore,

$$P(\eta) = C_1P_1(\eta) + C_2P_2(\eta) \tag{17}$$

where

$$P_1(\eta) = Ai(\eta) + \Omega[F_{Ai}(\eta)Bi(\eta) - F_{Bi}(\eta)Ai(\eta)] \int_{\eta_1}^{\eta_2} Ai(\xi)d\xi,$$
$$P_2(\eta) = Bi(\eta) + \Omega[F_{Ai}(\eta)Bi(\eta) - F_{Bi}(\eta)Ai(\eta)] \int_{\eta_1}^{\eta_2} Bi(\xi)d\xi, \tag{18}$$

where $$\Omega = \frac{\pi}{\alpha h} \cdot (\frac{D}{\Lambda_1})^{1/3} \cdot \{1 - \frac{\pi}{\alpha h \Lambda_1} \int_{\eta_1}^{\eta_2} [F_{Ai}(\xi)Bi(\xi) - F_{Bi}(\xi)Ai(\xi)]d\xi\}^{-1}$$

$$\eta_1 = (D/\Lambda_1)^{2/3} \cdot \tfrac{1}{D}(\Lambda_2 + \lambda), \eta_2 = (D/\Lambda_1)^{2/3} \cdot \tfrac{1}{D}(\Lambda_2 + \lambda + \Lambda_1 h)$$

And $F_{Ai}(\eta)$, $F_{Bi}(\eta)$ are original functions of $Ai(\eta)$, $Bi(\eta)$, respectively.

Substitution of equation (17) into the first one of boundary conditions (14) yields a relation between the constant $C_1$ and $C_2$ as follows:

$$C_2 = \gamma C_1 \tag{19}$$

where $$\gamma = [\frac{1}{\delta} P_2(\eta_1) - (\frac{\Lambda_1}{D})^{1/3} P_2'(\eta_1)]^{-1} [(\frac{\Lambda_1}{D})^{1/3} P_1'(\eta_1) - \frac{1}{\delta} P_1(\eta_1)] \tag{20}$$

where $P_1'(\eta_1)$, $P_2'(\eta_1)$ are derivatives with respect to $\eta$ at $\eta = \eta_1$, respectively. Therefore solution (17) becomes,

$$P(\eta) = C_1[P_1(\eta) + \gamma P_2(\eta)] \tag{21}$$

Substitution of equation (21) into the second boundary condition of equation (14) yields

$$P_1'(\eta_2) + \gamma P_2'(\eta_2) = -\frac{1}{\delta}[P_1(\eta_2) + \gamma P_2(\eta_2)] \tag{22}$$

To obtain the Curie temperature, one can set the eigenvalue $\lambda = 0$ and $-\omega = A(T - T_{c0}) + 1/\alpha$, by solving $\omega$ versus the thickness of the film numerically, the Curie temperature can be obtained as

$$T_c = T_{c0} - \frac{1}{A\alpha} - \frac{\omega_{\min}}{A} \tag{23}$$

where $\omega_{\min}$ means the minimum solution of $\omega$.

Within the present formulation, if the substrate thickness $H \to \infty$, equation (13) gives,

$$\Lambda_1 = 0, \quad \Lambda_2 = A(T - T_{c0}) + 1/\alpha + 4QG\varepsilon^0 \tag{24}$$

and equation (15) reduces to the same form as discussed in Wang and Woo's paper [1]. For a free standing thin film, and the polarization distribution is asymmetric with respect its center axis, the misfit stress is zero in such cases, one can determine the Curie temperature in the same way as shown in Ref. (1). These two cases represent the two limiting cases for the substrate effects. With a substrate, usually one cannot get rid of the misfit stress in the fabrication process. Depending the combination of the film and the substrate, the misfit stress may help the ferroelectric transformation or suppress the transformation, which means the misfit stress may increase or decrease the Curie temperature of the film. Therefore one can tune the Curie temperature of the film by changing these external factors such as the thicknesses of the film and substrate in addition to change the internal factor, such as the material's compositions of the film.

To consider the effect of the substrate on the Curie temperature of the ferroelectric thin film, the Curie temperature of $PbTiO_3$ thin film on $SrTiO_3$ substrate was calculated using equation (23) as a function of the thickness of the film and substrate, respectively, and the polarization is assumed to be in perpendicular with the film surfaces. The materials' constants are shown in Table 1. And the calculation result of the Curie temperature versus the thickness of the film is shown in Fig.2 under different thickness of the substrate.

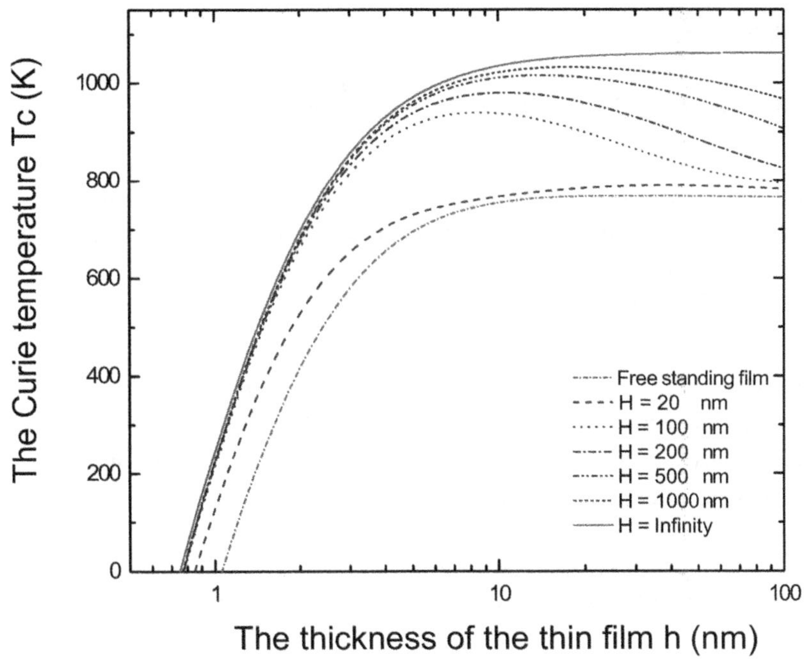


*Figure 2.* The Curie temperature of PbTiO3 film on SrTiO3 substrate versus the film thickness under different substrate thicknesses.

It can be found that the Curie temperature is changing between the two limiting cases: the upper limit is for the totally constrained film with largest misfit stress, whereas the lower limit is for the free standing film with almost zero misfit stress. One can tune the Curie temperature in such a range just by controlling the substrate thickness. With the consideration that the Curie temperature will generally decrease with decrease the film thickness, by adjusting the thickness of the film and substrate, one can tune the Curie temperature in quite a large range.

## 4. Critical Behavior of Ferroelectric thin Films near Phase Transition Point

At steady state, equation (8) satisfies $\partial P/\partial t = 0$, thus giving

$$L_c P = -(\lambda_1 z + \lambda_{2c})P + D\frac{d^2 P}{dz^2} + \frac{\phi}{\alpha h}\int_0^h P dz$$
$$= A(T - T_c)P + \Lambda_3 P^3 + CP^5 + (4g_2\Delta_2 + 2g_3\Delta_1)zP + [(4g_1 - 2ag_3)\Delta_1 + (2g_3 - 4ag_2)\Delta_2]P \quad (25)$$

where

$$\lambda_1 = -\frac{2QG\varepsilon^0}{I}(h^2 - 2ah),$$
$$\lambda_{2c} = A(T_c - T_{c0}) + \frac{1}{\alpha} - 4QG\varepsilon^0\left(\frac{h}{H+h} - \frac{ah^2 - 2a^2 h}{2I} - 1\right),$$

where $T_c$ is the Curie temperature of the film, in the neighborhood of which the solution $P$ can be assumed small, due to continuity. Following Nicholis and Prigogine[7], both $P$ and $\gamma = T - T_c$ can be expanded in a power series in terms of a small perturbation $\varepsilon$ from the critical point:

$$P = \varepsilon P_1 + \varepsilon^2 P_2 + \cdots$$
$$\gamma = T - T_c = \varepsilon\gamma_1 + \varepsilon^2\gamma_2 + \cdots \quad (26)$$

This expansion is more flexible than the seemingly more natural one in which $P$ is expanded in a power series of $(T - T_c)$. More importantly, it allows fractional power dependence of P on $(T - T_c)$. By substituting the expansion (26) into equation (25), and equating coefficients of equal powers of $\varepsilon$, a set of relations of the following form can be obtained:

$$L_c P_k = a_k, \quad k = 1, 2, \cdots \quad (27)$$

which have to be satisfied together with the boundary conditions (11). The first several coefficients $\alpha_k$ are:

$$a_1 = 0$$
$$a_2 = A\gamma_1 P_1$$
$$a_3 = A\gamma_1 P_2 + A\gamma_2 P_1 + (B + 4GQ^2)P_1^3 \quad (28)$$
$$+(4g_1 - 2ag_3 + 2g_3 z)P_1\int_0^h P_1^2 dz + (2g_3 - 4ag_2 + 4g_2 z)P_1\int_0^h P_1^2(z-a)dz$$

It is easy to check that under the boundary condition (11), the operator $L_c$ is self-adjoint, and

$$\int_{-h/2}^{h/2} P_k L_c P^* dz = (P^*, L_c P_k) = (P^*, a_k) = \int_{-h/2}^{h/2} P^* a_k dz = 0 \quad (29)$$

where $P^*$ is the solution of the following homogeneous equation:

$$L_c P^* = 0;$$
$$\frac{dP^*}{dz} = \mp\frac{P^*}{\delta}; \quad z = \pm\frac{h}{2} \quad (30)$$

Equation (29) can be used to determine the coefficients $\gamma_i$. Then from the second relation of (26), one can determine $\varepsilon$ as a function of $(T-T_c)$. Substituting the resulting $\varepsilon$ into the first relation of (26) and solving the inhomogeneous equations (27) results in an explicit expression for the solution $P$.

The equation (30) is simplified with the changed variable $\eta=(\frac{D}{\lambda_1})^{2/3}\frac{1}{D}(\lambda_1 z+\lambda_{2c})$,

$$\frac{d^2P^*}{d\eta^2}-\eta P^*=-R\lambda_1^{-2/3}D^{-1/3} \tag{31}$$

The general solution of equation (31) is a linear combination of the two Airy functions $Ai(\eta)$, $Bi(\eta)$, therefore,

$$P^*(\eta)=C_1P_1(\eta)+C_2P_2(\eta) \tag{32}$$

where

$$P_1(\eta)=Ai(\eta)+R_0,\quad P_2(\eta)=Bi(\eta)+R_0, \tag{33}$$

and $R_0$ is a constant, give by

$$R_0=\frac{\pi}{\alpha h}(\frac{\lambda_1}{D})^{-1/3}\int_{\eta_1}^{\eta_2}Ai(\eta)d\eta\cdot[Ai(\eta)\int_{\eta_1}^{\eta_2}Bi(\xi)d\xi-Bi(\eta)\int_{\eta_1}^{\eta_2}Ai(\xi)d\xi] \tag{34}$$

where $\eta_1=(D/\lambda_1)^{2/3}\cdot\frac{1}{D}\lambda_{2c}$, $\eta_2=(D/\lambda_1)^{2/3}\cdot\frac{1}{D}(\lambda_{2c}+\lambda_1 h)$

Substitution of equation (32) into the boundary conditions (11) yields a relation between the constant $C_1$ and $C_2$ as follows:

$$C_2=\gamma C_1 \tag{35}$$

where

$$\gamma=[\frac{1}{\delta}P_2(\eta_1)-(\frac{\Lambda_1}{D})^{1/3}P_2'(\eta_1)]^{-1}[(\frac{\Lambda_1}{D})^{1/3}P_1'(\eta_1)-\frac{1}{\delta}P_1(\eta_1)] \tag{36}$$

where $P_1'(\eta_1)$, $P_2'(\eta_1)$ are derivatives with respect to $\eta$ at $\eta=\eta_1$, respectively.

By using the orthogonal condition (29), we obtain

$$\gamma_1=0 \tag{37}$$

$$\gamma_2=-\frac{1}{A}\frac{\int_{-h/2}^{h/2}P_1^4dz}{\int_{-h/2}^{h/2}P_1^2dz}(B+4GQ^2)-\frac{1}{A\int_{-h/2}^{h/2}P_1^2dz}[\int_0^h(4g_1-2ag_3+2g_3z)P_1^2dz\int_0^hP_1^2dz+\int_0^h(2g_3-4ag_2+4g_2z)P_1^2dz\int_0^hP_1^2(z-a)dz] \tag{38}$$

Substitution into equation (26b) and neglecting the higher order terms, we obtain $\varepsilon\simeq\pm\sqrt{\frac{T-T_c}{\gamma_2}}$. From equation (26a), for $P$ to be real and physical,

$\gamma_2$ and $(T-T_c)$ must have the same sign. Thus, in the supercritical regime ($T<T_c$), a continuous solution exists only if $\gamma_2<0$. Equivalently, if $\gamma_2>0$, a real and physical solution cannot be continuous in the supercritical regime.

In the case $\gamma_2=0$, one has to take higher order terms of the expansions, and thus the critical exponents will differ from their bulk counterpart. Indeed, the critical system behavior obtained from the foregoing analysis can also be understood in another way. When $\gamma_2<0$, a real and continuous solution only exists for $T<T_c$, and bifurcation analysis[8] shows that both branches of the solutions are asymptotically stable (theorem 3.1). The corresponding phase diagram is shown in Fig. 3a. When $\gamma_2>0$, real and continuous solutions only exist in the $T>T_c$ regime, but both branches of the solution found to be unstable, and are therefore physically unimportant. Nevertheless, it is reasonable to argue, from physical considerations, that away from the neighborhood of the critical point, there should be a maximum value of $T_{\max}$, above which only one solution (the trivial $P=0$ solution) is admissible. Therefore, the subcritical branches should turn in the direction of decreasing temperature at some value $T_{\max}$ (Fig. 3b), and join the discontinuous solution in the $T<T_c$ regime.

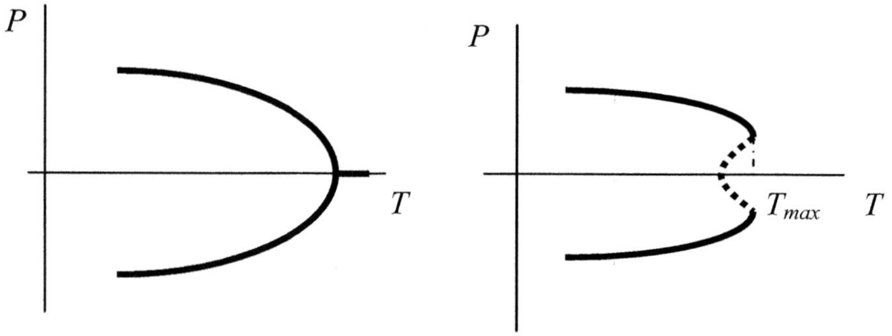


*Figure 3.* Schematic of bifurcation diagrams for ferroelectric thin films.

Thus, the critical system behavior of a thin film is determined by the sign of $\gamma_2$, and is therefore not necessarily the same as that of the bulk material, which is completely governed by the sign of $B$. The sign of $\gamma_2$ depends, in a complicated way, on the bulk material parameters $B,G,Q$, film thickness h, substrate thickness and the surface conditions, through equation (38). Some general observations can be made. Thus, if $B>0$, then $\gamma_2<0$, and hence the transition of a thin film is second order with a critical exponent of ½, if the bulk material is second-order. This is independent of the film thickness, substrate thickness and surface conditions considered here.

The interesting case is when $B<0$, $\gamma_2$ may then take either signs for different film and substrate thickness and surface conditions. Thus, $\gamma_2$ is negative, and the ferroelectric transition of the film is second-order and continuous, despite being first-order and discontinuous in the bulk. An example can be found in a well-known first-order ferroelectric material PbTiO3, the materials parameters of which are listed in Table 1. Plotted in Fig. 4 is $\gamma_2/C_1^2$ VS the ratio of the film thickness over the substrate thickness. One can see that if one can control the ratio $h/H < 0.156$, one can change the first-order transition into the second-order ones.

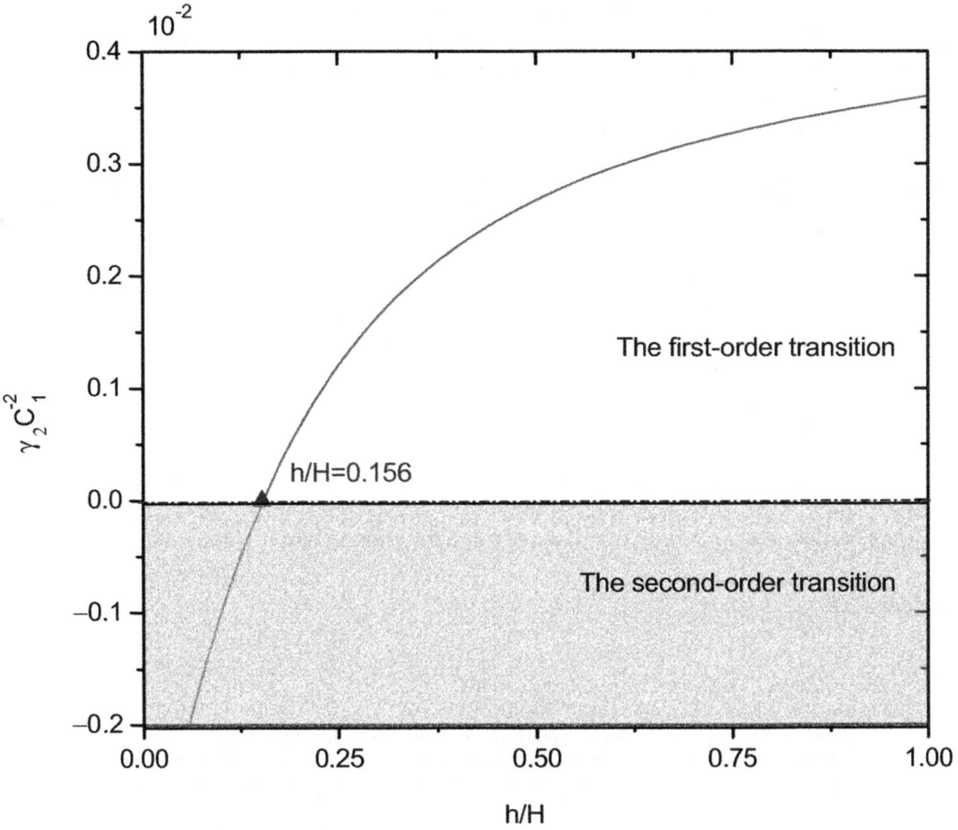


*Figure 4.* $\gamma_2/C_1{}^2$ VS the ratio of the film thickness over the substrate thickness of PbTiO3/SrTio3 system.

## 5. Concluding Remarks

Ferroelectric thin films generally exhibit characteristics that are different from their bulk counterparts, and their behaviors also depend on the surface conditions. By considering ferroelectric thin films on a compliant substrate, we found that the characteristics of the phase transition, such as, the Curie temperature, the order of the phase transition, etc., can be tuned simply by changing the relative thickness of the film and substrate. For PbTiO3 film/SrTiO3 substrate system, the Curie temperature of the film can be adjusted in the range from $0\,^{\circ}K$ to $1000\,^{\circ}K$, and the ferroelectric phase transition can also be tuned from the first-order as their bulk counterpart to the second-order.

Table 1. Materials' constant of the film and substrate (SI) [4,8-10]

| $T_{c0}$ $(K)$ | $A(10^5)$ $JmC^{-2}K^{-1}$ | $D(10^{-9})$ $Jm^5C^{-2}$ | $\delta$ $nm$ | $\alpha/\alpha_0$ | $G$ $10^{11}Pa$ | $G_0$ $10^{11}Pa$ | $Q$ $SI$ | $\varepsilon_0$ |
|---|---|---|---|---|---|---|---|---|
| 763 | 7.6 | 2.7 | 0.5 | 210 | 1.82 | 3.5 | –0.026 | 0.012 |

## Acknowledgments

This project was supported by grants from the Research Grants Council of the Hong Kong Special Administrative Region (PolyU 5173/01E, 5309/03E, 5312/03E), and the National Science Foundation of China (50232030, 10172030).

## References

1. Wang, B. and Woo, C. H., Curie temperature and critical thickness of ferroelectric thin films, *Journal of Applied Physics, (submitted for publication), 2004.*
2. J. Junquera and Ph. Ghosez, *Nature,* **422**, *506 (2003).*
3. W. L. Zhong, Y. G. Wang, P. L. Zhang, and B. D. Qu, *Phys. Rev,* **B50**, *698 (1994).*
4. N. A. Pertsev, A. G. Zembilgotov, and A. K. Tagantsev, *Phys. Rev. Lett.* **80**,*1988 (1998).*
5. A. G. Zembilgotov, N. A. Pertsev, H. Kohlstedt, and R. Waser, *J. Appl. Phys.,* **91**, *2247 (2002).*
6. R. Kretschmer and K. Binder, *Phys. Rev.* **B20**, *1065 (1979).*
7. G. Nicolis and I. Prigogine, Self-organization in nonequilibrium systems, **85** *(1977) by John Wiley & Sons, Inc.*
8. B. S. Kwak, A. Erbil, J. D. Budai, M. F. Chisholm, and L. A. Boatner, *Phys. Rev.* **B49**, *14865, (1994).*
9. J. S. Speck and W. Pompe, *J. Appl. Phys.,* **76**, *466, (1994).*
10. S. Li, J. A. Eastman, J. M. Vetrone, C. H. Foster, R. E. Newnham, and L. E. Cross, *Jpn. J. Appl. Phys.,* **36**, *5169 (1997).*

# MULTI-SCALE CHARACTERIZATIONS FOR DUCTILE THIN FILM DELAMINATION

YUEGUANG WEI[1] and GUANSHUI XU[2]
[1]*LNM, Institute of Mechanics, Chinese Academy of Sciences, Beijing 100080, China*
[2]*Department of Mechanical Engineering, University of California at Riverside, CA 92521*

## 1. Introduction

The interfacial fracture behavior of a structured material, such as a thin film/substrate system, is governed by processes occurring over a diverse range of length scales. Most models of the fracture processes, however, usually focus on the process at a limited range of length scales. Continuum models based on elastic, elastic-plastic, and recently developed strain gradient constitutive relations are appropriate for the analysis of the macro-/microscopic mechanical response of solids, but they do not accurately represent the underlying fundamental deformation mechanism of crystal defects such as nucleation and motion of discrete dislocations and slip of grain boundaries. On the other hand, the discrete/continuum models, such as the discrete dislocation theory, may be capable of accounting for the fundamental dislocation mechanisms in the deformation and fracture process, but at the present stage of the development, this kind of model is limited to the consideration of nominally elastic behavior (i.e., relatively small or non-existent dislocation densities) at very small scale. The multiple-scale approaches toward materials modeling have led to a wealth of understanding of materials mechanical behavior within each domain of model's applicability. However, it is still of central importance to address the linkage between the models at the different length scales in order to develop mechanism based modeling of mechanical behavior of materials.

In this paper, we generalize our previous multiscale model (Wei and Xu, 2005) that is built upon the elastic-plastic and strain gradient models for plastic deformation at the macro-scale and the discrete dislocation for plastic deformation at the submicron scale



*Q. P. Sun and P. Tong (eds.), IUTAM Symposium on Size Effects on Material and Structural Behavior at Micron- and Nano-Scales,* 13–22.

to an analysis for the thin film delamination. In our model, we attempt to establish the linkage between these two models to address the large disparity between the relevant length scales involved in the ductile thin film delaminating processes at the interfacial crack tip.

For the macroscale analysis model for ductile thin film delamination, we shall use a two-parameter criterion to characterize the ductile fracture process as usually for elastic-plastic fracture analyses (Betegon and Hancock, 1991; O'Dowd and Shih, 1991; Xia et al., 1993; Tvergaard and Hutchinson, 1993; Wei and Wang, 1995a; 1995b). The two-parameter criterion adopted here will be a modified elastic-core model (or dislocation-free zone model, or called SSV model, refs. Suo et al., 1993 and Beltz et al., 1996). The ductile thin film will be treated with a strain gradient plasticity material (Fleck and Hutchinson, 1997; Gao et al., 1999), while the substrate is an elastic material. On the discrete dislocation model, there have been many researches related with the subject (Rice and Thomson, 1974; Lin and Thomson, 1986; Rice, 1992; Hsia et al., 1994; Xu and Argon, 1995; 1997; Wang, 1998; Mao and Li, 1999; Yang et al., 2001). We shall investigate dislocation shielding effect on the thin film delamination considering an elasticity mixed mode K-field including several discrete dislocations. The K-field and its radius will be related to the macroscopic crack-tip fracture toughness.

## 2. Model Descriptions

A linkage model for the thin film delamination is presented, as shown in figure 1. The entire description of thin film delamination should consist of both the macroscopic interface fracture process and the microscopic interface fracture process. These interface fracture characteristics can be described by using the continuum model and the discrete dislocation model, respectively, as sketched in figures 1(a), 1(b) and 1(c). Figure 1(a) and 1(b) are the thin film peeling macroscale model and delaminating macroscale model, respectively. Both macroscale models corresponding to the scale level which is larger than a micron or a sub-micron can be characterized using the conventional continuum model, i.e., the conventional elastic-plastic theory or the strain gradient theory. In this scale, with increasing loading there exist three regions around the interface crack tip: the elastic zone far away from the tip, the strain gradient dominated zone very near the crack tip and the plastic zone between the elastic zone and the strain gradient zone. In order to describe the multi-scale problem clearly, an elastic core model, or a modified SSV model is adopted here, as shown in figure 1(a) and 1(b). This model is an improvement on the conventional SSV model (Suo et al., 1993), which supposed that near the growing crack tip and the crack surface a dislocation-free zone strip exists with an infinite length. In the elastic core model or the modified SSV model, an elastic core

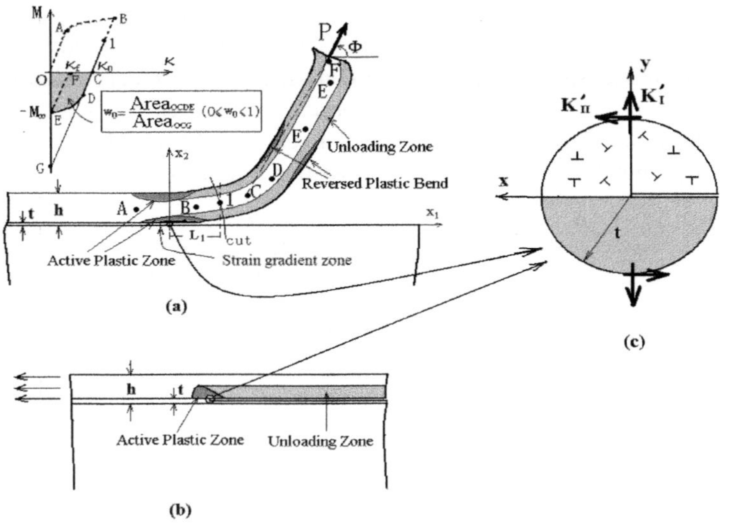


*Figure 1.* Multi-scale model for thin film delemination.

always surrounds crack tip during crack advance. Under steady-state crack growth condition, a semi-infinite elastic strip around the crack surface is left behind the crack tip. The radius of the elastic core (or thickness of the elastic layer), **t**, can be taken as a model parameter (Wei and Hutchinson, 1997, 1999). Within a much small scale which is smaller than a micron or sub-micron, the interface fracture behaviors will be predicted by using the discrete dislocation theory, as shown in Figure 1(c). With increasing load, dislocation nucleates and emits from the crack tip, and discrete dislocations exist (keep in equilibrium) within the region. The key problems are noted here: what is the discrete dislocation number in equilibrium or in limit equilibrium? What is the effective size of the discrete dislocation region, t? In addition, what are the macro-/microscale linkage conditions, i.e., the outer boundary conditions of the discrete dislocation model, or the inner boundary conditions of the continuum model? A treatment in the present study, the macroscopic analysis results will be taken as the outer boundary conditions for microscopic problem. The key point is to determine the intersection radius between microscopic problem and macroscopic problem. We shall study the problems in the present research.

## 3. Thin Film Delaminating (peeling) Analysis Using the Continuum Model

Specifically, we shall focus our attention on the thin film peeling analyses, and corresponding model has been given in figure 1(a). Using the modified SSV model and the mechanism-based strain gradient (MSG) plasticity flow theory (Gao et al., 1999; Huang et al., 2000), the relations of the normalized total energy release rate (or the normalized peel force) with the material parameters and the model parameter under steady-state delamination of thin film can be written through dimensional analysis,

$$\frac{P(1-\cos\Phi)}{G_0'} = f\left(\frac{E}{\sigma_Y}, \nu, N, \frac{h}{R_0}, \frac{l}{R_0}, \frac{R_0}{t}\right) \tag{1}$$

Where the length parameter $l$ describes the strain gradient effect, $t$ is the elastic core size, or intersection radius of microscopic with macroscopic fields to be determined; a length parameter $R_0$ is defined as follows

$$R_0 = \frac{EG_0'}{3\pi(1-\nu^2)\sigma_Y^2} \tag{2}$$

which is the plastic zone size in small scale yielding. $G_0'$ is the macroscopic fracture toughness at tip. $E$ is the Young's modulus and $\nu$ is Poisson's ratio. For simplicity, we consider that the Young's modulus and Poisson's ratio of substrate material are the same as those of thin film. Other parameters are defined in figure 1(a). Through finite element numerical simulation by using the MSG strain gradient theory, the detailed parameter relation (1) is given in figure 2 for several parameter values. The numerical process is similar to that by Wei et al. (2004). From figure 2, the variation of the normalized energy release rate is very sensitive to the elastic core size and length parameter $R_0$. For the typical metal materials, the value of $R_0$ is about one micron. Therefore, when elastic core size of the macroscopic model is taken as around submicron, the normalized energy release rate is much sensitive to the value of the elastic core size.

In order to determine the remote boundary condition for microscopic fracture analysis, the macro-scale crack tip solution is equivalently expressed into a standard interface K-field,

$$G_0' = \tfrac{1}{2}(1-\beta_D^2)\left(\frac{1-\nu^2}{E} + \frac{1-\nu_s^2}{E_s}\right)(K_I'^2 + K_{II}'^2) \tag{3}$$

$$\tan\Psi = \frac{\sigma_{12}}{\sigma_{22}} = \frac{\mathrm{Im}[(K_I' + iK_{II}')L_0^{i\varepsilon}]}{\mathrm{Re}[(K_I' + iK_{II}')L_0^{i\varepsilon}]} \tag{4}$$

where $\Psi$ is the mode mixity, $L_0$ is a reference length, and

$$\sigma_{22} + i\sigma_{12} = (K_I' + iK_{II}')(2\pi r)^{-1/2} r^{-\varepsilon} \tag{5}$$

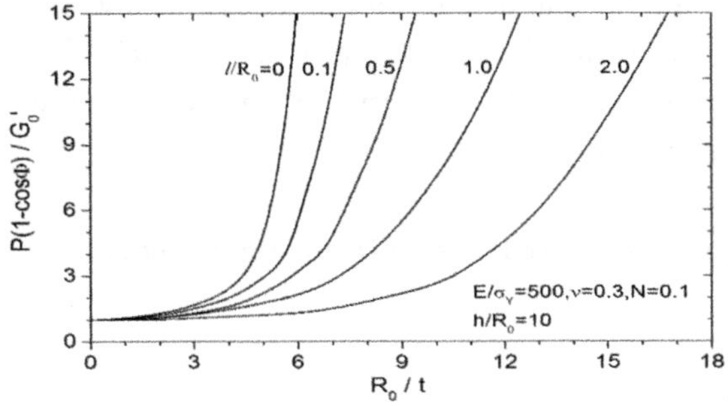


*Figure 2.* Macron scale model solution of the normalized energy release rate vs. thin film thickness.

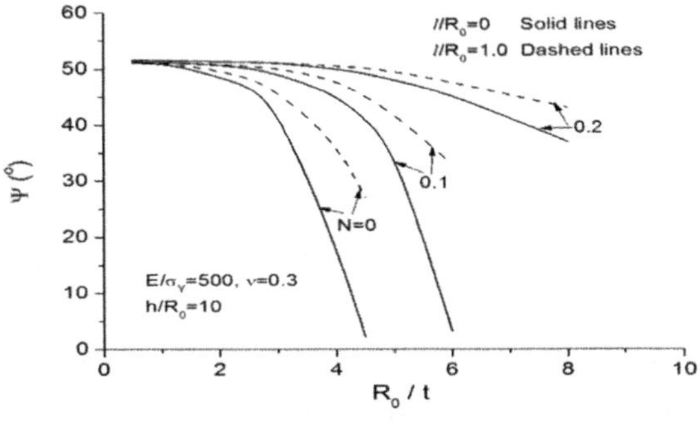


*Figure 3.* Variation of mode mixity with normalized thin film thickness.

$$\varepsilon = \tfrac{1}{2}\ln\left(\frac{1-\beta_D}{1+\beta_D}\right), \quad \beta_D = \tfrac{1}{2}\frac{\mu(1-2\nu_s) - \mu_s(1-2\nu)}{\mu(1-\nu_s) + \mu_s(1-\nu)} \tag{6}$$

For $E_s = E$ and $\nu_s = \nu$, then $\beta_D = 0$ and $\varepsilon = 0$, one has

$$G_0' = \frac{(1-\nu^2)(K_I'^2 + K_{II}'^2)}{E}, \quad \tan\Psi = \frac{K_{II}'}{K_I'} \tag{7}$$

The mode mixity variation with other parameters is calculated and the result is shown in figure 3. From figure 3, the mode mixity tends to mode I case with increasing the value of $R_0 / t$. Especially for conventional metals which are with weak hardening property. Therefore, we can approximately consider a Mode I crack field $(K_I', K_{II}' << K_I')$ exerted on the remote boundary of the micron scale model (the elastic interface fracture including discrete dislocations).

## 4. Fracture Analyses using the Discrete Dislocation Model

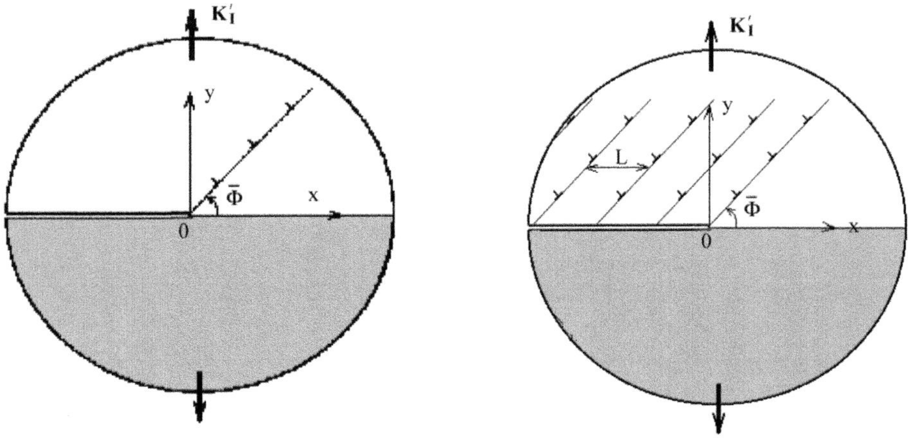


(a) Stationary crack (b) Steady-state growing crack

*Figure 4.* Dislocation slip models for interface crack.

As discussed above, the macroscopic fracture process is accompanied by a microscopic fracture process within the submicron zone near the crack tip. For the microscopic fracture analysis, the discrete dislocation theory is adopted here. The simplification model has been presented previously, as shown in figure 1(c), and as discussed above, we shall approximately consider a mode I field exerted on the remote boundary. The problem is characterized by figure 4 (a) and (b) for stationary and steady-state crack growing, respectively. The continuum solutions discussed above, a $K_I'$ field, can be obtained based on the strain gradient theory and formula (7). The key problem here is how to properly select the radius of the outer boundary, t. The t value will be determined through bridging the continuum result with the micro-scale analysis result after the discrete dislocation analysis. In order to analyze crack growth behavior in micro-scale influenced by the discrete dislocations, as usual, a typical kind of the discrete dislocation arrangements will be considered, as shown in figure 4 (a), or (b). The possibility of putting the greatest numbers of dislocations within the region $0 < r < t$, will be investigated (corresponding to the limit equilibrium state for each dislocation), where $r$ is polar coordinate. The arrangement of the discrete dislocations is according to the dislocation equilibrium status: $-1 \le f_d / f_d^c \le 1$, where $f_d$ is dislocation force, $f_d^c = \sigma_f b$ is referred to as the lattice frictional resistance, $\sigma_f$ and $b$ are the critical shear strength along the slip plane and Burgers vector, respectively. For $f_d = f_d^c$, dislocation is in the limit equilibrium state. The limit equilibrium state will be considered in the present research. It is worth noting that Lin and Thomson (1986) obtained the dislocation force formulations and the dislocation shielding effect formulations (solutions), so we can directly base on the Lin and Thomson's formulations to analyze the interaction of crack with discrete dislocations.

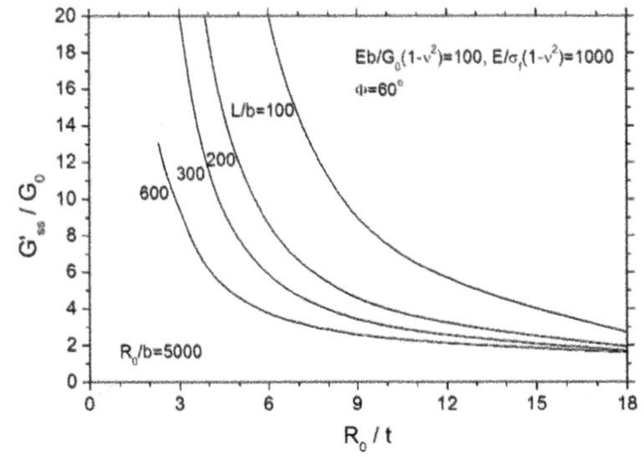


(a)

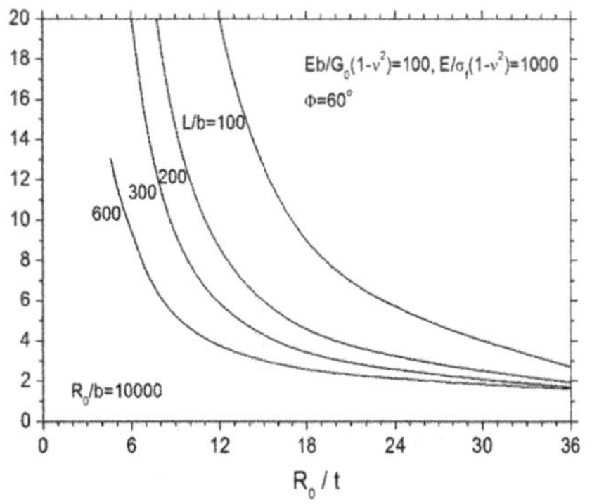


(b)

*Figure 5*. Ratio of the macron crack tip toughness to the micron crack tip toughness.

The macroscopic fracture solutions for crack steady-state growth have analyzed and obtained in section 3. Here we shall focus our attention on the microscopic fracture analysis. We have considered the possibility of existing discrete dislocations along a slip plane within the submicron scale in above sections. Let's further investigate the existing possibility of dislocation distributed pattern as crack grows under the steady-state condition, i.e., investigate whether the dislocation pattern can be kept around the crack surface, accompanied by the new slip plane produced near crack tip (see figure 4(b)).

Here suppose that the space between slip planes is equal to L. There exists a question: what is the size of the slip plane space L? According to the experimental observation and measurement by Mao and Evans (1997), L is about 1 micron, or submicron. By calculating the dislocation forces for each dislocation or the dislocation pattern, one can examine the stability of the dislocation pattern (figure 4(b)), i.e., check whether or not $| f_d / f_d^c | \leq 1$ is met as crack grows. After the condition has been confirmed, we obtain several results about toughness ratio. Figures 5(a) and (b) show the toughness ratio variations with the normalized elastic core radius, t, for several normalized material parameters $Eb / G_0(1-\nu^2)$, $R_0 / b$, $L / b$, $E / \sigma_f(1-\nu^2)$. The toughness ratio $G'_{ss} / G_0 = G'_0 / G_0$ describes the ratio of the macroscopic crack tip toughness to the micron crack tip toughness.

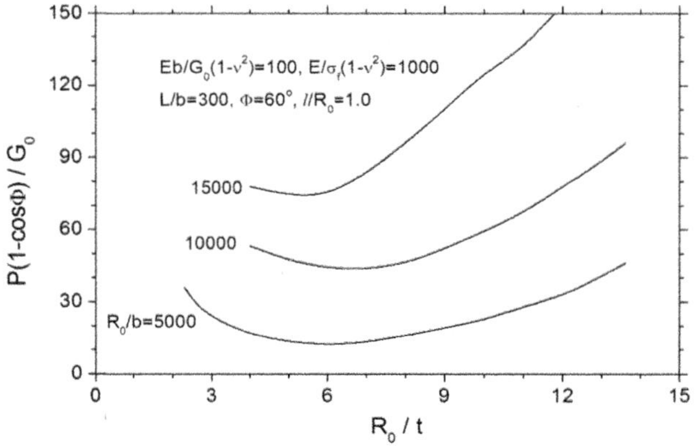


*Figure 6.* Total energy release rate normalized by the micron scale crack tip toughness vs. the normalized "elastic core" size. There exists the stationary value region.

## 5. The selection of "elastic core" size

We can directly calculate the energy release rate changing with the selection of the "elastic core" size t for the steady-state delamination. By increasing the discrete dislocation number along a slip plane, the corresponding "elastic core" size increases. From Figure 5, the toughness ratio from microscopic theory increases with increasing the "elastic core" size, however, it is just inverse feature from macroscopic theory (from figure 2). In order to conveniently compare both microscopic and macroscopic cases, the results shown in figure 5 and figure 2 will be used to plot the normalized total energy release rate:

$$\frac{P(1-\cos\Phi)}{G_0} = \frac{P(1-\cos\Phi)}{G'_{ss}}\frac{G'_{ss}}{G_0} = \frac{P(1-\cos\Phi)}{G'_0}\frac{G'_{ss}}{G_0} \tag{8}$$

where $G'_0 = G'_{ss}$ from present analysis. Figure 6 shows the normalized total energy release rate varying with the "elastic core" size for three parameter values of $R_0 / b$. Curves in figure 6 are come from both microscopic analysis results and macroscopic analysis results shown in figure 2 and figure 5 by using formula (8). From figure 6, around the value, $R_0 / t \approx 6$, the variation of the total energy release rate takes stationary value, or insensitive to the selection of t value, so the proper selection of the elastic core size is about $t \approx R_0 / 6$. For conventional metal materials, $R_0 \approx 1\mu$m, so the elastic core size is at about submicron scale.

## 6. Conclusion

A complete delaminating process for ductile thin film has been presented in the present research. The macroscopic delaminating process has been analyzed based on the strain gradient plasticity theory. The thin film delaminating characteristics, such as the interface fracture toughness, the plastically shielding effect etc. have been investigated. By using the discrete dislocation theory, microscopic fracture process has been analyzed, and a bridging model has been developed. In the bridging model, the macroscopic fracture solution (equivalent interface K field) has been taken as the outer boundary condition of the microscopic fracture problem. The shielding effect of discrete dislocations on the crack growth has been studied. The selection of the elastic core size t, the intersection radius between macron scale and micron scale, has been determined from a stationary value theorem that the total energy release rate should take the stationary value. According to present analysis, we have obtain elastic core size $t \approx R_0 / 6$, in submicron level.

**Acknowledgements**

The work is supported by National Science Foundations of China through Grants 10432050 and 10428207.

**References**

Beltz, G.E., Rice, J.R., Shih, C.F., Xia, L., 1996. A self-consistent model for cleavage in the presence of plastic flow. Acta Mater. 44, 3943-3954.

Betegon, C., Hancock, J.W., 1991. Two-parameter characterization of elastic-plastic crack-tip fields. J. Appl. Mech. 113, 104-110.

Fleck, N.A., Hutchinson, J.W., 1997. Strain Gradient Plasticity. Advances in Applied Mechanics, Hutchinson JW, Wu TY, editors. New York: Academic, 33, 295-361.

Gao, H., Huang, Y., Nix, W.D., Hutchinson, J.W., 1999. Mechanism-based strain gradient plasticity -I. Theory. J. Mech. Phys. Solids 47, 1239-1263.

Hsia, K.J., Suo, Z., Yang, W., 1994. Cleavage due to dislocation confinement in layered materials. J. Mech. Phys. Solids 42, 877-896.

Huang, Y., Gao, H., Nix, W.D., Hutchinson, J.W., 2000. Mechnism-based strain gradient plasticity-II. Analysis. J. Mech. Phys. Solids 48, 99-128.

Lin, I.H., Thomson, R., 1986. Cleavage, dislocation emission, and shielding for cracks under general loading. Acta Metall. 34, 187-206.

Mao, S.X., Evans, A.G., 1997. The influence of blunting on crack growth at oxide/metal interfaces. Acta Materialia 45, 4263-4270.

Mao, S.X., Li, M.Z., 1999. Effects of dislocation shielding on interface crack initiation and growth in metal/ceramic layered materials. J. Mech. Phys. Solids 47, 2351-2379.

O'Dowd, N.P., Shih, C.F., 1991. Family of crack tip fields characterized by a triaxiality parameter-I. Structure of fields. J. Mech. Phys. Solids 39, 989-1015.

Rice, J.R., 1992. Dislocation nucleation from a crack tip: an analysis based on the Peierls concept. J. Mech. Phys. Solids 40, 239-271.

Rice, J.R., Thomson, R., 1974. Ductile versus brittle behaviour of crystals. Phil. Mag. 29, 73-97.

Suo, Z., Shih, C.F., Varias, A.G., 1993. A theory for cleavage cracking in the presence of plastic flow. Acta Metall. Mater. 41, 151-1557.

Tvergaard, V., Hutchinson, J.W., 1993. The influence of plasticity on mixed mode interface toughness. J. Mech. Phys. Solids 41, 1119-1135.

Wang, T.C., 1998. Dislocation behaviours ahead of crack tip. Int. J. Solids Structures 35, 5033-5050.

Wei, Y., Wang, T.C., 1995. Fracture criterion based on the higher-order asymptotic fields. Int. J. Fracture 73, 39-50.

Wei, Y., Wang, T.C., 1995. Characterization of elastic-plastic fields near stationary crack tip and fracture criterion. Engin. Fract. Mech. 51, 547-553.

Wei, Y., Hutchinson, J.W., 1997. Nonlinear delamination mechanics for thin films. J. Mech. Phys. Solids 45, 1137-1159.

Wei, Y., Hutchinson, J.W., 1999. Models of interface separation accompanied by plastic dissipation at multiple scales. Int. J. Fracture 95, 1-17.

Wei, Y., Qiu, X., Hwang, K.C., 2004. Steady-state crack growth and fracture work based on the theory of mechanism-based strain gradient plasticity. Engineering Fracture Mechanics 71, 107-125.

Wei, Y., Xu, G., 2005. A multiscale model for the ductile fracture of crystalline materials. Int. J. Plasticity 21, 2123-2149.

Xia, L., Wang, T.C., Shih, C.F., 1993. Higher-order analysis of crack tip fields in elastic power-law hardening materials. J. Mech. Phys. Solids 41, 665-687.

Xu, G., Argon, A.S., Ortiz, M., 1995. Nucleation of dislocations from crack tips under mixed modes of loading: implications for brittle against ductile behaviour of crystals. Phil. Mag. A. 72, 415-451.

Xu, G., Argon, A.S., 1997. Critical configurations for dislocation nucleation from crack tips. Phil. Mag. A. 75, 341-367.

Yang, W., Tang, J.C., Ing, Y.S., Ma, C.C., 2001. Transient dislocation emission from a crack tip. J. Mech. Phys. Solids 49, 2431-2453.

# FRACTAL AND SCALING PHENOMENA ON FRACTURE AT MICRO-SCALES

CHUNSHENG LU[1,2], YIU-WING MAI[1], and YILONG BAI[2]
[1]*Centre for Advanced Materials Technology (CAMT), School of Aerospace, Mechanical and Mechatronic Engineering J07, The University of Sydney, NSW 2006, Australia*
[2]*State Key Laboratory of Nonlinear Mechanics (LNM), Institute of Mechanics, Academia Sinica, Beijing 100080, P. R. China*

## Abstract

A simple numerical model is proposed to investigate the fracture owing to the coalescence of numerous microcracks, in which the spatial nucleation of a microcrack is simulated through a random distribution, and the coalescence of two microcracks is determined using a critical condition in accordance with the load-sharing principle. As the number density of nucleated microcracks increases, stochastic coalescence occurs, and finally, a newly nucleated microcrack triggers a catastrophic failure. Simulation results show that fracture profiles exhibit fractal (self-affine) characteristics with a universal roughness exponent, but the critical damage threshold is sensitive to details of the model. The spatio-temporal distribution of nucleated microcracks in the vicinity of critical failure exhibits a power-law behaviour, which implies that the system is in a self -organized critical state.

## 1. Introduction

Most failure of inhomogeneous materials, such as rock, ceramics and composites, involves a variety of processes occurring on a range of length scales from nano-, micro-, to macro-scales (Curran *et al.*, 1987). Thus, the fracture process due to damage evolution (nucleation, growth and coalescence) initiated at micro- or nano-scales is usually one of typical complex phenomena. Generally speaking, the features of complex phenomena are adaptation, self-organization and emergence (Mandelbrot, 1982; Takayasu, 1990). Fracture, as an important nonlinear dynamic phenomenon, has attracted much attention from physicists in many decades (Fineberg and Marder, 1999). Recently, several remarkably universal features in the fracture process have been discovered. For example, the fracture surface of metals is observed to show fractal characteristics, and its topology is usually self-affine with a universal roughness exponent (Mandelbrot *et al.*, 1984; Bouchaud *et al.*, 1990). Acoustic emission studies of



*Q. P. Sun and P. Tong (eds.), IUTAM Symposium on Size Effects on Material and Structural Behavior at Micron- and Nano-Scales*, 23–30.

micro-fracture processes reveal power-law behaviour and self-organized criticality (Zapperi *et al.*, 1997; Sethna *et al.*, 2001). All these discoveries in the study of fracture at micro-scales can be viewed as a hallmark of complex systems. To understand the underlying physical mechanisms in such a complex process, one must understand not only the threshold conditions that trigger this process as we did in traditional fracture mechanics, but also the dynamics by which it proceeds. So the methods used in the study of complex systems are instructive to the investigation of multi-scale fracture. Some simple models related to critical phenomena, such as fractals or multi-fractals, percolation and renormalization group, have been applied to fracture studies (Herrmann and Roux, 1990; Barenblatt, 1993; Diao and Mai, 1999).

Although significant progress has been achieved, many aspects associated with fracture itself, especially cases where disorder is important and even governs the finally catastrophic fracture, are still unclear. In this paper, a two-dimensional (2D) model is proposed to describe fracture due to the coalescence of numerous microcracks. Here, we will focus attention on the statistical properties of collective evolution of numerous microcracks, and spatio-temporal activities in the vicinity of critical failure.

## 2. Numerical Simulation Model

To study the interplay between disorder and dynamical effects at a mesoscopic scale, we have developed a 2D evolution-induced catastrophe model from experimental findings (Bai *et al.*, 1994; Lu *et al.*, 1995). In the model, two main processes (microcrack nucleation and coalescence) are involved. The nucleated microcracks are simulated through a random distribution function which is often closely related to the distribution of microstructures, such as second-phase particles, inclusions, and flaws etc. The coalescence condition between two neighbouring microcracks can be approximately determined by mechanical analysis. As an example, we consider the simple dynamics of damage evolution, that is, a load-sharing principle where the load supported by a broken part is shared by its two neighbouring intact parts. The average stress on the segment with distance $d$ between two neighbouring microcracks is: $\bar{\sigma} = (1 + \bar{c}/d)\sigma_0$, where $\sigma_0$ is nominal stress and $\bar{c}$ is average length of two microcracks (Figure 1). Provided two neighbouring microcracks will coalesce when the average stress $\bar{\sigma}$ achieves a critical value (strength) $\sigma_c$, the critical coalescence condition can be written as:

$$r = \frac{d}{\bar{c}} \leq r_c = \frac{\sigma_0}{\sigma_c - \sigma_0} \tag{1}$$

Here, the probability of microcrack coalescence is independent of the angle $\theta$ between the orientation of two cracks and the line joining their tips (see Figure 1). The length of a coalesced crack is approximately represented by its projection normal to the loading direction. The stress level can be indirectly calculated from the crack density evolution *versus* time relationship.

More details on the model were discussed in Bai *et al.* (1994) and Lu *et al.* (1995, 1999). The dynamics used in the numerical simulations can be briefly summarized as follows:

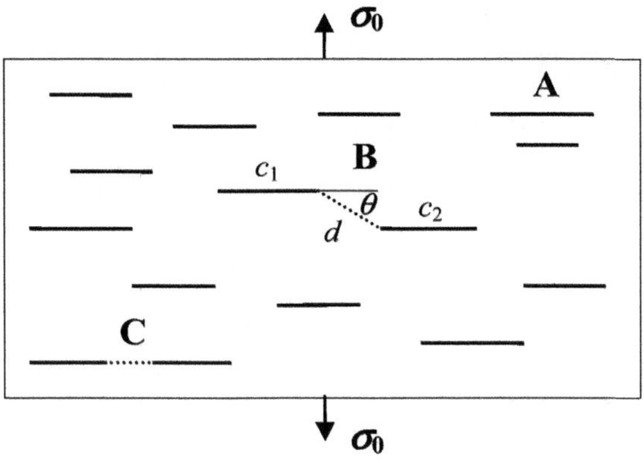


*Figure 1.* Illustration of numerous stacked microcracks horizontally and randomly nucleated in a two-dimensional space, in which two microcracks shield and attract each other in cases A and C, respectively.

(1) A newly nucleated microcrack is added randomly in the lattice, with horizontal orientation and length chosen from a given size distribution.

(2) Check coalescence condition for the newly nucleated crack with all pre-existing cracks, in turn, starting from the closest pair. If the left tip of the new crack lies below or above the right tip of an existing crack (or *vice versa*), it is supposed that the new crack is shielded by the old one, and coalescence will not occur.

(3) Repeat step (2) until there are no more cracks satisfying the condition and, at the same time, there is not a crack spanning the whole lattice. Then, return to step (1).

## 3. Simulation Results and Discussion

A typical evolution process in a 1000 × 1000 lattice is given in Figure 2, where the critical coalescence condition $r_c = 1$ and a uniform distribution of nucleated microcracks with a mean length equal to 2 were used in the simulation (Lu *et al.*, 1999). It is obvious that there are three different phases (stages) in the process. (i) Initial phase: new (small) events are too sparse to cause coalescence, and the number of microcracks increases linearly with time. (ii) Stable phase: a significant proportion of new events fall close enough to an existing crack-tip to cause coalescence. Such coalescence has two effects: (a) it increases the zone of influence around the crack-tip and so increases the chance of a coalescence event, perhaps a large one, and (b) it increases the area shielded by the coalescence. In this case, if a new event appears in the shielded region, it does not coalesce. These two effects balance each other out producing a near-stable regime. (iii) Critical phase: after some time, the zones of inference extend and overlap to such an extent that a newly nucleated microcrack (small fluctuation) will trigger a cascade coalescence of microcracks and result in a catastrophic failure (avalanche). We cannot predict, *a priori*, the position where the final fracture will appear (Figure 2).

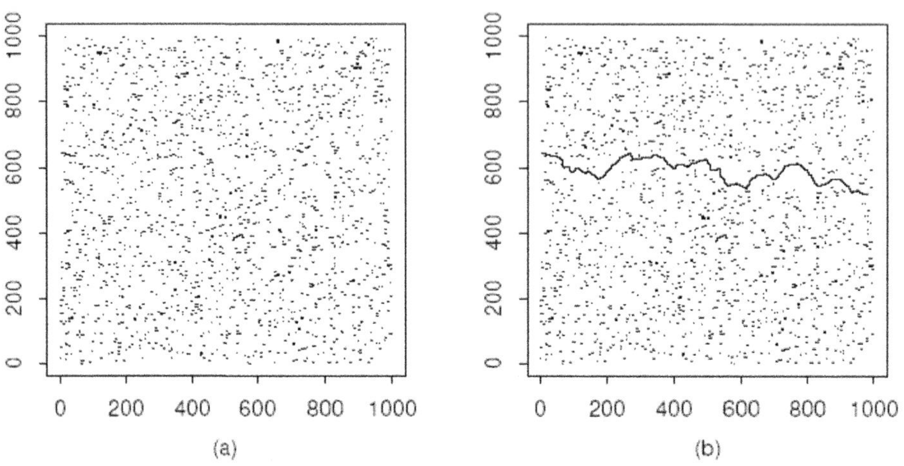


*Figure 2.* Patterns of microcrack coalescence in a 1000 × 1000 lattice: (a) just before fracture; and (b) a newly nucleated microcrack triggers catastrophic failure.

### 3.1. ROUGHNESS OF FRACTURE PROFILES

Fractography is a useful tool to understand the failure mechanisms in heterogeneous materials since the fracture surface, more or less, records some features or events that are created during a fracture process. In their pioneering work, Mandelbrot *et al.* (1984) showed that fracture surfaces of metals are fractal. Fractal dimension has been used as a measure of toughness; however, this is still a somewhat controversial topic. It was conjectured that the value of roughness exponent $\zeta \approx 0.8$ (usually between 0.7 and 0.9 in most experiments) is a universal exponent in the micro-scale range, whereas $\zeta \approx 0.5$ in nano-scale range (Bouchaud *et al.*, 1990; Bouchaud, 1997).

For a given fracture surface, the roughness index (Hurst exponent) $\zeta$ of a fracture profile is defined as

$$w = \langle \max h_i(x') - \min h_i(x') \rangle \sim L^{\zeta}, \tag{2}$$

where $h_i(x')$ is the height of the fracture profile in the window with length $L$, that is, $x' \in (x, x + L)$, and $\zeta$ lies between the values 0 and 1, and is related to the fractal dimension through the relation, $d_f = D - \zeta$, where $D$ is the Euclidean dimension. As shown in Figure 3, it is clearly seen that there is a power-law scaling, $w \sim L^{\zeta}$ with the exponent $\zeta \approx 0.8$, which is independent of the details used in the model, such as the distribution functions of nucleated microcracks (whether uniform, normal, Weibull, power-law, exponential functions with the same characteristic length), and the critical coalescence conditions (Figure 3). Hence, the fracture surface obtained from simulation has self-affine fractal characteristics. Several models have been suggested to describe this universality, but there is still no clear picture for why the roughness exponent should be about 0.8 in the micro-scale range.

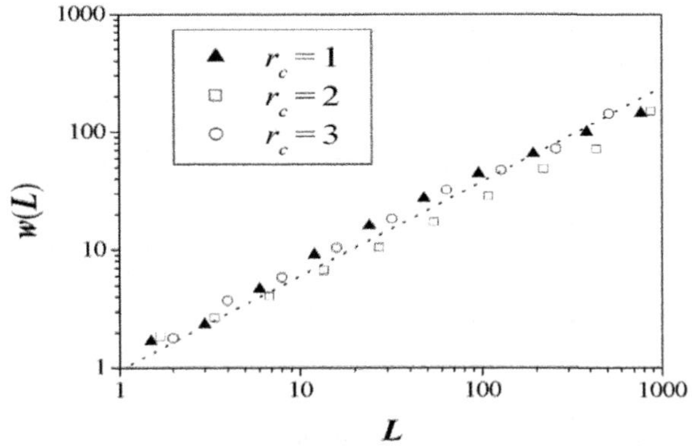


*Figure 3*. Log-log plot of fracture profile roughness $w(L)$ *versus* size $L$ for 3 cases of different critical coalescence conditions: $r_c = 1$, 2, and 3, where the slope of dotted line is 0.8.

### 3.2. CRITICAL DAMAGE THRESHOLD

A microcrack evolution process is intuitively very similar to percolation. The final catastrophic failure corresponds to the occurrence of infinitive cluster in percolation. Here, another interesting parameter is the accumulative damage at critical failure, which can be defined by the accumulated number, length, area, or volume of nucleated microcracks. For simplicity, we define this value as the accumulated number of nucleated microcracks. As shown in Figure 4, the threshold value is not a constant like that in percolation, but is sensitive to the details of the model. However, the smaller the coalescence condition $r_c$, the narrower the variance of the normal distribution becomes. Thus, percolation can be seen as an approximation of the fracture process in extreme disordered media (i.e., $r_c = 0$).

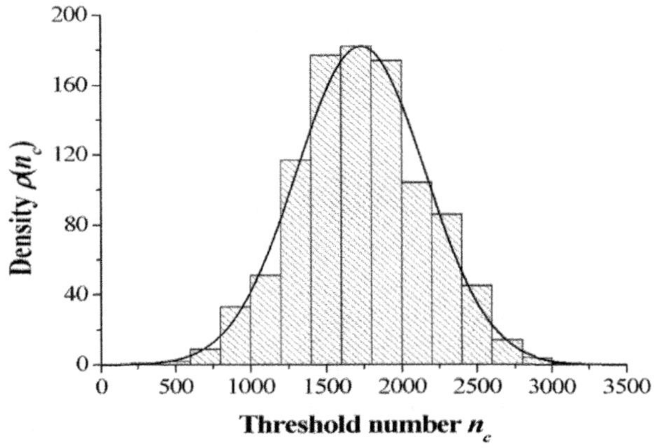


*Figure 4*. Density distribution of the threshold number of nucleated microcracks in 1000 simulations ($r_c = 1$).

### 3.3. SPATIO-TEMPORAL SCALING AND PREDICTABILITY

Next we will focus on the description of the final stage of damage evolution, since our main motivation is to examine if the fracture is a critical phenomenon. Conversely, the predictability of final catastrophic failure is closely dependent on the sensitivity of the system to disorder or stochastic fluctuation (Xia *et al.*, 1997). Here, we just randomly change the spatial position of the nucleated microcrack which triggers the catastrophic failure, and study their spatio-temporal distribution. That is, we try to find the waiting time to the next critical point, and the new spatial position of a microcrack which triggers the final avalanche. We suppose the waiting time is proportional to the number of steps (or nucleated microcracks). As we usually see and expect, the cumulative distribution of waiting time follows a modified power-law distribution:

$$N(>\Delta t) \sim (\Delta t)^{-\gamma} \exp(-\varepsilon \Delta t) \tag{3}$$

where $\varepsilon$ is small and the exponential term is important for large time intervals, as shown in Figure 5.

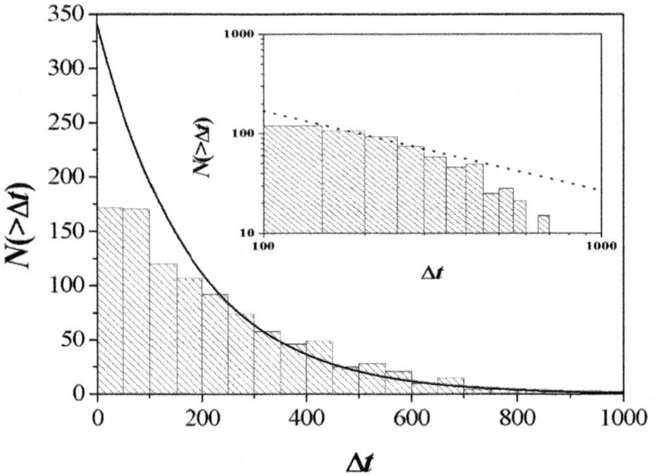


*Figure 5.* The cumulative distribution of number of nucleated microcracks $\Delta n$ (the difference from original critical number) in 1000 simulations with $r_c = 1$ (here, suppose $\Delta t \sim \Delta n$). As $\Delta t$ increases, the distribution decays by a power law (see inset, where the slope of the dotted line is 0.8) for small $\Delta t$, but follows an exponential law (solid line) for large $\Delta t$.

In the vicinity of critical failure, we have $N(>\Delta t) \sim (\Delta t)^{-\gamma}$ with $\gamma \approx 0.8$ (see inset in Figure 5). At the same time, it can be seen that the spatial distribution of next critical microcracks is not uniform (Figure 6). The clustering distribution follows the power-law, $N(r) \sim r^{-\alpha}$ with $\alpha \approx 1$, where $r$ is the coarse-grained size and $N(r)$ is the number in which there is at least one critical microcrack. Here, the spatio-temporal power-law scaling strongly implies that the microcrack evolution system is indeed at the self-organized critical state (Bak, 1996; Lu *et al.*, 1999).

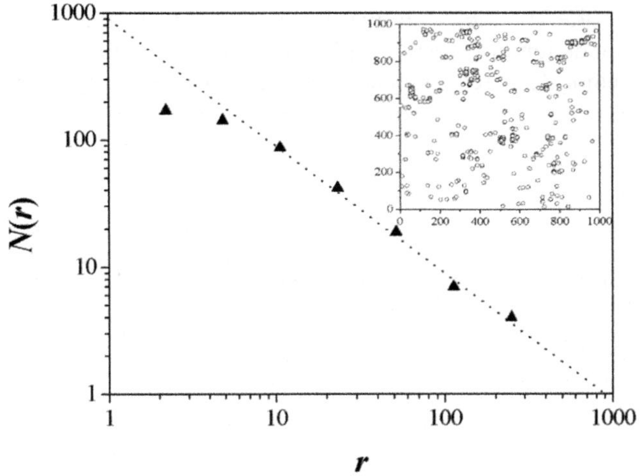


*Figure 6*. The spatial distribution of critical nucleated microcracks *versus* coarse-grained length $r$ in 1000 simulations with $r_c = 1$. The inset shows the spatial pattern of nucleated microcracks that trigger the catastrophic failure, where the slope of dotted line is –1.

As is well known, a self-organized critical system is exactly at the verge of chaos, where the uncertainty grows algebraically rather than exponentially. We gradually lose our ability to prediction, but much more slowly than for chaotic systems (Bak, 1996). This seems to shed some light on the predictability of fracture in disordered media. In fact, some precursors could be found from the analysis of acoustic emission signals in the vicinity of critical failure. It should also be noted that these scaling behaviours imply that material properties could be insensitive to the details at micro-scales. Recent study indeed has found that some materials become insensitive to flaws at the micro- or nano-scales (Lu *et al.*, 2002; Gao *et al.*, 2003).

## 4. Conclusion

In this paper, a simple two-dimensional simulation model is proposed to describe the fracture process due to collective evolution of numerous microcracks. Simulation results show that fracture profiles exhibit fractal (self-affine) characteristics, and the roughness exponents of fracture profiles seem to be a universal value, which are independent of the details used in the model, such as random (size and space) distributions of nucleated microcracks, coalescence conditions, etc. However, in contrast to classical percolation theory, the critical damage threshold is sensitive to the model details, and the spatial position triggering final fracture cannot be predicted *a priori.* The spatio-temporal distribution of nucleated microcracks in the vicinity of critical failure follows power-law scaling, which implies that the system may lie in a self-organized critical state and also sheds light on the predictability of catastrophic fracture.

## Acknowledgements

This work was supported by the Australian Research Council (ARC) and the National Natural Science Foundation of China. YWM and CL are, respectively, ARC Federation Fellow and ARC Senior Research Associate tenable at the University of Sydney.

## References

Bai, Y., Lu, C., Ke, F., and Xia, M. (1994) Evolution induced catastrophe, *Phys. Lett.* **A185**, 196-200.

Bak, P. (1996) *How Nature Works: The Science of Self-Organized Criticality*, Springer, New York.

Barenblatt, G.I. (1993) Micromechanics of fracture, in S.R. Bodner, J. Singer, A. Solan, and Z. Hashin (eds.), *Theoretical and Applied Mechanics*, North-Holland, Amsterdam, pp. 25-52.

Bouchaud, E., Lapasset, G., and Planès, J. (1990) Fractal dimension of fractured surfaces: a universal value? *Europhys. Lett.* **13**, 73-79.

Bouchaud, E. (1997) Scaling properties of cracks, *J. Phys.: Condens. Matter* **9**, 4319-4344.

Curran, D.R., Seaman, L., and Shockey, D.A. (1987) Dynamic failure of solids, *Phys. Rep.* **147**, 253-388.

Diao, X. and Mai, Y.-W. (1999) Self-similarity of crack propagation in inhomogeneous materials, *Phil. Mag. Lett.* **79**, 187-193.

Fineberg, J. and Marder, M. (1999) Instability in dynamic fracture, *Phys. Rep.* **313**, 1-108.

Gao, H., Ji, B., Jäger, I.J., Arzt, E., and Fratzl, P. (2003) Materials become insensitive to flaws at nanoscale: Lessons from nature, *Proc. Natl. Acad. Sci. USA* **100**, 5597-5600.

Herrmann, H.J. and Roux, S. (eds.) (1990) *Statistical Models for the Fracture of Disordered Media*, North-Holland, Amsterdam.

Lu, C., Danzer, R., and Fischer, F.D. (2002) Fracture statistics of brittle materials: Weibull or normal distribution, *Phys. Rev.* **E65**, 067102.

Lu, C., Ke, F., Bai, Y., and Xia, M. (1995) Numerical simulation of evolution induced catastrophe, *Sci. China* **A38**, 462-471.

Lu, C., Vere-Jones, D., and Takayasu, H. (1999) Avalanche behaviour and statistical properties in a microcrack coalescence process, *Phys. Rev. Lett.* **82**, 347-350.

Mandelbrot, B.B. (1982) *The Fractal Geometry of Nature*, Freeman, San Francisco.

Mandelbrot, B.B., Passoja, D.E., and Paullay, A.J. (1984) Fractal character of fracture surfaces of metals, *Nature* **308**, 721-722.

Sethna, J.P., Dahmen, K.A., and Myers, C.R. (2001) Crackling noise, *Nature* **410**, 242-250.

Takayasu, H. (1990) *Fractals in the Physical Sciences*, Manchester University Press, Manchester.

Xia, M., Ke, F., Bai, J., and Bai, Y. (1997) Threshold diversity and trans-scales sensitivity in a finite nonlinear evolution model of materials failure, *Phys. Lett.* **A236**, 60-64.

Zapperi, S., Vespignani, A., and Stanley, S. (1997) Plasticity and avalanche behaviour in microfacturing phenomena, *Nature* **388**, 658-660.

# NUMERICAL SIMULATION OF SIZE EFFECTS OF MICROCRACKED ELASTIC SOLIDS

X.Q. FENG
*Department of Engineering Mechanics, Tsinghua University, Beijing 100084, China*

**Abstract**

A numerical scheme of micromechanics is presented by combining the effective stress field concept for estimating effective elastic moduli and Kachanov's method for calculating strong interaction of microcracks. Some applications of this method are illustrated, including: (1) calculation of the effective elastic moduli of microcracked solids, (2) determination of the strengths of brittle specimens under uniaxial tension or bending, and (3) simulation of the failure process associated with interaction and propagation of distributed microcracks. The size effects of these mechanical properties induced by stochastic microcrack distributions are examined.

## 1. Introduction

Microcracking is one of the most typical damage mechanisms in engineering materials (e.g., concrete, ceramics, polycrystalline rocks, some metals and composites) and biomaterials (e.g., bones, teeth and shells). The dependences of material properties on interacting microcracks can be divided into two classes, weak (or indirect) and strong (or direct). The micromechanics approaches dealing with the two types of interaction problems are different. Such properties as the effective elastic moduli and thermal conductivities depend basically upon the statistical distributions of orientations, sizes and positions of microcracks as well as the statistically averaged effects of interaction, but are insensitive to the concrete orientations, sizes and locations of individual microcracks. Self-consistent method (SCM), differential method (DM) and many other methods based on the concept of effective (equivalent) medium or effective stress field have been established for estimating the elastic moduli of microcracked solids [1–4]. However, the accuracy and the validation scopes of these established methods are yet to be evaluated further. There is a lack of experimental data available in the literature for effective moduli of microcracked solids, especially when the microcrack density is high. Therefore, it seems promising to calculate the effective moduli for complex microcrack distributions and to evaluate the accuracy of these micromechanics schemes via numerical approaches.

As far as strong interaction of microcracks is concerned, some approximate micromechanical discrete schemes (e.g., pseudo-traction method, complex potential



*Q. P. Sun and P. Tong (eds.), IUTAM Symposium on Size Effects on Material and Structural Behavior at Micron- and Nano-Scales,* 31–40.

method and double potential method) and finite element methods have been developed. Interaction of multiple cracks is often reduced to a set of integral equations, which need to be solved by the series expansion method, perturbation method, collocation method, and some other approximate methods. However, calculation of interaction among microcracks of a large number is time-consuming even for numerical schemes since the number of equations increases very rapidly with the increase in the microcrack number.

In addition, the failure process of brittle or quasi-brittle materials is generally preceded by coalescence of interacting microcracks to form a fatal crack that will propagate unstably. Simulation of material failure characterized by microcrack growth and linkage is a problem of considerable interest both in theoretical analysis and in engineering applications. Due to the inherent prohibitive complexity and difficulties in calculation, nevertheless, little attention has been paid on the stochastic coalescence process of distributed microcracks. In the present paper, a numerical method recently developed by Feng *et al.* [5] is used to calculate the effective elastic moduli, strength and failure process of microcracked solids.

## 2. Effective Stress-Subregion Method

Consider a plate of area $A$ containing many randomly distributed, planar microcracks and subjected to a uniform stress $\boldsymbol{\sigma}_\infty$ in the far field, as shown in Fig. 1(a). Let us consider a microcrack in it, say the $\alpha$th one, whose length is denoted as $2l_\alpha$. Refer to a global Cartesian coordinate system $(o-x_1x_2)$ and a local one $(o-x_1'x_2')$, as shown in Fig. 1(c), where the $x_2'$-axis is aligned with the normal $\mathbf{n}^\alpha$ of this microcrack. The microcrack orientation is then expressible in terms of an angle, $\theta_\alpha$, measured from $x_2$ to $\mathbf{n}^\alpha$. Assume that the statistical distribution of orientations and sizes of microcracks satisfies a probability density function, $\Phi(l,\theta)$.

To calculate the SIFs of the $\alpha$th crack which interacts with all the others, some simplifications are made to yield an efficient numerical method. On one hand, the local stress field around a microcrack is sensitive to the positions, orientations and sizes of its neighboring cracks. Therefore, we define a subregion $\Omega$ around the $\alpha$th microcrack and remove all the microcracks outside $\Omega$, as shown in Fig. 1(a) and (b). The characteristic size of $\Omega$ is much larger than that of microcracks (e.g., 10–20 times the average length of microcracks), while the shape of $\Omega$ may be specified according to the microcrack orientation distribution. A circular shape is generally appropriate for isotropic or weakly anisotropic distributions.

In spite of that each microcrack outside $\Omega$ has weak interaction with the $\alpha$th one, the total contribution of all the outside microcracks to the SIFs of $\alpha$ may be important because of their large number. If all the microcracks both inside and outside the subregion were considered, evidently, the average stress $\boldsymbol{\sigma}_\Omega$ over $\Omega$ should equal approximately to the far-field stress $\boldsymbol{\sigma}_\infty$. Since all the microcracks outside $\Omega$ have been "removed" in the approximate model (Fig. 1b), $\Omega$ exists in the plate as a weaker "inclusion" with stiffness lower than the pristine matrix, and thus the average stress $\boldsymbol{\sigma}_\Omega$

over $\Omega$ is different from $\boldsymbol{\sigma}_\infty$. Therefore, we replace the far-field stress $\boldsymbol{\sigma}_\infty$ by $\boldsymbol{\sigma}_0$ in our calculation model such that $\boldsymbol{\sigma}_\Omega = \boldsymbol{\sigma}_\infty$. According to Eshelby's inclusion theory, the modified far-field stress $\boldsymbol{\sigma}_0$ should equal to

$$\boldsymbol{\sigma}_0 = \mathbf{B}^{-1} : \boldsymbol{\sigma}_\infty = \left[\mathbf{I} + \mathbf{P} : (\mathbf{M} - \mathbf{M}_0)\right] : \boldsymbol{\sigma}_\infty, \tag{1}$$

where the fourth-order tensor $\mathbf{B} = [\mathbf{I} + \mathbf{P} : (\mathbf{M} - \mathbf{M}_0)]^{-1}$ is the average stress-concentration tensor, $\mathbf{P} = \mathbf{M}_0^{-1} : (\mathbf{I} - \mathbf{S})$, $\mathbf{I}$ is the fourth-order identity tensor, $\mathbf{S}$ is the Eshelby tensor, $\mathbf{M}_0$ and $\mathbf{M}$ denote the elastic compliance tensors of the matrix and the subregion $\Omega$, respectively. To modified far-field stress $\boldsymbol{\sigma}_0$ in Eq. (1), one needs to estimate first the tensor $\mathbf{M}$ by using a micromechanics method [1].

In summary, the central idea of the present method is that all the microcracks in the specimen are divided into two sets, which are treated in different manners in calculation of their contributions to the SIFs of the $\alpha$th crack. The interacting microcracks inside $\Omega$ are calculated directly from a direct micromechanics method (e.g., Kachanov's method [3, 6]) considering their concrete sizes, locations and orientations, while the influence of those cracks outside is reflected merely by modifying the far-field stress. For more discussion of this method, the reader is referred to Ref. [5].

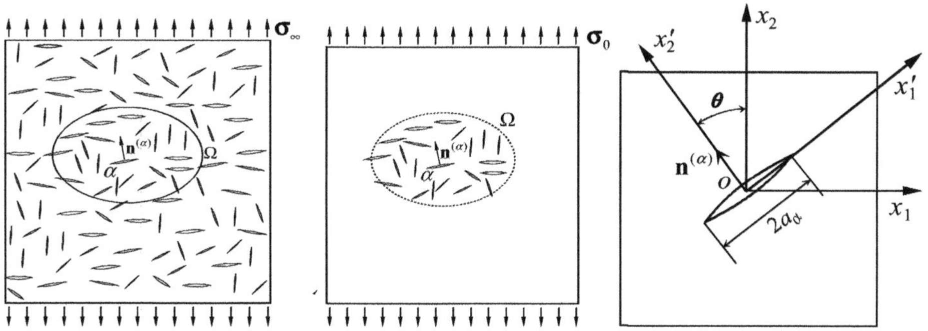


(a) (b) (c)

*Figure 1.* (a) A microcracked specimen, (b) the simplified calculation model, and (c) coordinate systems.

## 3. Effective Elastic Moduli

### 3.1. CALCULATION METHOD

The first application of the above presented model is to calculate directly the Young's moduli of plane specimens containing many microcracks with arbitrary distributions of sizes, positions and orientations. A uniform tensile stress $\sigma_{ij}^\infty$ is applied on its boundary. For a concise discussion, we mainly consider open microcracks in this paper though closed cracks can also be considered by this method [5]. The average strain tensor $\varepsilon_{ij}$ is decomposed additively as

$$\varepsilon_{ij} = \varepsilon^0_{ij} + \varepsilon^c_{ij}, \tag{2}$$

where $\varepsilon^0_{ij}$ and $\varepsilon^c_{ij}$ denote the average strain tensor of the elastic matrix and its increase due to the presence of microcracks, respectively. Assume that the matrix is isotropic. $\varepsilon^0_{ij}$ and $\varepsilon^c_{ij}$ can be obtained respectively by [1, 2]

$$\varepsilon^0_{ij} = M_{ijkl}\sigma^\infty_{kl} = \frac{1}{2E_0}\left[(1+\nu_0)(\delta_{ik}\delta_{jl} + \delta_{il}\delta_{jk}) - 2\nu_0\delta_{ij}\delta_{kl}\right]\sigma^\infty_{kl}, \tag{3}$$

$$\varepsilon^c_{ij} = \frac{1}{A}\sum_{\alpha=1}^{N} l_\alpha\left[\bar{b}_i^{(\alpha)} n_j^\alpha + \bar{b}_j^{(\alpha)} n_i^\alpha\right], \tag{4}$$

where $\bar{b}_i^{(\alpha)}$ are the average opening displacements of the $\alpha$th crack. It is calculated by

$$\bar{b}_1^{(\alpha)} = \frac{\pi l_\alpha}{E}\left[\cos\theta_\alpha\left\langle\tau^\alpha\right\rangle - \sin\theta_\alpha\left\langle p^\alpha\right\rangle\right],$$
$$\bar{b}_2^{(\alpha)} = \frac{\pi l_\alpha}{E}\left[\sin\theta_\alpha\left\langle\tau^\alpha\right\rangle + \cos\theta_\alpha\left\langle p^\alpha\right\rangle\right], \tag{5}$$

where $p^\alpha$ and $\tau^\alpha$ are the pseudo-tractions on the $\alpha$th crack, determined from the method in Section 2. The angle brackets denote an average over the crack surfaces.

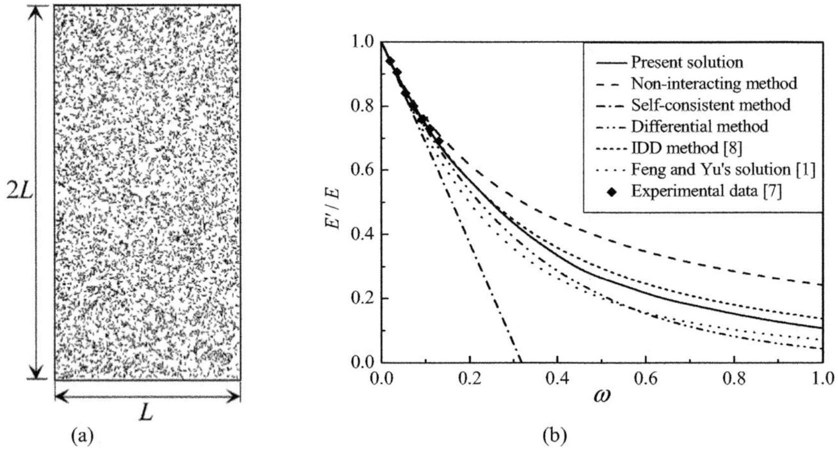


*Figure 2.* (a) Specimen containing many microcracks, and (b) effective Young's modulus.

## 3.2. EFFECTIVE YOUNG'S MODULUS

For instance, we calculate the effective Young's modulus of a rectangular specimen with length $2L$ and width $L$ and containing many microcracks (e.g., 6000), as shown in Fig. 2(a). The half-lengths of the microcracks satisfy a normal distribution law $\Phi(l)$ with the mathematical expectation being 5 mm and the variance being 1 mm. The material parameters of the matrix are taken as $E = 0.35\times10^5$ MPa and $\nu = 0.17$. In the case of completely random orientations, the effective Young's modulus is shown in Fig. 2(b) in comparison with the experimental data in [7], the numerical results in [3], as well as the analytical results obtained from the non-interacting approximation, SCM, DM, IDD [8] and Feng and Yu's solution in [1]. Because each specimen has a very large number of microcracks, no evident dispersion in the calculation results has been observed for different specimens. As expected, the effective Young's modulus exhibits almost no size effect provided that the specimen is large enough to contain a sufficient number of microcracks. In addition, it is seen that the numerical estimates agree well with the experimental data [7], and the results of DM, IDD [8], and Feng and Yu's method [1]. Therefore, the present method provides a useful tool to evaluate various micromechanics methods for the effective elastic moduli of microcracked solids as well as to examine the influences of various factors of microcrack distributions.

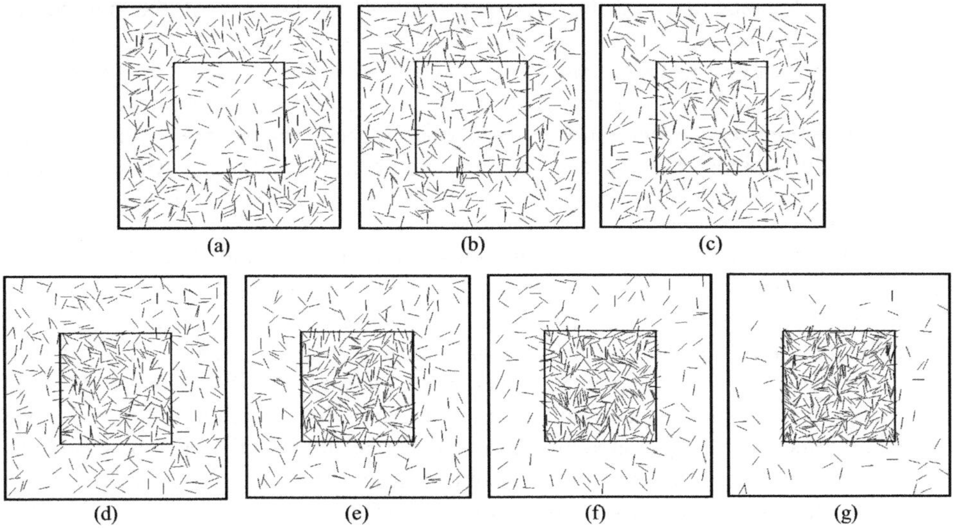


*Figure 3.* Seven different spatial distributions of randomly oriented microcracks.

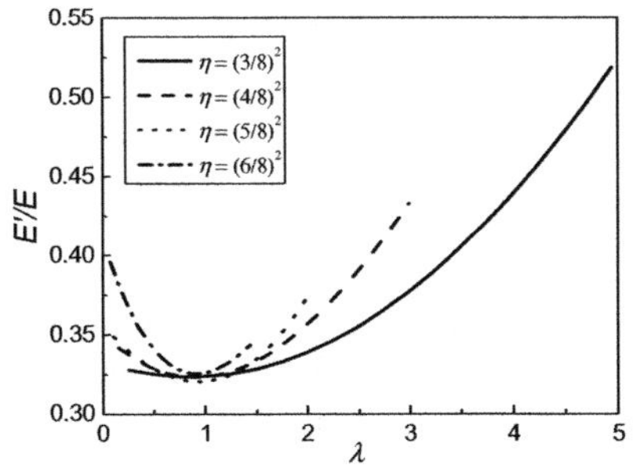


*Figure 4.* Effect of nonuniform spatial crack distribution on Young's modulus.

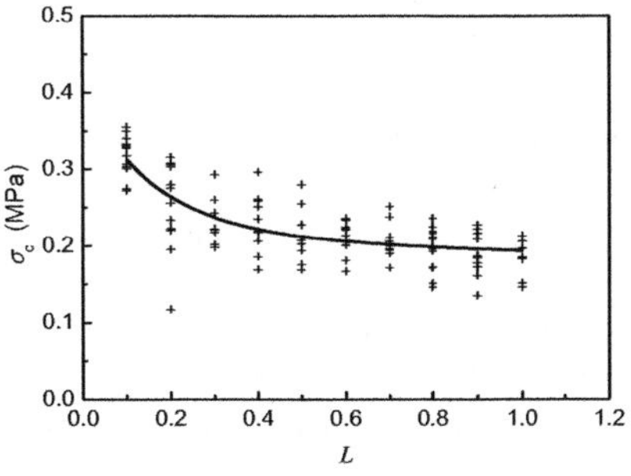


*Figure 5.* Size effect of the tensile strength of microcracked solids under tension.

### 3.3. AGGLOMOERATION EFFECT

In actual materials, the position distribution and damage evolution of microcracks are usually non-uniform in space. The conventional estimation methods of effective elastic moduli are applicable only for statistically homogeneous materials. Here we examine how non-uniform distributions of microcrack locations influence the elastic moduli.

Assume that microcracks of identical size are distributed non-uniformly in a solid. The average density $\omega$ of microcracks over the specimen is specified, but the microcrack density $\omega_c$ in some local regions, denoted as $\Omega_c$, is different from $\omega$. Introduce two parameters $\eta$ and $\lambda$ to describe the non-uniform distribution of microcracks, which are defined as

$$\eta = A_c / A, \quad \lambda = \omega_c / \omega, \tag{6}$$

where $A$ is the total area of the specimen, and $A_c$ is the area of the microcrack concentration zones $\Omega_c$. The parameter $\eta$ stands for the relative fraction of $\Omega_c$ with respect to the whole specimen, and $\lambda$ the concentration magnitude of microcracks in $\Omega_c$. $\lambda=1$ corresponds to a uniform distribution. $\lambda>1$ and $\lambda<1$ represent two different modes of microcrack concentrations: the former means that there are some weaker regions or "inclusions" that have a higher microcrack concentration, while the latter indicates that the material has some stronger regions or "inclusions" with a lower microcrack density.

Figure 3 shows seven different distributions of randomly oriented microcracks. Each specimen has 400 microcracks, but the crack numbers in the central zones $\Omega_c$ in Fig. 3(a–g) are 50, 100, 150, 200, 250, 300 and 350, respectively. The calculated effective Young's moduli as a function of the concentration parameter $\lambda$ are given in Fig. 4 for four values of $\eta$, with the average microcrack density $\omega = 0.4$. Interestingly and reasonably, it is found that the more uniform the spatial distribution of microcracks, the lower the effective Young's modulus. In other words, the effective Young's modulus has its minimum when the distribution of microcracks is uniform, i.e., at $\lambda = \omega_c / \omega = 1$. Since microcracks are usually non-uniformly distributed in most actual solids, the assumption of uniform distribution leads to an underestimation of elastic moduli. We have also examined by the same method the influences of some other microcrack distribution parameters on the macroscopic elastic moduli.

## 4. Tensile Strength

### 4.1. CALCULATION METHOD

The elastic energy release rate concept developed initially by Griffith is adopted here as the controlling indicator of crack growth in brittle solids. Accordingly, the mixed-mode fracture criterion is written in a simplified form

$$G = (K_I / K_{IC})^2 + (K_{II} / K_{IIC})^2 = 1, \tag{7}$$

where $K_{\rm I}$ and $K_{\rm II}$ denote the mode-I and II SIFs, $K_{\rm IC}$ and $K_{\rm IIC}$ their intrinsic critical values, respectively. For simplicity, $K_{\rm IC}$ and $K_{\rm IIC}$ are regarded here as material constants without dependence upon crack propagation. To determine the load-bearing capacity of a specimen, the far-field stress is denoted as $\boldsymbol{\sigma}_\infty = m\boldsymbol{\sigma}_\infty^0$, where $\boldsymbol{\sigma}_\infty^0$ is a tri-axial reference stress tensor, and $m$ is a non-dimensional load factor. Thus, the $\alpha$th crack will propagate at one of its two tips when the factor $m$ reaches the following value

$$m_\alpha = \min\left\{\left[\left(\frac{K_{\rm I0}^\alpha(l_\alpha)}{K_{\rm IC}}\right)^2 + \left(\frac{K_{\rm II0}^\alpha(l_\alpha)}{K_{\rm IIC}}\right)^2\right]^{-1/2}, \left[\left(\frac{K_{\rm I0}^\alpha(-l_\alpha)}{K_{\rm IC}}\right)^2 + \left(\frac{K_{\rm II0}^\alpha(-l_\alpha)}{K_{\rm IIC}}\right)^2\right]^{-1/2}\right\}, \tag{8}$$

where $K_{\rm I0}^\alpha(\pm l_\alpha)$ and $K_{\rm II0}^\alpha(\pm l_\alpha)$ denote the mode-I and II SIFs at the two tips of the $\alpha$th crack subjected to the reference stress $\boldsymbol{\sigma}_\infty^0$.

The tensile strength of a perfectly-brittle specimen is defined as the applied stress $\boldsymbol{\sigma}_{\rm c}$ at which any one of the cracks starts to propagate. That is,

$$\boldsymbol{\sigma}_{\rm c} = m_{\rm c}\boldsymbol{\sigma}_\infty^0, \tag{9}$$

where the critical load factor $m_{\rm c}$ is the minimum value of $m_\alpha$ among all the cracks.

## 4.2. SIZE EFFECT OF TENSILE STRENGTH

The strength of specimens of a brittle or quasi-brittle material usually exhibits a significant size effect, i.e., the strength measured decreases as the specimen size increases. The size effect of such engineering materials as concrete and rocks are caused mainly by their heterogeneous microstructures due to the stochastic distribution of constituent phases and microdefects. Using the effective field-subregion method presented above, the size effect of material strength associated with interacting microcracks is examined here for otherwise homogeneous brittle solids.

Two-dimensional rectangular plate specimens with length $2L$ and width $L$ are taken as an example, as shown in Fig. 2(a). Specimens of different sizes are considered by varying the parameter $L$ in order to examine the size effect. Each specimen may contain microcracks of a great number, e.g., 5000. If the direct interaction of all these microcracks is calculated, one has to solve a system of 10000 equations, each of which has interactive terms (or transmission factors) of a very large number (about $1.2\times10^7$). Evidently, such a calculation is very difficult, and, in fact, unnecessary. By using our present method, however, if each subregion contains 100 microcracks, one needs to solve systems of only 200 equations, each equation having less than 5000 interactive terms. As the total number of microcracks increases, the computation time increases proportionally, instead of exponentially.

In the case of completely random orientations of microcracks (Fig. 2a), the fracture stress averaged from multiple samples of the same size is plotted in Fig. 5 as a function of the specimen size. A pronounced size effect of the material strength can be observed. It is seen that the particular statistics of microcracks dictates the failure and the associated size effect of brittle solids. The critical condition may fluctuate significantly depending upon the actual arrangement of the crack array.

4.3. SIZE EFFECT OF A BENDING BEAM

The above method is also applied to calculate the strength of a microcracked beam subjected to bending, as shown in the inset of Fig. 6. A bending beam possesses some basic feature of engineering structures, and has non-uniform, both tensile and compressive, stresses. With the same distribution rule of microcracks as in Section 3.2, the tensile strengths calculated from the present numerical approach are given in Fig. 6, which exhibits also a significant size effect.

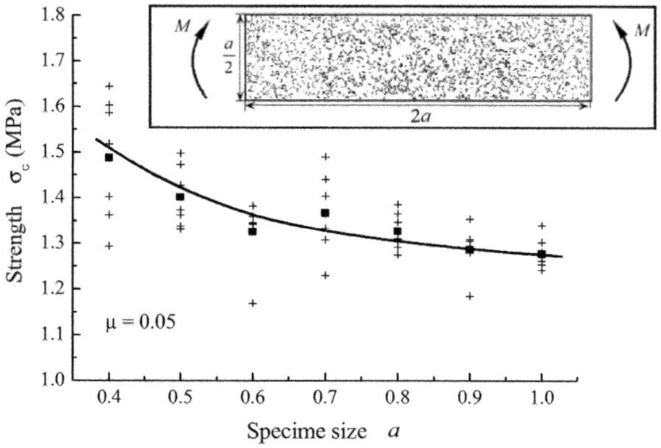


*Figure 6.* Size effect of the bending strength of a microcracked beam.

## 5. Microcrack Coalescence and Failure Process

Rupture of brittle materials is the outcome of a process which in most cases involves successive coalescence of interacting microcracks, formation and unstable propagation of a main crack. Due to the prohibitive complexity of mathematical solution, however, there seems a lack of efficient methods for simulating the failure process associated with interaction and evolution of numerous disordered microcracks. To clarify some of the fundamental aspects of the physics of brittle deformation process, the above numerical method is extended to study coalescence of microcracks and the resulted facture process of elastic solids containing many microcracks.

An essential issue for simulating the failure process associated with microcrack coalescence is to specify appropriate criteria to determine at each linking step: (i) when crack propagation and coalescence will happen, (ii) which crack will grow, (iii) with which crack it will coalesce, and (iv) the linking path [9]. In fact, it is very difficult to give a unified criterion for linking of randomly distributed microcracks. The actual propagation and coalescence path of interacting cracks is usually curved, even for two collinear cracks. As a result of the complexity of the problem and the contradicting requirements of rigor and simplicity, however, we assume that two cracks are connected always along a straight path and that the linked microcrack, though zigzag, is treated as a planar one in the next coalescing step. These two assumptions were also made by Li and Yang [10], who discussed their acceptable reasonability and accuracy.

Two criteria in terms of the energy release ratio are introduced in our simulation, though other criteria can also be implemented in the present program. First, the SIFs $K_{\rm I}$ and $K_{\rm II}$ and the non-dimensional energy release rates $G$ of all the cracks in the considered specimen are calculated in each step. The crack, denoted as $\alpha$, that has the maximum value of $G$ will propagate in the following step. The critical load for propagation of $\alpha$ is obtained by using the criterion $G_\alpha = 1$ defined in Eq. (7).

Besides Eq. (7), another criterion is required to determine which nearby microcrack will coalesce with $\alpha$. The coalescence is dictated here by an energy ratio defined as [10]

$$R = \Delta \Pi / (2c\gamma_{\rm s}), \tag{10}$$

where $\Delta \Pi$ denotes the release of potential energy because of the linkage of two neighboring microcracks, $c$ the ligament size, and $\gamma_{\rm s}$ the surface energy per area of the brittle matrix. The parameter $R$ represents the ratio between the released potential energy and the energy required to create two surfaces along the broken ligament during the coalescence. The larger the ratio $R$, the bigger the driving force of coalescence. Microcracks $\alpha$ and $\beta$ will be connected if $R_{\alpha\beta}$ is the maximum among $R_{\alpha I}$, where $I$ stands for any of the microcracks neighboring to $\alpha$.

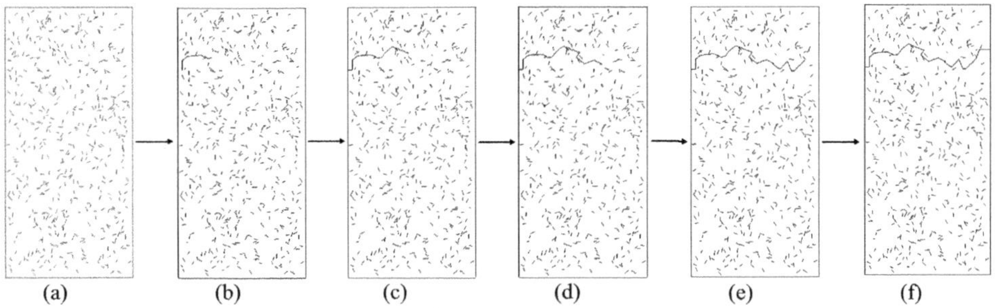

*Figure* 7. Microcrack coalescence process of a specimen.

To calculate $\Pi$, the normal and shear pseudo-tractions transferred to the $\alpha$th microcrack, $p^\alpha(\xi)$ and $\tau^\alpha(\xi)$, need to be determined first from the method in Section 2. The opening and sliding displacements of the crack can be given by [3]

$$b_2^\alpha(\xi) = \frac{4l_\alpha}{E}\sigma_n\left(1+\alpha_n\frac{\xi}{l_\alpha}+\beta_n\frac{\xi^2}{l_\alpha^2}\right)\sqrt{1-\frac{\xi^2}{l_\alpha^2}},$$

$$b_1^\alpha(\xi) = \frac{4l_\alpha}{E}\sigma_\tau\left(1+\alpha_\tau\frac{\xi}{l_\alpha}+\beta_\tau\frac{\xi^2}{l_\alpha^2}\right)\sqrt{1-\frac{\xi^2}{l_\alpha^2}}. \tag{11}$$

with $\sigma_n$, $\sigma_\tau$, $\alpha_n$, $\alpha_\tau$, $\beta_n$ and $\beta_\tau$ being given in Ref. [3, 9]. Then the potential energy contributed from the $\alpha$th microcrack is given by

$$\Pi_\alpha = \frac{1}{2}\int_{-l_\alpha}^{l_\alpha}\left[p^\alpha(\xi)b_2^\alpha(\xi)+\tau^\alpha(\xi)b_1^\alpha(\xi)\right]\mathrm{d}\xi. \tag{12}$$

Summing over $\Pi_\alpha$ before and after the microcrack linkage, we obtain the potential energy release $\Delta \Pi$, from which the energy ratio $R$ follows from Eq. (10).

Before the final rupture of a specimen, there are many coalescence steps. Such a complicated process can be simulated by a step-by-step numerical method. For convenience, only one coalescence is considered in each step of calculation. The failure process of a specimen is schematized, for instance, in Fig. 7, which includes 27 steps of crack linkage prior to the final rupture. The macroscopic averaged stress and strain history with the fracture process is recorded in the program, which shows that the material ruptures along a zigzag path basically normal to the tensile direction.

## 6. Conclusions

An approximate method is presented to calculate the interaction of microcracks of a large number. Its central idea is that all the microcracks in a specimen are divided into two groups, which are dealt with in different ways. The interacting microcracks within a subregion surrounding the considered crack are calculated directly by using Kachanov's interaction method, while the influence of other microcracks is reflected by modifying the far-field stress. This method can be used to analyze various problems associated with interaction and propagation of numerous microcracks.

## Acknowledgements

The supports from the National Natural Science Foundation (Grant Nrs. 10121202 and 10028204) and the Education Ministry of China (National Excellent Doctoral Theses Funding Grant Nr. 199926) are acknowledged.

## References

1. Feng, X.Q. and Yu, S.W. (2002) Damage Micromechanics of Quasi-Brittle Materials, Higher Education Press, Beijing.
2. Krajcinovic, D. (1996) Damage Mechanics, Elsevier, Amsterdam.
3. Kachanov, M. (1993) Elastic solids with many cracks and related problems, *Advances in Applied Mechanics* **30**, 259–445.
4. Yu, S.W. and Feng, X.Q. (1997) Damage Mechanics, Tsinghua University Press, Beijing.
5. Feng, X.Q., Li, J.Y. and Yu, S.W. (2003) A simple method for calculating interaction of numerous microcracks and applications, *Int. J. Solids Structures* **40**, 447–464.
6. Kachanov, M. (1987) Elastic solids with many cracks: A simple method of analysis, *Int. J. Solids Structures* **23**, 23–43.
7. Carvalho, F.C.S. and Labuz, J.F. (1996) Experiments on effective elastic modulus of two-dimensional solids with cracks and holes. Int. J. Solids Structures **33**, 4119–4130.
8. Zheng, Q.S. and Du, D.X. (2001) An explicit and universally applicable estimate for the effective properties of multiphase composites which account for inclusion distribution. *J. Mechanics and Physics of Solids* **49**, 2765–2788.
9. Ma, L. Wang, X.Y., Feng, X.Q. and Yu, S.W. (2005) Numerical analysis of interaction and coalescence of numerous microcracks, *Engineering Fracture Mechanics* **72**.
10. Li, T. and Yang, W. (2002) Critical linkages of coalescing microcracks under stress loading, *Fatigue and Fracture in Engineering Materials and Structures* **25**, 499–508.

# MORPHOLOGICAL INSTABILITY AND KINETICS OF AN ELASTIC FILM ON A VISCOELASTIC LAYER

S.H. IM and R. HUANG
*Research Center for Mechanics of Solids, Structures & Materials*
*Department of Aerospace Engineering and Engineering Mechanics*
*The University of Texas, Austin, TX 78712, USA*

## Abstract

This paper develops a theoretical model for wrinkling of an elastic film on a viscoelastic layer. The film is elastic and subjected to a compressive residual strain. The viscoelastic layer is sandwiched between the film and a rigid substrate. The nonlinear von Karman plate theory is employed to model the film, and a thin-layer approximation is adopted for the viscoelastic layer. The stability of the system and the kinetics of structural evolution are studied first by a linear perturbation analysis and then by numerical simulations, both for plane strain deformation only. Three stages of evolution are identified: initial growth of the fastest growing mode, intermediate growth with mode transition, and finally an equilibrium wrinkled state. The results qualitatively agree with experimental observations of wrinkling in metal/polymer bilayer films.

## 1. Introduction

Complex wrinkle patterns have been observed in thin film systems with integrated compliant layers. The wrinkles are a nuisance in some applications (Iacopi *et al.*, 2003; Yin *et al.*, 2003), but may be used as stretchable interconnects for flexible electronics (Watanabe *et al.*, 2002; Lacour *et al.*, 2003), or biological assays (Cerda and Mahadevan, 2003). The wrinkle pattern can be controlled by engineering the surface structures or chemistry for micro to nanoscale fabrication (Bowden *et al.*, 1998; Chua *et al.*, 2000). It is also possible to extract mechanical properties (e.g., Young's modulus and residual stress) of thin film materials from wrinkle patterns (Martin *et al.*, 2000; Stafford *et al.*, 2004).

The underlying mechanism of wrinkling has been understood as a stress-driven instability, similar to Euler buckling of an elastic column under compression. For a solid film bonded to a substrate, however, the instability is constrained. If the substrate is elastic, there exists a critical compressive stress, beyond which the film wrinkles with a particular wavelength selected by minimizing the total elastic energy in the film and the substrate (Groenewold, 2001; Chen and Hutchinson, 2004; Huang *et al.*, 2004). If the substrate is viscous, wrinkling becomes a kinetic process (Sridhar *et al.*, 2001;



*Q. P. Sun and P. Tong (eds.), IUTAM Symposium on Size Effects on Material and Structural Behavior at Micron- and Nano-Scales,* 41–50.

Huang and Suo, 2002A). Since the viscous substrate does not store elastic energy, a compressed blanket film is always unstable energetically. The viscous flow in the substrate controls the kinetics, selecting a fastest growing wavelength. More generally, if the substrate is viscoelastic, both energetics and kinetics play important roles. A spectrum of evolving wrinkle patterns has been observed experimentally in metal/polymer bilayers (Yoo and Lee, 2003), exhibiting a peculiar kinetic process. A linear perturbation analysis has shown that the viscoelastic property of the substrate significantly influences the wrinkling kinetics, even at the initial stage (Huang, 2005). This paper goes beyond the linear perturbation analysis and simulates the evolution of wrinkles in an elastic film on a thin viscoelastic layer.

## 2. Model Formulation

Figure 1 shows a model structure, an elastic film of thickness $h_f$ lying on a viscoelastic layer of thickness $H$, which in turn lies on a rigid substrate. At the reference state (Fig. 1a), both the elastic film and the viscoelastic layer are flat, and the film is subjected to an in-plane biaxial compressive stress $\sigma_0$ (i.e, $\sigma_0 < 0$). Upon wrinkling (Fig. 1b), the film undergoes both in-plane and out-of-plane displacements, and the viscoelastic layer deforms concomitantly, with the interface assumed to remain bonded.

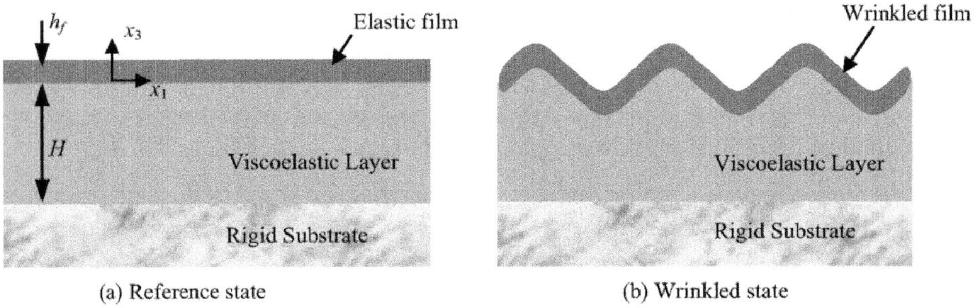


*Figure 1.* Schematic of a model structure

### 2.1. DEFORMATION OF AN ELASTIC FILM

Let $w$ be the lateral deflection of the film, $u_\alpha$ the in-plane displacement ($\alpha = 1,2$), $p$ the pressure at the interface, and $\tau_\alpha$ the shear tractions. Employing the nonlinear von Karman plate theory, equilibrium requires that

$$p = D_f \frac{\partial^4 w}{\partial x_\alpha \partial x_\alpha \partial x_\beta \partial x_\beta} - N_{\alpha\beta} \frac{\partial^2 w}{\partial x_\alpha \partial x_\beta} - \tau_\alpha \frac{\partial w}{\partial x_\alpha}, \tag{1}$$

$$\tau_\alpha = \frac{\partial N_{\alpha\beta}}{\partial x_\beta}, \tag{2}$$

where

$$D_f = \frac{E_f h_f^3}{12(1-\nu_f^2)}, \tag{3}$$

$$N_{\alpha\beta} = \sigma_0 h_f \delta_{\alpha\beta} + \frac{E_f h_f}{1-\nu_f^2}\left[(1-\nu_f)\varepsilon_{\alpha\beta} + \nu_f \varepsilon_{\gamma\gamma}\delta_{\alpha\beta}\right], \tag{4}$$

$$\varepsilon_{\alpha\beta} = \frac{1}{2}\left(\frac{\partial u_\alpha}{\partial x_\beta} + \frac{\partial u_\beta}{\partial x_\alpha}\right) + \frac{1}{2}\frac{\partial w}{\partial x_\alpha}\frac{\partial w}{\partial x_\beta}. \tag{5}$$

Here $E_f$ is the Young's modulus of the film, $\nu_f$ the Poisson's ratio, $N_{\alpha\beta}$ the in-plane membrane force, $\varepsilon_{\alpha\beta}$ the in-plane strain, $\delta_{\alpha\beta}$ the Kronecker delta. The Greek subscripts $\alpha$ and $\beta$ take on the values of in-plane coordinates 1 and 2, and a repeated Greek subscript implies summation over 1 and 2. Note that, both the in-plane displacements and the deflection contribute to the relaxation of the compressive stress. The nonlinear term in Eq. (5) accounts for moderately large deflection of the film.

## 2.2. DEFORMATION OF A VISCOELASTIC THIN-LAYER

The underlayer material is assumed to be isotropic, linear viscoelastic, with a relaxation modulus $\mu(t)$ and Poisson's ratio $\nu(t)$. The layer is stress free initially ($t = 0$) and subjected to a normal and a shear traction at the top surface for $t > 0$, namely

$$\sigma_{31} = S_1(x_1, t) \text{ and } \sigma_{33} = S_3(x_1, t) \text{ at } x_3 = 0. \tag{6}$$

At the lower surface, the displacement is fixed:

$$u_1 = u_3 = 0 \text{ at } x_3 = -H. \tag{7}$$

Consider a periodic traction in the form of

$$S_1(x_1, t) = A(t)\sin kx_1, \tag{8}$$

$$S_3(x_1, t) = B(t)\cos kx_1, \tag{9}$$

where $k$ is the wave number, $A(t)$ and $B(t)$ are the amplitudes. The plane-strain deformation of a viscoelastic layer subjected to such a periodic surface traction has been solved by Huang (2005) using Laplace transform and the correspondence principle of viscoelasticity. The Laplace transform of the displacements at the top surface are

$$\bar{u}_1(x_1, s) = \frac{1}{2ks\bar{\mu}(s)}\left[\gamma_{11}(s\bar{\nu}, kH)\bar{A}(s) + \gamma_{12}(s\bar{\nu}, kH)\bar{B}(s)\right]\sin(kx_1), \tag{10}$$

$$\bar{u}_3(x_1, s) = \frac{1}{2ks\bar{\mu}(s)}\left[\gamma_{21}(s\bar{\nu}, kH)\bar{A}(s) + \gamma_{22}(s\bar{\nu}, kH)\bar{B}(s)\right]\cos(kx_1), \tag{11}$$

where

$$\gamma_{11} = \frac{1+\kappa}{4}\frac{\kappa\sinh(2kH) + 2kH}{\kappa\cosh^2(kH) + (kH)^2 + \left(\frac{1-\kappa}{2}\right)^2}, \tag{12}$$

$$\gamma_{22} = \frac{1+\kappa}{4}\frac{\kappa\sinh(2kH) - 2kH}{\kappa\cosh^2(kH) + (kH)^2 + \left(\frac{1-\kappa}{2}\right)^2}, \tag{13}$$

$$\gamma_{12} = \gamma_{21} = -\frac{\frac{\kappa(1-\kappa)}{2}\sinh^2(2kH) + (kH)^2}{\kappa\cosh^2(kH) + (kH)^2 + \left(\frac{1-\kappa}{2}\right)^2}, \tag{14}$$

and $\kappa = 3 - 4s\bar{\nu}(s)$. A bar over a variable designates its Laplace transform, and $s$ is the transform variable.

In general, the surface of the viscoelastic layer undergoes both out-of-plane and in-plane displacements and they are coupled. The coupling is broken in two special cases, one for an infinitely thick ($kH \to \infty$) and incompressible ($\nu = 0.5$) layer and the other for a very thin layer ($kH \to 0$). The former has been considered previously (Huang, 2005). This paper considers the thin layer limit, for which Eqs. (10) and (11) reduce to

$$\bar{u}_1(x_1, s) = \frac{1}{2ks\bar{\mu}(s)}\left[2kH\bar{A}(s) - \frac{1-4\nu}{2(1-\nu)}(kH)^2\bar{B}(s)\right]\sin(kx_1), \tag{15}$$

$$\bar{u}_3(x_1, s) = \frac{1}{2ks\bar{\mu}(s)}\left[\frac{1-2\nu}{1-\nu}(kH)\bar{B}(s) - \frac{1-4\nu}{2(1-\nu)}(kH)^2\bar{A}(s)\right]\cos(kx_1), \tag{16}$$

where only the leading terms in $kH$ are retained and the Poisson's ratio has been assumed to be a constant independent of time. In addition, $\nu \neq 0.5$ is assumed. As will be shown in a subsequent study, the kinetics is quite different if the underlayer is incompressible (i.e., $\nu = 0.5$).

The surface displacement as a function of time is obtained by the inverse Laplace transform of Eqs. (15) and (16). To be specific, consider the Kelvin model of viscoelasticity with a spring and a dashpot in parallel, for which the relaxation modulus is,

$$\mu(t) = \mu_\infty + \eta\delta(t), \tag{17}$$

where $\mu_\infty$ is the elastic shear modulus at the rubbery limit and $\eta$ is the viscosity. The Laplace transform of the relaxation modulus is

$$\bar{\mu}(s) = \frac{\mu_\infty}{s} + \eta. \tag{18}$$

After substituting (18) into Eqs. (15) and (16), the inverse Laplace transform leads to

$$\frac{\partial u_1}{\partial t} = \frac{H}{\eta}S_1 - \frac{1-4\nu}{4(1-\nu)}\frac{H^2}{\eta}\frac{\partial S_3}{\partial x_1} - \frac{\mu_\infty}{\eta}u_1, \tag{19}$$

$$\frac{\partial u_3}{\partial t} = \frac{1-2\nu}{2(1-\nu)}\frac{H}{\eta}S_3 + \frac{1-4\nu}{4(1-\nu)}\frac{H^2}{\eta}\frac{\partial S_1}{\partial x_1} - \frac{\mu_\infty}{\eta}u_3. \tag{20}$$

Note that, in case of incompressible materials ($\nu = 0.5$), the first term at the right hand side of Eq. (20) vanishes. Using the next leading term of $kH$ in Eq. (16), Eq. (20) becomes

$$\frac{\partial u_3}{\partial t} = -\frac{H^3}{3\eta}\frac{\partial^2 S_3}{\partial x_1^2} - \frac{H^2}{2\eta}\frac{\partial S_1}{\partial x_1} - \frac{\mu_\infty}{\eta}u_3. \tag{21}$$

Equations (19) and (21) are similar to those for a thin liquid layer by Reynold's lubrication theory (Huang and Suo, 2002A), but with an additional term accounting for the elastic limit.

Assuming $\nu \neq 0.5$ and neglecting the $H^2$ terms in Eqs. (19) and (20) as a thin-layer approximation, we obtain that

$$\frac{\partial u_1}{\partial t} = \frac{H}{\eta} S_1 - \frac{\mu_\infty}{\eta} u_1, \tag{22}$$

$$\frac{\partial u_3}{\partial t} = \frac{1-2\nu}{2(1-\nu)} \frac{H}{\eta} S_3 - \frac{\mu_\infty}{\eta} u_3. \tag{23}$$

Equation (22) is a combination of the shear lag model for elastic and viscous layers (Xia and Hutchinson, 2000; Huang *et al.*, 2002), and Eq. (23) is similar to the Winkler model for elastic foundation (Allen, 1969) but includes a viscous term. The two equations are uncoupled under the thin-layer approximation.

In the development of Eqs. (22) and (23), plane strain deformation and periodic tractions have been assumed. The restriction of periodic tractions have been relaxed by linear superposition of all Fourier components in the form of (8) and (9). The restriction of plane strain deformation can also be relaxed by including the other in-plane component in the same form as Eq. (22), which leads to

$$\frac{\partial u_\alpha}{\partial t} = \frac{H}{\eta} S_\alpha - \frac{\mu_\infty}{\eta} u_\alpha, \tag{24}$$

for $\alpha = 1,2$.

## 2.3. COUPLED EVOLUTION EQUATIONS

The interface between the elastic film and the viscoelastic layer is assumed to remain bonded during deformation. Consequently, the displacements and the tractions are continuous across the interface. Therefore, the equilibrium equations (1) and (2) are coupled with Eqs. (23) and (24), leading to

$$\frac{\partial w}{\partial t} = \frac{1-2\nu}{2(1-\nu)} \frac{H}{\eta} \left( -D_f \frac{\partial^4 w}{\partial x_\alpha \partial x_\alpha \partial x_\beta \partial x_\beta} + N_{\alpha\beta} \frac{\partial^2 w}{\partial x_\alpha \partial x_\beta} + \frac{\partial N_{\alpha\beta}}{\partial x_\beta} \frac{\partial w}{\partial x_\alpha} \right) - \frac{\mu_\infty}{\eta} w, \tag{25}$$

$$\frac{\partial u_\alpha}{\partial t} = \frac{H}{\eta} \frac{\partial N_{\alpha\beta}}{\partial x_\beta} - \frac{\mu_\infty}{\eta} u_\alpha. \tag{26}$$

Equations (25) and (26) are coupled, nonlinear evolution equations, which can be solved numerically to simulate the evolution of two-dimensional wrinkles. In the remainder of this paper, however, we consider one-dimensional plane-strain wrinkles only, for which some analytical solutions can be obtained to provide useful insights. The reduced equations for the plane-strain wrinkles are summarized as below:

$$\frac{\partial w}{\partial t} = \frac{1-2\nu}{2(1-\nu)} \frac{H}{\eta} \left( -D_f \frac{\partial^4 w}{\partial x^4} + N \frac{\partial^2 w}{\partial x^2} + \frac{\partial N}{\partial x} \frac{\partial w}{\partial x} \right) - \frac{\mu_\infty}{\eta} w, \tag{27}$$

$$\frac{\partial u}{\partial t} = \frac{H}{\eta} \frac{\partial N}{\partial x} - \frac{\mu_\infty}{\eta} u, \tag{28}$$

$$N = \sigma_0 h_f + \frac{E_f h_f}{1-\nu_f^2} \left[ \frac{\partial u}{\partial x} + \frac{1}{2} \left( \frac{\partial w}{\partial x} \right)^2 \right]. \tag{29}$$

### 3. Linear Perturbation Analysis

Assume a small deflection from the reference state in the form of

$$w(x,t) = A(t)\cos kx . \tag{30}$$

The in-plane displacement is uncoupled from the lateral deflection in the linear perturbation analysis and therefore ignored. Inserting (30) into Eq. (27) and retaining only the leading order terms in $A$, we obtain that

$$\frac{dA}{dt} = \frac{\alpha E_f - \mu_\infty}{\eta} A(t) , \tag{31}$$

where,

$$\alpha = \frac{(1-2\nu)k^2 H h_f}{24(1-\nu)(1-\nu_f^2)} \left[ -k^2 h_f^2 - \frac{12(1-\nu_f^2)\sigma_0}{E_f} \right] . \tag{32}$$

Solving Eq. (31) leads to

$$A(t) = A_0 \exp(s\frac{t}{\tau}) , \tag{33}$$

where $\tau = \eta / E_f$ is the time scale and $s = \alpha - \mu_\infty / E_f$ is the dimensionless growth rate. Figure 2 plots the growth rate as a function of the wavelength ( $L = 2\pi / k$ ) of the perturbation for various ratios between the rubbery modulus of the viscoelastic layer and the elastic modulus of the film. For a given residual stress in the film, there exists a critical ratio between the moduli, above which the growth rate is negative for all wavelengths and the film is stable against any perturbation. By setting the maximum growth rate to be zero, we find the critical ratio

$$\left( \frac{\mu_\infty}{E_f} \right)_c = \frac{3(1-\nu_f{}^2)(1-2\nu)}{2(1-\nu)} \frac{H}{h_f} \left( \frac{\sigma_0}{E_f} \right)^2 . \tag{34}$$

In figure 2, the critical modulus ratio is 0.00025. For smaller modulus ratios, however, the growth rate is positive for intermediate wavelengths within a window bounded by two critical values; the film becomes unstable and a small perturbation grows to form wrinkles. The wavelength of the fastest growing mode is

$$L_m = \pi h_f \sqrt{-\frac{2E_f}{3(1-\nu_f^2)\sigma_0}} , \tag{35}$$

which is indicated by the dashed line ($L_m = 26.9h_f$) in Figure 2.

The effects of the viscoelastic layer on the stability of the elastic film can be drawn from the above analysis. The critical condition for the flat film to remain stable is controlled by the rubbery modulus of the underlayer, as given by Eq. (34). In cases that the film is unstable, the growth time scale is proportional to the viscosity and the growth rate increases as the rubbery modulus decreases. However, the wavelength of the fastest growing mode is independent of the underlayer properties, as given in Eq. (35). A similar analysis was conducted for a film on an incompressible viscous layer (Huang and Suo, 2002B), which led to a slightly smaller fastest growing wavelength; the difference is due to the incompressibility of the viscous liquid.

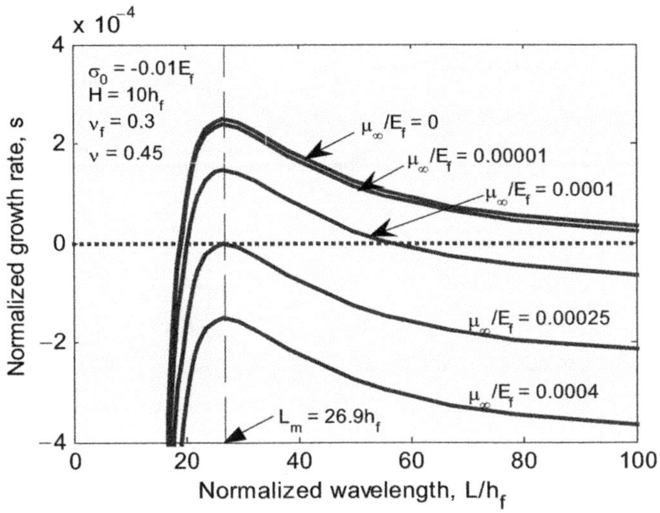


*Figure 2.* The dimensionless growth rate by the linear perturbation analysis.

## 4. Equilibrium Wrinkles

Setting $\partial/\partial t = 0$ in Eqs. (27) and (28) leads to two coupled nonlinear ordinary differential equations, from which one can solve for the equilibrium state. An approximate solution has been obtained by neglecting the in-plane displacement and the shear traction (Huang, 2005), from which the equilibrium amplitude of a sinusoidal wrinkle is

$$A_{eq} = \frac{2\sqrt{1-\nu_f^{\,2}}}{k}\left[-\frac{\sigma_0}{E_f} - \frac{(kh_f)^2}{12(1-\nu_f^{\,2})} - \frac{2(1-\nu)}{1-2\nu}\frac{\mu_\infty}{E_f}\frac{1}{k^2 H h_f}\right]^{1/2}. \tag{36}$$

It can be confirmed that only when the film is unstable does there exist a nontrivial solution corresponding to the equilibrium wrinkled state.

Furthermore, at the equilibrium state, the wrinkle selects a wavelength that minimizes the elastic energy, namely,

$$L_{eq} = \pi h_f\left[\frac{2(1-2\nu)}{3(1-\nu)(1-\nu_f^2)}\frac{E_f}{\mu_\infty}\frac{H}{h_f}\right]^{1/4}. \tag{37}$$

The equilibrium state is identical to that for a film on a thin elastic layer with the shear modulus $\mu_\infty$ (Huang *et al.*, 2004). For a viscoelastic layer with zero rubbery modulus, however, the equilibrium wavelength approaches infinity. Consequently, the wrinkle wavelength coarsens indefinitely.

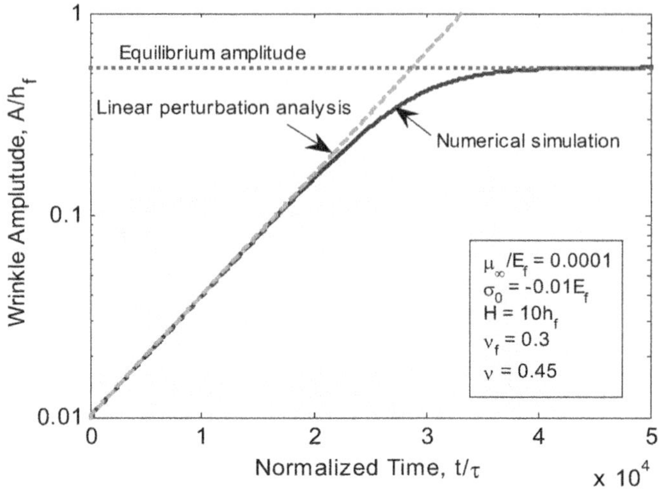


*Figure 3.* Growth of the amplitude of a sinusoidal wrinkle with wavelength $L = 30h_f$.

## 5. Numerical Simulations

In this section, we simulate the wrinkling process by integrating Equations (27) and (28) numerically with a finite difference method. First, a sinusoidal perturbation is assumed for the initial deflection. The wavelength was selected to be close to the fastest growing wavelength and it remains constant for the entire period of simulation. The amplitude grows over time, first exponentially and then approaching a constant corresponding to the equilibrium amplitude, as shown in Figure 3. The time is normalized by the time scale $\tau = \eta/E_f$. The exponential growth predicted by the linear perturbation analysis is plotted as the dashed line in Figure 3, and the equilibrium amplitude corresponding to the selected wavelength ($L = 30h_f$, $A_{eq} = 0.537h_f$) is indicated by the horizontal dotted line. The numerical simulation confirms the validity of the linear perturbation analysis at the initial stage and the equilibrium wrinkle amplitude of a particular wavelength.

Next, a random perturbation is generated for the initial deflection. Figure 4 shows the evolving deflection curves and their Fourier transforms. Figure 5 plots the evolution of the dominant wavelength (maximum intensity in the Fourier transform), and Figure 6 shows the growth of the RMS (i.e., the root mean square value). While many wavelengths co-exist in the initial perturbation, only those of intermediate wavelengths grow and the fastest growing wavelength ($L_m = 26.9h_f$) dominates at the initial stage. Three stages of the wrinkle evolution are identified: initial growth of the fastest growing mode, intermediate growth with mode transition, and finally an equilibrium wrinkled state. For $\mu_\infty / E_f = 0.00001$, the equilibrium wrinkle has a wavelength $L_{eq} = 60.0h_f$ and an amplitude $A_{eq} = 1.63h_f$ (RMS = $1.15h_f$). Such behavior qualitatively agrees with the experimental observations in a metal/polymer bilayer film (Yoo and Lee, 2003).

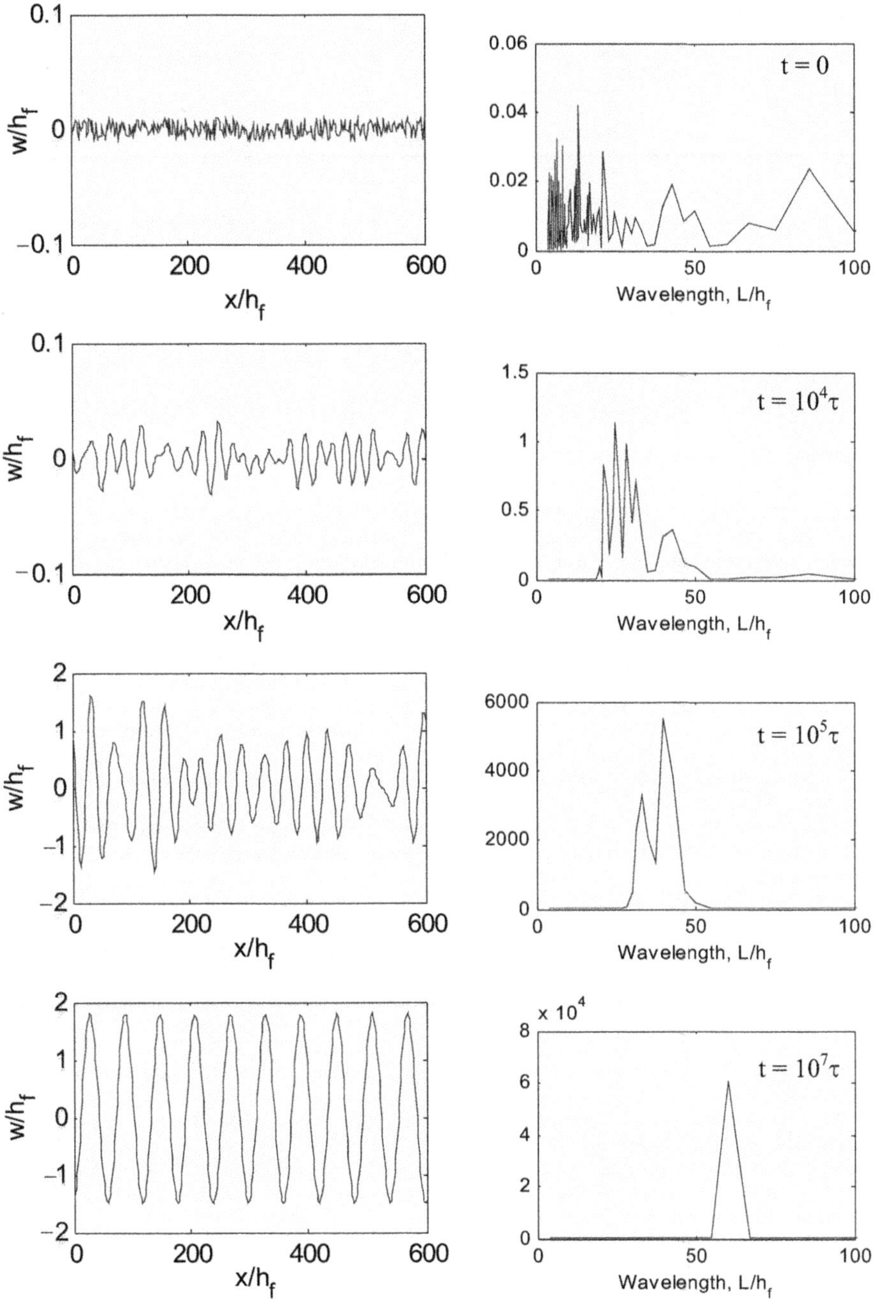


*Figure 4.* Simulated evolution of wrinkles (left) and their Fourier transform (right).

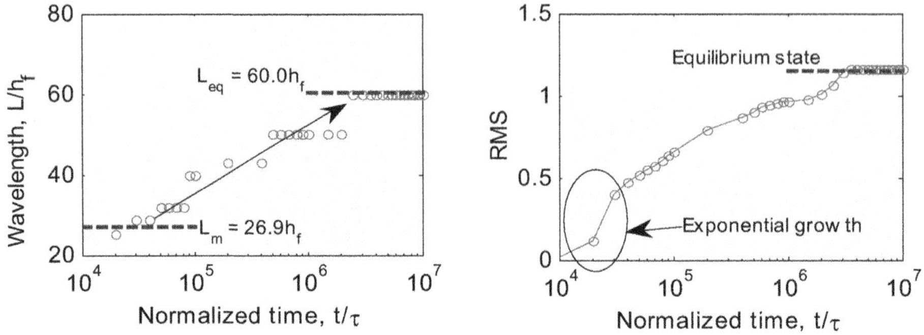


*Figure 5.* Simulated evolution of the dominated wavelength (left) and the RMS of a wrinkled film on a viscoelastic layer.

## References


Allen, H.G., 1969. Analysis and Design of Structural Sandwich Panels, Pergamon, New York.

Bowden, N., Brittain, S., Evans, A.G., Hutchinson, J.W., Whitesides, G.M., 1998. Spontaneous formation of ordered structures in thin films of metals supported on an elastomeric polymer. Nature 393, 146-149.

Cerda, E. and Mahadevan, L., 2003. Geometry and physics of wrinkling. Phys. Rev. Lett. 90, 074302.

Chen, X. and Hutchinson, J.W., 2004. Herringbone buckling patterns of compressed thin films on compliant substrates. J. Appl. Mech. in press.

Chua, D.B.H., Ng, H.T., Li, S.F.Y., 2000. Spontaneous formation of complex and ordered structures on oxygen-plasma-treated elastometric polydimenthylsiloxane. Applied Physics Letters 76, 721-723.

Groenewold, J., 2001. Wrinkling of plates coupled with soft elastic media. Physica A 298, 32-45.

Huang, R., 2005. Kinetic Wrinkling of an Elastic Film on a viscoelastic Substrate. J. Mech. Phys. Solids 53, 63-89.

Huang, R., Prevost, J.H., Suo, Z., 2002. Loss of constraint on fracture in thin film structures due to creep. Acta Mater. 50, 4137-4148.

Huang, R. and Suo, Z., 2002A. Wrinkling of an elastic film on a viscous layer. J. Appl. Phys. 91, 1135-1142.

Huang, R. and Suo, Z., 2002B. Instability of a compressed elastic film on a viscous layer. Int. J. Solids Struct. 39, 1791-1802.

Huang, Z.Y., Hong, W., Suo, Z., 2004. Nonlinear analysis of wrinkles in films on soft elastic substrates. Submitted.

Iacopi, F., Brongersma, S.H., Maex, K., 2003. Compressive stress relaxation through buckling of a low-k polymer-thin cap layer system. Appl. Phys. Lett. 82, 1380-1382.

Lacour, S.P., Wagner, S., Huang, Z.Y., Suo, Z., 2003. Stretchable gold conductors on elastomeric substrates. Appl. Phys. Lett. 82, 2404-2406.

Martin, S.J., Godschalx, J.P., Mills, M.E., Shaffer II, E.O., Townsend, P.H., 2000. Development of a low-dielectric-constant polymer for the fabrication of integrated circuit interconnect. Advanced Materials 12, 1769-1778.

Sridhar, N., Srolovitz, D.J., Suo, Z., 2001. Kinetics of buckling of a compressed film on a viscous substrate. Appl. Phys. Lett. 78, 2482-2484.

Stafford, C.M., Harrison, C., Beers, K.L., Karim, A., Amis, E.J., Vanlandingham, M.R., Kim, H.-C., Volksen, W., Miller, R.D., Simonyi, E.E., 2004. A buckling-based metrology for measuring the elastic moduli of polymeric thin films. Nature Materials 3, 545-550.

Watanabe, M., Shirai, H., Hirai, T., 2002. Wrinkled polypyrrole electrode for electroactive polymer actuators. J. Appl. Phys. 92, 4631-4637.

Xia, Z.C. and Hutchinson, J.W., 2000. Crack patterns in thin films. J. Mech. Phys. Solids 48, 1107-1131.

Yin, H., Huang, R., Hobart, K.D., Liang, J., Suo, Z., Shieh, S.R., Duffy, T.S., Kub, F.J., Sturm, J.C., 2003. Buckling suppression of SiGe islands on compliant substrates. J. Appl. Phys. 94, 6875-6888.

Yoo, P.J. and Lee, H.H., 2003. Evolution of a stress-driven pattern in thin bilayer films: spinodal wrinkling. Phys. Rev. Lett. 91, 154502.

# A MULTI-SCALE MODEL-BASED SIMULATION OF THIN FILM DELAMINATION

LUMING SHEN and ZHEN CHEN
*Department of Civil and Environmental Engineering*
*University of Missouri-Columbia*
*Columbia, MO 65211-2200, USA*

## Abstract

To bridge the different spatial scales involved in the process of tungsten (W) film delaminating from silicon (Si) substrate, a multi-scale model-based simulation procedure is proposed. In the proposed procedure, a bifurcation-based multi-scale decohesion model, which represents the link between molecular and continuum levels, is first formulated within the framework of the continuum mechanics. The proposed model is then calibrated by using molecular dynamics simulation of single crystal W block under tensile loading and available experimental data, with a power scaling law to account for the spatial size effect. A multi-scale model-based simulation of W film delamination from Si substrate is performed by using the proposed procedure within the framework of the material point method. The simulated results have provided new insights into the mechanisms of the thin film delamination process.

**Keywords:** Multi-Scale Simulation, Film Delamination, Molecular Dynamics, Size Effect

## 1. Introduction

Thin films usually develop high residual stresses during the deposition process and may fail by delaminating from the substrates in the working environment. The delamination of compressed films has been studied by many researchers in both academia and industry, as shown by representative papers (Gioia and Ortiz, 1997; Hutchinson and Suo, 1991; among others). Existing approaches are mainly based on conventional elastic stability theory and interfacial fracture mechanics, with a focus on the stability of blisters (Audoly, 1999). Recently, much research has been conducted to model and simulate pattern formation during the delamination process, based on the buckling-driven mechanism (Huang and Suo, 2002). However, a systematic study of the complete process from the formation of films to their eventual delamination from substrates is not yet available from the open literature, due to the complexity of multi-physics and multi-scales involved.



*Q. P. Sun and P. Tong (eds.), IUTAM Symposium on Size Effects on Material and Structural Behavior at Micron- and Nano-Scales,* 51–65.

Based on the experimental observation of the transition from tensile to compressive stress as a function of the argon gas pressure in sputtered tungsten (W) film onto silicon (Si) substrate (Shen et al., 2000), Chen et al. (2003) proposed that the delamination of compressed films is essentially due to the interaction between geometrical and material instabilities which results in the formation and evolution of localization, depending on different stress states in the domain of influence. In the proposed procedure, a discrete constitutive model was formulated by using the bifurcation analysis to predict decohesion of continuum, since the discontinuous bifurcation analysis could be performed without too much computational cost if an associated von Mises elastoplasticity model is used with a linear hardening and softening law (Chen and Fang, 2001). As a result, rigorous and mesh-objective results could be obtained since the location and orientation of the cohesive surfaces are determined via the discontinuous bifurcation analysis (Shen and Chen, 2004). By implementing the bifurcation-based decohesion model into the material point method (MPM), which is one of the "meshfree" methods (Chen et al., 2002), a numerical effort has been made to investigate the transition from continuous to discontinuous failure modes involved in the W film delaminating from the Si substrate. The numerical study of the effects of aspect ratio on the evolution of failure patterns under different boundary conditions provides a better understanding on the physics behind the film delamination process (Chen et al., 2003).

However, the bifurcation-based simulation of the film delamination process is based on a phenomenological framework, which could only provide a qualitative analysis of film failure mechanism. The constitutive relation of decohesion-traction, $\tau(\bar{u})$, as the key part of the decohesion model, cannot be calibrated via existing experimental techniques. Since $\tau(\bar{u})$ is directly related to the interatomic binding energy and film microstructure formed during the film deposition process, atomistic simulation might provide required information for establishing the decohesion model. Since the size of the film-substrate structure usually ranges from nano/micro to macro scale, the conventional continuum mechanics alone is handicapped for film delamination problems. Therefore, a thorough understanding of the physics behind the film delamination process necessitates a multi-scale investigation ranging from atomistic simulation to continuum mechanics.

For the purpose of simplicity, a multi-scale sequential approach is being proposed in this study, as illustrated in Figure 1, to simulate the film delamination process. When bifurcation occurs in the MPM discretization of macroscopic responses, the proposed procedure would zoom in to the desired scale level and obtain the corresponding $\tau(\bar{u})$ curve. By inputting the obtained multi-scale $\tau(\bar{u})$ constitutive relation into the bifurcation-based decohesion model within the framework of the MPM, the multi-degree discontinuous failure modes involved in the film delamination process could then be simulated.

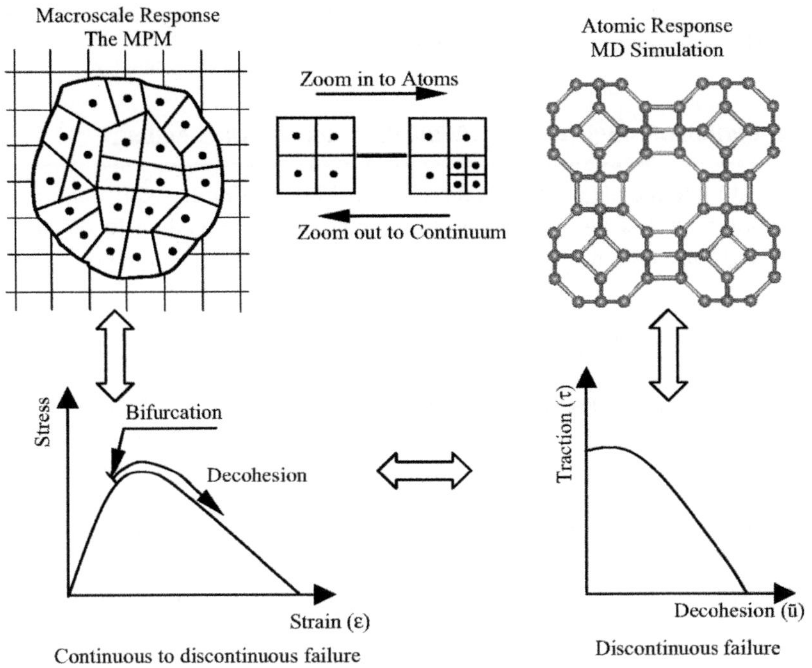


*Figure 1.* Link between continuum mechanics and atomistic simulation.

With the rapid development of micro-electronic mechanical system (MEMS) and nano-technology in recent years, atomistic simulation of mechanical behavior of metals has drawn considerable attention in recent years (Horstemeyer et al., 2001; Liang and Zhou, 2003; Li and Yip, 2002). Size and strain rate effects arise out of several factors and play important roles in determining the response of nanostructures. The behavior and properties of micro/nanostructure are size-dependent due to the discreteness of atomic system, crystal orientation and boundary condition. The dynamic inertia effect and the finite speeds at which lattice waves propagate also introduce length scales to the problem and contribute to size-dependence of atomic behavior. The inertia effect and finite wave speeds also cause the response of nanostructures to be deformation-rate dependent. A part of the study therefore focuses on the size and rate effects on the constitutive law of single crystal W block under tensile loading.

To determine material properties for elasto-plasticity and decohesion at different size scales, a power scaling law is proposed to account for the effect of structure size. In this approach, MD simulation of single crystal W block under tensile loading is combined with available experimental data to develop a multi-scale $\tau(\bar{u})$ model for W based on a scaling law. The multi-scale decohesion-traction model is then employed to simulate the film delamination process.

The remaining sections of the paper are arranged as follows. The MD simulation of single crystal W block under tensile loading is conducted in Section 2, which is followed by the establishment of the multi-scale decohesion-traction relation for W in Section3.

The multi-scale simulation of W film delamination from Si substrate is conducted in Section 4. The conclusions are then made in the last section.

## 2. MD Simulation of Single Crystal W block under Tensile Loading

In order to calibrate the parameters in the proposed bifurcation-based decohesion model and to develop a multi-scale decohesion-traction constitutive relation for W at different size scales, MD simulation of single crystal W block under tension is conducted at the atomic level.

### 2.1. SIMULATION MODEL

The computational set-up for MD simulation of single crystal W block under uniaxial tensile loading is shown in Figure 2. The 3-D simulation super-cell consists of two parts. One part is referred to as the active zone in which the atoms move according to the interactions among the neighboring atoms; the other part, wrapped by the boxes as presented in Figure 2, is referred to as the boundary zone where the atoms are assigned a fixed rigid body velocity. The size of active zone is indicated by $H$, $D$ and $W$, while the size of each boundary zone in the $z$-direction is $2a_o$ with $a_o$ being the lattice parameter of W.

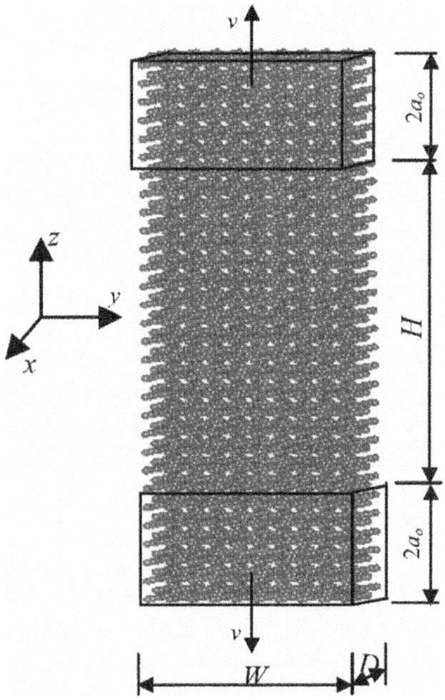


*Figure 2.* Model of single crystal W block under uniaxial tensile load.

Many studies (Milstein and Chantasiriwan, 1998; Roundy et al., 2001; among others) have identified the <100> axes as the weak directions in tension and the {100} planes as the cleavage planes for bcc W. The crystal orientation of ($x$[1 0 0], $y$[0 1 0], $z$[0 0 1]) will be investigated since the decohesion of single crystal W is the interest of current research.

The preliminary simulations have shown that neither decohesion nor cleavage of {100} planes will occur with the use of free surface in either x- or y-direction because the lack of constrain in that direction results in the shrinking of the x-y cross sectional area in the active zone and the subsequent rearrangement of the crystal structure. Thus, the periodic boundary condition (PBC) will be imposed along both the *x*- and y-directions.

In the simulation, initially all atoms are placed at their equilibrium positions at the temperature of 298 K. Those atoms in the boundary zone are then fixed. After the system has equilibrated for a certain period, constant velocities with the same magnitude and opposite direction are assigned to the atoms in the top and bottom boundary zones, respectively, to simulate a displacement-controlled uniaxial tensile loading in the z-direction. A velocity scaling technique (Allen and Tildseley, 1990) is employed through the simulation to maintain a constant temperature of 298 K. The Embedded Atom Method (EAM) developed by Daw and Baskes (1984) is employed to model the interatomic potential among W atoms. The corresponding model parameters for W are based on the work of Zhang and Ouyang (1993). The method applied to integrate the equations of motion is the 6-value Gear predictor-corrector algorithm. The largest time step size that can keep the total system energy remaining a constant in the adiabatic simulations for W atoms with the EAM potential is used for the numerical study. A time step size of 0.5 fs is chosen after several adiabatic simulations for W at different initial temperatures up to 2000 K.

## 2.2. THE CALCULATION OF STRESS AND STRAIN

Stress calculation in MD simulations has been the focus of many studies over the past years (Cheung and Yip, 1991; Cormier, et al., 2001; Irving and Kirkwood, 1950; Zhou, 2003; among others). In this study, the formulations employed to calculate atomic-level stresses are motivated by the work of Zhou (2003). At each atom, the local stress tensor, $\boldsymbol{\beta}$, is given by

$$\boldsymbol{\beta}_i = -\frac{1}{\Omega_i}\sum_{j>i}^{N^n} \mathbf{f}_{ij} \otimes \mathbf{r}_{ij} \tag{1}$$

where i refers to the atom considered and j refers to the neighboring atom, $\mathbf{r}$ij is the position vector between atoms i and j, Nn is the number of neighboring atoms surrounding atom i, Ωi is the volume of atom i, and $\mathbf{f}$ij is the force vector on atom i due to atom j. The global continuum stress tensor is defined by,

$$\boldsymbol{\sigma} = \frac{1}{N^*}\sum_{i}^{N^*} \boldsymbol{\beta}_i \tag{2}$$

in which $N^*$ represents the total number of atoms in a representative volume of continuum.

Since the W block is under very large deformation, true strain, ε, is used in this study with $\varepsilon = \ln(L/L_o)$ where $L_o$ and $L$ are the original and deformed lengths of the specimen, respectively.

## 2.3. SIMULATION RESULTS

To study the effect of specimen size and strain rate on the dynamic responses of single crystal W block under tensile loading, seven MD simulation cases are designed. The details of the numerical experimental configurations are listed in Table 1. By introducing different number of defects in Simulation 2, the effect of the vacancies in W specimen on the stress-strain relation is also investigated.

TABLE 1. Description of the MD simulation of W block under tensile loading.

| Simulation # | Specimen Size $D \times W \times H$ ($nm^3$) | Number of Active Atoms | Initial Strain Rate ($s^{-1}$) | Peak Stress (GPa) |
|---|---|---|---|---|
| 1 | 1.27 × 1.27 × 3.83 | 384 | $2 \times 10^9$ | 39.0 |
| 2 | 1.59 × 1.59 × 4.78 | 750 | $2 \times 10^9$ | 39.1 |
| 3 | 2.54 × 2.54 × 7.65 | 3,072 | $2 \times 10^9$ | 28.6 |
| 4 | 3.19 × 3.19 × 9.56 | 6,000 | $2 \times 10^9$ | 24.8 |
| 5 | 6.38 × 6.38 × 19.13 | 48,000 | $2 \times 10^9$ | 18.5 |
| 6 | 3.19 × 3.19 × 9.56 | 6,000 | $2 \times 10^8$ | 22.2 |
| 7 | 3.19 × 3.19 × 9.56 | 6,000 | $2 \times 10^{10}$ | 28.5 |

### *2.3.1. The Effect of Specimen Size*

To study the effect of specimen size on the tensile deformation of single crystal W, Simulations 1-5 are performed. Figure 3 shows the corresponding stress-strain curves of different W blocks under tensile loading. As can be seen from the figure, the initial elastic modulus of W is almost independent on the specimen size. However, the peak stress increases as the specimen size decreases, which is mainly due to the fact that larger specimens offer more opportunities for dislocation to occur. In addition to the constraints due to the boundary zones and the PBC used in the simulations, the single crystal W specimen is a discrete system with atoms distributed only at certain positions. The bonds among atoms are the relatively weak parts while the individual atoms are the strong parts of the system. The increase of specimen size increases the number of relatively weak bonds, which in turn offers more opportunities for bond breaking and dislocation to occur and thus decreases the strength of the specimen.

### *2.3.2. The Effect of Defects*

To verify the argument that the decrease of the W strength with the increase of specimen size is mainly due to the fact that larger specimens offer more opportunities for dislocation to occur, the effect of artificially introduced defects in single crystal W block on the stress-strain relation is investigated. Different numbers of vacancies are implemented into Simulation 2 with other simulation conditions being kept the same. The stress-strain curves of W blocks with zero, one, two and four vacancies distributed in the same *x-y* plane, which is initially located 2.152 nm away from the top end of the active zone, are presented in Figure 4. As can be seen from the figure, all stress-strain curves are initially the same until failure occurs. The strength of W decreases as the number of vacancies increases, since more vacancies offer more opportunities for bond breaking and dislocation to occur.

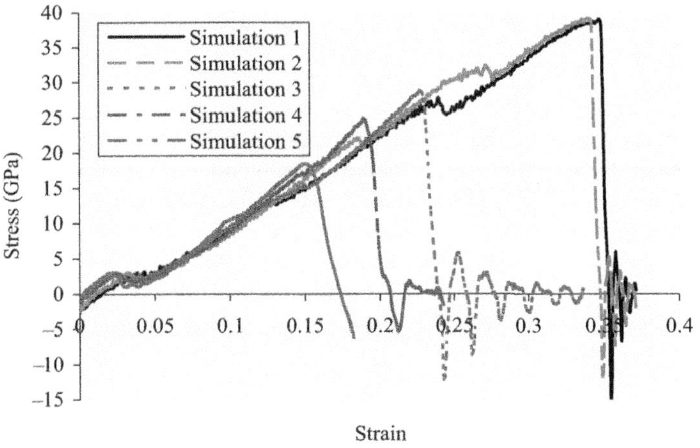


*Figure 3.* The effect of specimen size on the stress-strain curves of W under tensile strain rate of $2 \times 10^9$ $s^{-1}$.

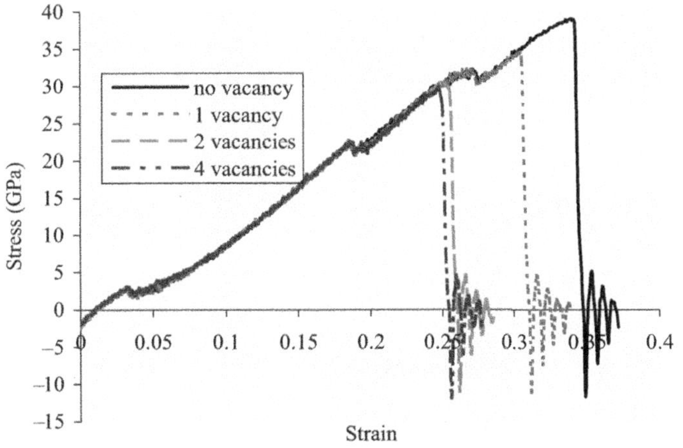


*Figure 4.* Stress-strain curves of W blocks with different number of vacancies distributed in the same x-y plane.

To study the effect of the vacancy distribution position in the *z*-direction on the stress-strain relation of single crystal W block, four cases with zero, one, two and four vacancies, respectively, located at the center of *x-y* plane but different positions in the *z*-direction, are simulated with other simulation conditions being kept the same as those in Simulation 2. The corresponding stress-strain curves (not shown here) are initially the same until failure occurs. The strength of single crystal W decreases as the number of vacancies increases, which is similar to the previous simulation with vacancies distributed in the same *x-y* plane. It seems that the increase of the number of defects regardless of the distribution position in single crystal W block will increase the possibilities for dislocation to occur and thus reduce the strength of W, which is very similar to the effect of increasing specimen size on the strength of W. Therefore, it is reasonable to argue that larger specimens offer more opportunities for dislocation to occur which in turn reduces the strength of W.

*2.3.3. The Effect of Strain Rate*

To investigate the effect of the tensile strain rate on the stress-strain curve of single crystal W block at the atomic level, Simulations 4, 6 and 7 are performed with the initial strain rates of $2 \times 10^9$ s$^{-1}$, $2 \times 10^8$ s$^{-1}$ and $2 \times 10^{10}$ s$^{-1}$, respectively. Figure 5 shows the corresponding stress-strain curves. As can be seen from the figure, the initial elastic modulus of W is almost independent on the strain rate, but the peak stress increases with the strain rate. The dependence of decohesion initiation on the strain rate is mainly due to the dynamic wave effect that impedes the motion of dislocations (Liang and Zhou, 2003).

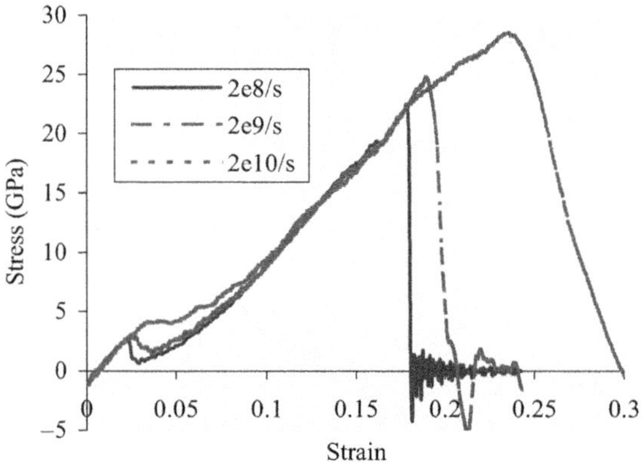


*Figure 5.* The effect of strain rate on the stress-strain curve of single crystal W under tension.

## 3. Multi-Scale Decohesion-Traction Model For W

In order to determine the strength and decohesion energy of W at different size scales and develop a multi-scale decohesion model for W, a power scaling law is proposed to combine the MD simulation of single crystal W block under tension and the available experimental data. By considering geometrically similar systems, the power scaling law takes the form of (Bazant, 2002):

$$f(\lambda) = \lambda^m \tag{3}$$

where $\lambda = D_1/D_2$ with $D_1$ and $D_2$ being the characteristic sizes of two similar structures, respectively, $f(\lambda) = Y_1/Y_2$ is a dimensionless function with $Y_1$ and $Y_2$ being the material properties at sizes $D_1$ and $D_2$, respectively, and exponent $m$ is an unknown constant.

Indentation tests on single crystal W have shown that there is dependence of material hardness on the size. However, indents on W with diagonals longer than about 100 μm cease to display any hardness-dependence on the size (Hutchinson, 2000). Since the material hardness is directly related to the strength, we have $m \neq 0$ when $D < 100$ μm and $m = 0$ when $D \geq 100$ μm. It seems that the exponent $m$ will change from nonzero to zero when the structure reaches the critical size of about 100 μm.

MD simulations of sheared single crystal metals conducted by Horstemeyer et al. (2001) have demonstrated some qualitative features of the rate effect on the material strength: a. strain rate independence at low strain rate, and b. an increase in the critical strain rate, under which rate-dependence disappears, with the decrease of specimen size. Their simulation results showed that there is no strain rate effect on nickel's yield stress with specimen size being smaller than 28 nm when the strain rate is in the order of $10^8$ s$^{-1}$. It is thus assumed that the strength of W obtained with the strain rate of $2 \times 10^8$ s$^{-1}$ is almost rate-independent since our specimen sizes are small than 28 nm. As can be seen from Figure 5, the strength of W in Simulation 4, decreases from 24.8 GPa at the rate of $2 \times 10^9$ s$^{-1}$ to 22.2 GPa at the rate of $2 \times 10^8$ s$^{-1}$ in Simulation 6 with a factor of 22.2/24.8 = 0.895. For the sake of simplicity, all the strengths obtained in Simulations 1-5 with the strain rate of $2 \times 10^9$ s$^{-1}$ will be multiplied by a factor of 0.895 to approximately get the rate-independent strengths at different structure sizes.

A multi-scale strength model for W is proposed here based on the assumption that the exponent $m$ in the power scaling law must change smoothly when the structure size increases from atomic-scale to macro-scale. Figure 6 presents the proposed multi-scale model for predicting W strength at the different structural sizes by fitting the available simulated and experimental data. The proposed multi-scale strength model for W takes the form of:

$$\begin{cases} \sigma_N = \sigma_P & D \le D_P \\ \dfrac{\lg\sigma_N - \lg\sigma_M}{\lg\sigma_P - \lg\sigma_M} = 1 - \sin\left[\dfrac{\pi}{2} \times \dfrac{\lg D - \lg D_P}{\lg D_M - \lg D_P}\right] & D_P < D < D_M \\ \sigma_N = \sigma_M & D \ge D_M \end{cases} \tag{4}$$

where $D$ is the characteristic size of W specimen, $D_P$ is the maximum atomic-scale size of W specimen at which the ultimate strength of W is reached, $D_M$ is the minimum macro-scale size of W specimen where the strength of W ceases its dependence on the size, $\sigma_N$ is the strength of W at size $D$, $\sigma_M$ is the strength of W at size $D_M$, and $\sigma_P$ is the ultimate strength of W. $\sigma_P = 34.9$ GPa at $D_P = 1.6$ nm, which is in a fair agreement with the maximum tensile strength of 29.5 GPa as reported by Roundy et al. (2001) using pseudopotential density functional theory, is applied based on the MD simulations of single crystal W block under tensile loading. According to Hutchison (2000), $D_M = 100$ μm is assumed. As a common material property for W, $\sigma_M = 1.5$ GPa is adopted in this study.

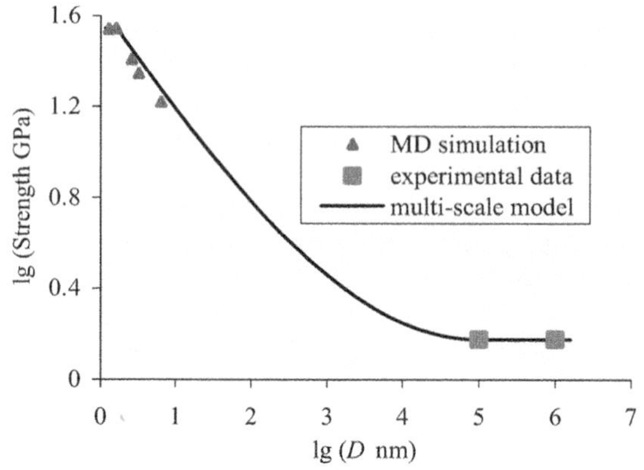


*Figure 6.* Multi-scale model prediction of the strength of W at different sizes.

Figure 7 presents the effect of specimen size on the decohesion energy of single crystal W with crystal orientation of (x[100], y[010], z[001]) under uniaxial tensile loading. The decohesion energy is defined by the area under the corresponding stress-strain curve within the softening regime in Figure 3, with a factor of 0.895 being multiplied to approximately account for the rate effect on the strength. The failure pattern of the deformed W block (not shown here) shows that the failure of the W is an averaged process since the breaking of the bonds across the failure surfaces cannot occur at the same time. When the specimen size increases, more bonds need to be broken before the complete separation of the material can occur, which mainly extends the whole failure evolution process and thus increases the simulated decohesion energy. It appears that the simulated data based on a smaller specimen could provide a more accurate estimation of the decohesion energy of W under tensile loading. Hence, the decohesion energy of W is chosen to be 0.835 N/m based on Simulation 1, as shown in Figure 3.

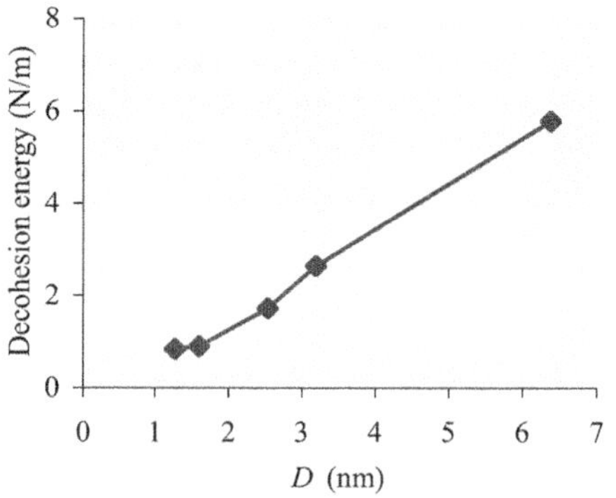


*Figure 7.* The size effect on the decohesion energy of single crystal W block under tension.

**4. Multi-Scale Simulation of Film Delamination**

Based on the proposed multi-scale simulation procedure, as shown in Figure 1, a plane-strain problem for simulating film delamination process is designed. The configuration of the problem is shown in Figure 8. The dimension of specimen is given as $L = 10$ μm, $h_t = 2.5$ μm and $h_s = 5$ μm, respectively. Due to the page limit, only the key parameters of the bifurcation-based decohesion model are discussed here. For the detailed information about the model, the reader is referred to Chen et al. (2003). The W strength is $\sigma_p \cong 2.3$ GPa based on Equation (4) by considering the effect of the film thickness. The corresponding yield strength of W is assumed to be $\sigma_y \cong 1.53$ GPa. Since elastic modulus is size-independent, $E = 411$ GPa is employed with Poisson's ratio $\nu = 0.28$ and mass density $\rho = 15{,}000$ kg/m$^3$. Before the discontinuous bifurcation occurs, the associated von Mises elasto-plasticity model with a linear hardening/softening function is used for W. After bifurcation occurs, the discrete constitutive model is active with reference decohesion energy $U_0 = 1.67$ N/m by considering $U_0$ as twice of the decohesion energy of W. Decohesion is not active inside Si due to the very high yield strength of Si. An elasto-perfectly-plastic von Mises model is employed for Si, with Young's modulus $E = 107$ GPa, Poisson's ratio $\nu = 0.42$, mass density $\rho = 3{,}200$ kg/m, and yield strength $\sigma_y = 8.0$ GPa. A step compressive stress of 1.8 GPa is uniformly applied along both ends of W film at the time $t = 0$ to simulate the dynamic response.

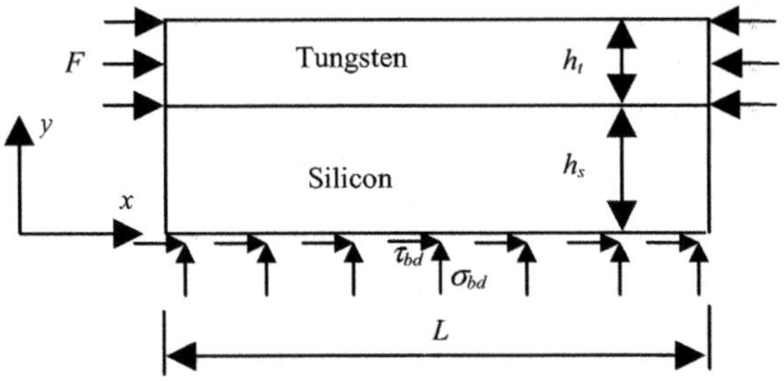


*Figure 8.* Geometry and boundary conditions for a plane strain problem.

Note that the designed substrate thickness is chosen to be much smaller than that in the real film-substrate problem in order to save computational costs. To reduce the effect of stress wave reflection from the bottom boundary of the substrate on the film failure pattern, a silent boundary is applied along the bottom surface of the substrate (Shen and Chen, 2004). The computational grid consists of square cells with each side being $1 \times 10^{-7}$ m long. Initially, one material point per cell is used to discretize both W and Si. To observe the deformation patterns clearly, the deformation fields are magnified by 10 times in the both x- and y-directions.

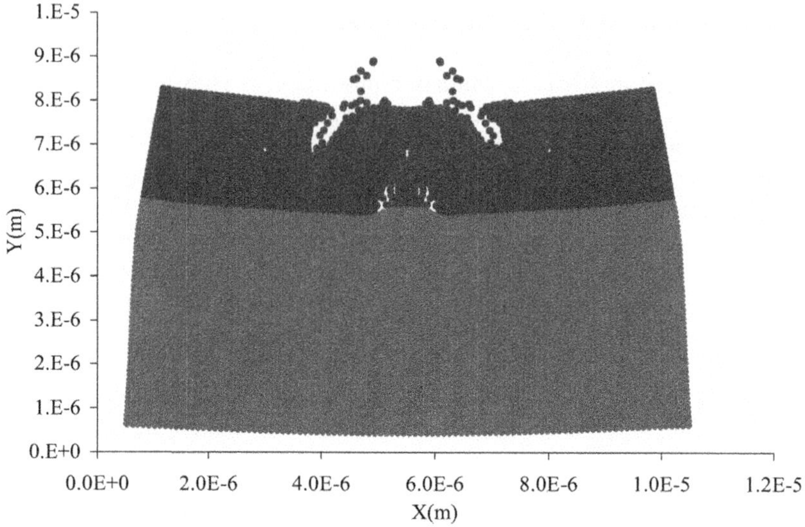


*Figure 9.* Failure pattern of film-substrate structure at t = 2.88 μs with mode I failure.

The effect of failure modes on the deformation pattern of the film-substrate structure is investigated by setting mode I, mode II and mixed mode failures, respectively, in the decohesion model for W. Figures 9-11 present the failure patterns of the W-Si structure at the time $t$ = 2.88 μs with mode I failure, at the time $t$ = 1.92 μs with mode II failure, and at the time $t$ = 2.88 μs with mixed mode failure, respectively. Since the film will be severely damaged at the time $t$ = 2.88 μs with mode II failure, only the deformation field at $t$ = 1.92 μs is being shown for mode II failure. As can be seen from Figures 9 and 11, the deochesion might initiate at the top film surface and evolves deeply into the film until reaching the film-substrate interface, with mode I and mixed model failures. However, the complete delamination of film might occur before the film decohesion reaches the film-substrate interface, with mode II failure, as shown in Figure 10. It appears that the decohesion evolves quicker and the damage is severer if mode II failure is dominant as compared with mode I and mixed mode failures. This might be due to the fact that von Mises model predicts mode II failure that governs the initiation and evolution of decohesion after the discontinuous bifurcation is identified.

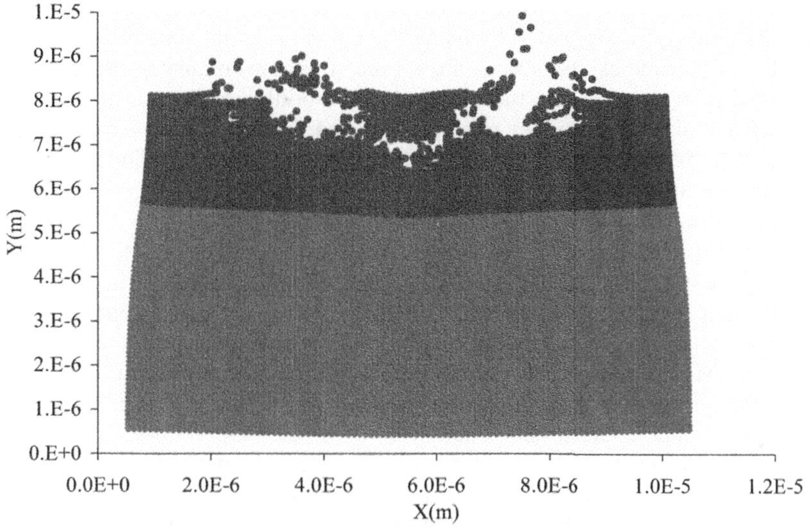


*Figure 10.* Failure pattern of film-substrate structure at t = 1.92 μs with mode II failure.

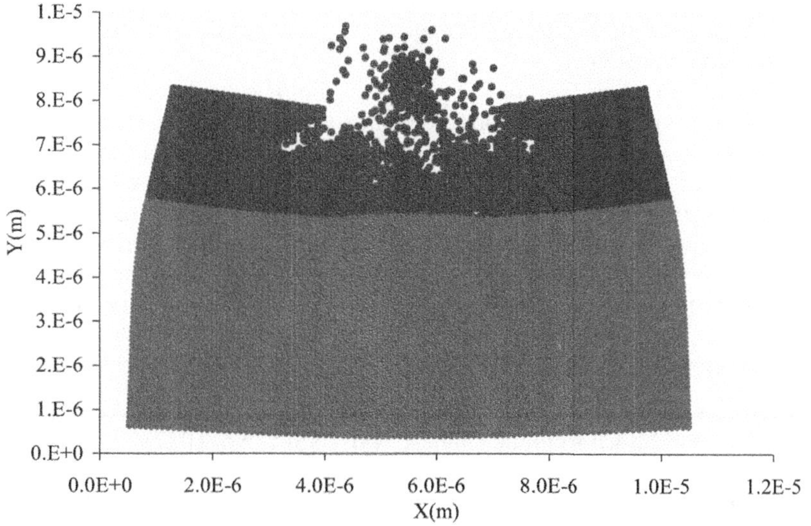


*Figure 11.* Failure pattern of film-substrate structure at t = 2.88 μs with mixed mode failure.

## 5. Conclusions

In this paper, a multi-scale model-based simulation procedure for film delamination process has been proposed. In the proposed procedure, a bifurcation-based decohesion model is first formulated within the framework of the MPM. To calibrate the key parameters, i.e., strength and decohesion energy, for the bifurcation-based decohesion model, MD simulation of W block with crystal orientation of (x[100], y[010], z[001]) under tensile loading is conducted by using the EAM potential. The effects of specimen size, number of vacancies and strain rate on the stress-strain curves are investigated at the atomic level. It appears that the initial elastic modulus of W is almost independent on the specimen size, the number of vacancies and the strain rate. However, the peak stress decreases as the specimen size increases, which is mainly due to the fact that larger specimens offer more opportunities for dislocation to occur, since further MD simulations with artificially introduced defects in the W block show the similar result that the strength of W decrease as the number of vacancies in W block increases regardless of the distribution position. The peak stress increases with the strain rate.

A multi-scale model for predicting W strength from atomic to continuum scales is formulated with the use of power scaling law. The model parameters are calibrated by combining the MD simulations of single crystal W block under tension and the macro-scale experimental results for W. The decohesion energy of W is obtained through the MD simulation of W block under tensile loading. By combining the multi-scale strength model and the decohesion energy for W, a multi-scale decohesion model is developed. The decohesion model is then employed to simulate the process of W film delaminating from Si substrate. It appears that the decohesion evolves quicker and the damage is severer if mode II failure is dominant as compared with mode I failure and mixed mode failure. Blistering usually initiates in the neighborhood of the film's edge and propagates towards the center of the specimen, although it might also initiate at discrete nucleation sites, as can be seen from Figures 9-11.

Since the size effect on the strength of thin W film is considered and the decohesion energy is obtained via MD simulation, the multi-scale simulation of W film delmination from Si substrate could provide better insight into the mechanisms of the film delamination process. However, the proposed multi-scale model for predicting W strength at different structure sizes is based on the MD simulation at the nano-scale and the experimental data for W at the macro-scale. Hence, an integrated experimental, analytical and numerical investigation on the structures with sizes ranging from nano to macro scales is required to verify the proposed multi-scale model-based simulation procedure.

## Acknowledgements

This work was sponsored in part by NSF-NIRT program, as well as by the Overseas Young Investigator Award from the Chinese NSF (NSFC No. 10228206).

## References

Allen, M.P., and Tildesley, D.J. (1990) *Computer Simulation of Liquids*, Oxford University Press, New York.

Audoly, B. (1999) Stability of Straight Delamination Blisters, *Physics Review Letters* **83**, 4124-4127.

Bazant, Z.P. (2002) *Scaling of Structural Strength*, Taylor and Francis, New York.

Chen, Z. and Fang, H.E. (2001) A Study on the Link between Coupled Plasticity/Damage and Decohesion for Multi-Scale Modeling, *J. Mechanical Engineering Science – Proceedings of the Institution of Mechanical Engineers Part C* **215**, 259-263.

Chen, Z., Hu, W., Shen, L., Xin, X., and Brannon, R. (2002) An Evaluation of the MPM for Simulating Dynamic Failure with Damage Diffusion, *Engineering Fracture Mechanics* **69**,1873-1890.

Chen, Z, Shen, L, Mai, Y-W and Shen, Y.G. (2003) A Bifurcation-Based Decohesion Model for Simulating the Transition from Localization to Decohesion with the MPM, *J. Applied Mathematics and Physics* (*ZAMP*), (in press).

Cheung, K.S. and Yip, S. (1991) Atomic-level Stress in an Inhomogeneous System, *J. Applied Physics* **70**, 5688-5690.

Cormier, J., Rickman, J. M. and Delph, T. J. (2001) Stress Calculation in Atomistic Simulations of Perfect and Imperfect Solids, *J. Applied Physics* **89**, 99-104.

Daw, M.S. and Baskes, M.I. (1984) Embedded-atom Method: Derivation and Application to Impurities, Surfaces, and Other Defects in Metals, *Physical Review Letter* **29,** 6443-6453.

Gioia, G. and Ortiz, M. (1997) Delamination of Compressed Thin Films, *Advances in Applied Mechanics* **33**, 119-192.

Horstemeyer, M.F., Baskes, M.I. and Plimpton, S.J. (2001) Computational Nanoscale Plasticity Simulations Using Embedded Atom Potentials, *Theoretical and Applied Fracture Mechanics* **37**, 49-98.

Huang, R. and Suo, Z. (2002) Wrinkling of a Compressed Elastic Film on a Viscous Layer, *J. Applied Physics* **91**, 1135-1142.

Hutchinson, J.W. and Suo Z. (1991) Mixed Mode Cracking in Layered Materials, *Advances in Applied Mechanics* **29**, 63-191.

Hutchinson, J.W. (2000) Plasticity at the Micron Scale, *International Journal of Solids and Structures* **37**, 225-238.

Irving, J.H. and Kirkwood, J.G. (1950) The Statistical Mechanical Theory of Transport Process IV The Equations of Hydrodynamics, *J. Chemical Physics* **18**, 817.

Li, J. and Yip, S. (2002) Atomistic Measures of Materials Strength, *Computer Modeling in Engineering & Sciences* **3**, 219-227.

Liang, W. and Zhou, M. (2003) Size and Strain Rate in Tensile Deformation of Cu Nanowires, *Nanotech* **2**, 452-455.

Milstein, F. and Chantasiriwan, S. (1998) Theoretical Study of the Response of 12 Cubic Metals to Uniaxial Loading, *Physical Review B* **58**, 6006–6018.

Roundy, D., Krenn, C.R., Cohen, M.L., and Morris, J.W. (2001) Ideal Strength of BCC Tungsten, *Philosophical Magazine A* **81**, 1725-1747.

Shen, L. and Chen, Z. (2004) A Silent Boundary Scheme with the Material Point Method for Dynamic Analyses Involving Different Material Phases, *Computer Modeling in Engineering and Sciences*, Submitted for Publication.

Shen, Y.G., Mai, Y.-W., Zhang, Q.C., McKenzie, D.R., Fall, W.D. and McBride, W.E. (2000) Residual Stress, Microstructure, and Structure of Tungsten Thin Films Deposited by Magnetron Sputtering, *J. Applied Physics* **87**, 177-187.

Zhang, B. and Ouyang, Y. (1993) Theoretical Calculation of Thermodynamics Data for BCC Binary Alloys with the Embedded-Atom Method, *Physical Review B* **48**, 3022-3029.

Zhou, M., (2003) A New Look at the Atomic Level Virial Stress: on Continuum-Molecular System Equivalence, *Proc. R. Soc. Lond. A* **459**, 2347-2392.

# INTERFACE FRACTURE: A STUDY BASED ON THE CONVENTIONAL THEORY OF MECHANISM-BASED STRAIN GRADIENT PLASTICITY

S. QU, Y. HUANG
*Department of Mechanical and Industrial Engineering, University of Illinois, Urbana, IL 61801*
C. LIU
*Materials Science and Technology Division, Los Alamos National Laboratory, Los Alamos, NM 87545*
K.C. HWANG
*FML, Department of Engineering Mechanics, Tsinghua University, Beijing 100084, China*

**Abstract**

Cleavage cracking along a bimaterial interface between Nb and sapphire was observed by Elssner et al. (1994) and Korn et al. (2002). The stress required for cleavage cracking is around the theoretical strength of the material. Classical plasticity models fall short to reach such a high stress level. We use the conventional theory of mechanism-based strain gradient plasticity (Huang et al., 2004), which has been established from the Taylor dislocation model, to investigate the stress level around the tip of an interface crack between Nb and sapphire. It is shown that the stress level significantly increases due to the strain gradient effect. At a distance of $0.1\mu m$ ahead of the interface crack tip, the tensile stress is high enough to trigger cleavage cracking.

## 1. Introduction

Elssner et al. (1994) and Korn et al. (2002) studied the behavior of Nb-sapphire interfaces using the notched bending tests. For an interface between single crystal niobium and sapphire, they measured both the macroscopic fracture toughness (from a four-point bend specimen) and the atomic work of separation (from the equilibrium shapes of microscopic pores on the interface). The macroscopic work of fracture is about 1000 times higher than the atomic work of separation. This large difference between macroscopic and atomic work of fracture is due to plastic dissipation in niobium, i.e., the significant plastic flow associated with dislocation activities around the crack tip. It is anticipated that the crack tip would be blunted by dislocations in niobium, and the maximum stress around the crack tip is no more than 4 to 5 times the tensile yield stress $\sigma_Y$ of niobium according to classical plasticity models (Hutchinson,



*Q. P. Sun and P. Tong (eds.), IUTAM Symposium on Size Effects on Material and Structural Behavior at Micron- and Nano-Scales*, 67–76.

1997). However, Elssner et al. (1994) and Korn et al. (2002) observed cleavage cracks in the niobium/sapphire system, i.e., the crack tip remained atomistically sharp without any blunting even though niobium had a large number of dislocations. The stress level around a cleavage crack tip for atomic decohesion of a lattice or strong interface is the theoretical strength of the metal, which is about one tenth of the shear modulus, $\mu/10$. For typical metals, it is approximately *E*/30 (*E* is the Young's modulus) or 10 times the yield stress, i.e., $10\sigma_Y$. This stress level of $10\sigma_Y$ required for cleavage cracks is more than twice the maximum stress around the crack tip predicted by classical plasticity models. It is clear that classical plasticity models fall short to explain Elssner et al.'s (1994) and Korn et al.'s (2002) experimental observation of cleavage cracks in presence of plastic flow.

There are significant efforts to develop continuum (but not classical) plasticity models that predict higher stress level around the crack tip than classical plasticity models do in order to explain the observed cleavage cracks in presence of significant plastic flow. Suo et al. (1993) introduced the so-called SSV model in which the crack tip was surrounded by a dislocation-free strip. The strip was elastic (since there are no dislocations inside), with the strip height on the order of dislocation spacing. The strip was in turn surrounded by the plastic zone. The SSV model indeed predicted much higher stress around the crack tip in the elastic strip. Using the self-consistent method, Beltz et al. (1996) determined the height of the dislocation-free strip. Wei and Hutchinson (1999) combined the SSV model with the cohesive zone model to investigate the stress level around a crack tip. Qu et al. (2004) found significant stress increase around the crack tip due to the strain gradient effect associated with geometrically necessary dislocations.

The present study extends Qu et al.'s (2004) analysis to investigate the stress level around the interface crack tip. It focuses on the effect of the intrinsic material length *l* associated with the strain gradient on the stress increase around the crack tip at the bimaterial interface between elastic and elastic-plastic solids. We use the conventional theory of mechanism-based strain gradient plasticity (Huang et al., 2004), CMSG, established from the Taylor dislocation model (1934, 1938). This lower-order theory of strain gradient plasticity is summarized in Section 2, along with the Taylor dislocation model. The interface crack between niobium and sapphire in Elssner et al.'s (1994) 4-point bend specimen is studied in Section 3 for different values of intrinsic material length *l*.

## 2. Taylor Dislocation Model and the Conventional Theory of Mechanism-based Strain Gradient Plasticity

Huang et al. (2004) developed a conventional theory of mechanism-based strain gradient plasticity (CMSG) based on the Taylor dislocation model (1934, 1938).

### 2.1. TAYLOR DISLOCATION MODEL

The shear flow stress $\tau$ is related to the dislocation density $\rho$ by (Taylor, 1934, 1938; Bailey and Hirsch, 1960)

$$\tau = \alpha\mu b\sqrt{\rho}\,, \tag{2.1}$$

where $\mu$ is the shear modulus, $b$ is the magnitude of the Burgers vector, and $\alpha$ is an empirical coefficient ranging from 0.3 to 0.5. The dislocation density $\rho$ is composed of the density $\rho_S$ for statistically stored dislocations, which accumulate by trapping each other in a random way (Ashby, 1970), and density $\rho_G$ for geometrically necessary dislocations, which are required for compatible deformation of various parts of the material (Nye, 1953; Cottrell, 1964; Ashby, 1970), i.e.,

$$\rho = \rho_S + \rho_G\,. \tag{2.2}$$

The density $\rho_G$ is related to the curvature of plastic deformation (Ashby, 1970; Nix and Gao, 1998), or effective plastic strain gradient $\eta^p$, by

$$\rho_G = \bar{r}\frac{\eta^p}{b}\,, \tag{2.3}$$

where $\bar{r}$ is the Nye factor introduced by Arsenlis and Parks (1999) to reflect the effect of crystallography on the distribution of geometrically necessary dislocations, and $\bar{r}$ is around 1.90 for face-centered-cubic (fcc) polycrystals (Arsenlis and parks, 1999; Shi et al., 2004).

The tensile flow stress $\sigma_{flow}$ is related to the shear flow stress $\tau$ by

$$\sigma_{flow} = M\tau\,, \tag{2.4}$$

where $M$ is the Taylor factor which acts as an isotropic interpretation of the crystalline anisotropy at the continuum level, and $M = 3.06$ for fcc metals (Bishop and Hill, 1951a, b; Kocks, 1970) as well as for body-centered-cubic metals that slip on $\{110\}$ planes (Kocks, 1970). The substitution of (2.1)-(2.3) into (2.4) yields

$$\sigma_{flow} = M\alpha\mu b\sqrt{\rho_S + \bar{r}\frac{\eta^p}{b}}\,. \tag{2.5}$$

For uniaxial tension, the flow stress can also be related to the plastic strain $\varepsilon^p$ by $\sigma_{flow} = \sigma_{ref} f(\varepsilon^p)$, where $\sigma_{ref}$ is a reference stress and $f$ is a nondimensional function determined from the uniaxial stress-strain curve. Since the plastic strain gradient $\eta^p$ vanishes in uniaxial tension, the density $\rho_S$ of statistically stored dislocations is determined from (2.5) as $\rho_S = \left[\sigma_{ref} f(\varepsilon^p)/(M\alpha\mu b)\right]^2$. The flow stress in (2.5) then becomes

$$\sigma_{flow} = \sqrt{\left[\sigma_{ref} f(\varepsilon^p)\right]^2 + M^2\bar{r}\alpha^2\mu^2 b\eta^p} = \sigma_{ref}\sqrt{f^2(\varepsilon^p) + l\eta^p}\,, \tag{2.6}$$

where

$$l = M^2\bar{r}\alpha^2\left(\frac{\mu}{\sigma_{ref}}\right)^2 b = 18\alpha^2\left(\frac{\mu}{\sigma_{ref}}\right)^2 b \tag{2.7}$$

is the intrinsic material length in strain gradient plasticity, $M = 3.06$ and $\bar{r} = 1.90$. This intrinsic material length represents a natural combination of the effects of elasticity (shear modulus $\mu$), plasticity (reference stress $\sigma_{ref}$) and atomic spacing (Burgers vector $b$). It is important to note that, even though this intrinsic material length $l$ depends on the choice of the reference stress $\sigma_{ref}$ in uniaxial tension, the flow stress $\sigma_{flow}$ in (2.6) does not because the strain gradient term inside the square root in (2.6) becomes $\sigma_{ref}^2 l\eta^p = 18\alpha^2\mu^2 b\eta^p$, and is independent of $\sigma_{ref}$. Here the term $18\alpha^2\mu^2 b\eta^p$ involves the plastic strain gradient $\eta^p$ to be given in Section 2.4, well-defined material properties such as shear modulus and Burgers vector, and the coefficient $\alpha$ in the Taylor dislocation model that varies from 0.3 to 0.5.

## 2.2. AN ALTERNATIVE EXPRESSION OF THE UNIAXIAL STRESS-STRAIN CURVE

Huang et al. (2004) suggested to rewrite the uniaxial stress-plastic strain relation $\sigma = \sigma_{ref} f(\varepsilon^p)$ to the following viscoplastic expression in order to pave the way for the establishment of a lower-order theory of CMSG

$$\dot{\varepsilon}^p = \dot{\varepsilon}_0 \left[ \frac{\sigma}{\sigma_{ref} f(\varepsilon^p)} \right]^m , \tag{2.8}$$

where $\dot{\varepsilon}^p$ is the rate of plastic strain, $\dot{\varepsilon}_0$ is a reference strain rate, and $m$ is the rate-sensitivity exponent which usually takes a large value (m $\geq$ 20). In the limit $m \to \infty$, (2.8) degenerates to $\sigma = \sigma_{ref} f(\varepsilon^p)$. Equation (2.8), however, displays a strain-rate sensitivity such that the resulting stress-strain relation depends on the normalized time $\dot{\varepsilon}_0 t$, even though this rate sensitivity is rather weak for a large $m$ ($\geq$ 20).

Following Kok et al. (2002a, b, c), Huang et al. (2004) proposed to replace the reference strain rate $\dot{\varepsilon}_0$ by the effective strain rate $\dot{\varepsilon}$ in order to eliminate the strain-rate sensitivity. Equation (2.8) then becomes

$$\dot{\varepsilon}^p = \dot{\varepsilon} \left[ \frac{\sigma}{\sigma_{ref} f(\varepsilon^p)} \right]^m . \tag{2.9}$$

In the limit $m \to \infty$, (2.9) also degenerates to the uniaxial stress-strain relation $\sigma = \sigma_{ref} f(\varepsilon^p)$. Since (2.9) has strain rate on both sides, the resulting stress-strain relation becomes independent of the strain rate, and the strains (including elastic and plastic strains) do not change once the stresses are fixed. Huang et al. (2004) showed that (2.9) with $m \geq 20$ is an excellent representation of the uniaxial stress-strain relation $\sigma = \sigma_{ref} f(\varepsilon^p)$.

## 2.3. THE CONSTITUTIVE MODEL IN CMSG

The volumetric strain rate $\dot{\varepsilon}_{kk}$ and deviatoric strain rate $\dot{\varepsilon}'_{ij}$ in CMSG plasticity are related to the stress rate in the same way as in classical plasticity, i.e.,

$$\dot{\varepsilon}_{kk} = \frac{\dot{\sigma}_{kk}}{3K} , \tag{2.10}$$

$$\dot{\varepsilon}'_{ij} = \frac{\dot{\sigma}'_{ij}}{2\mu} + \frac{3\dot{\varepsilon}^p}{2\sigma_e} \sigma'_{ij} , \tag{2.11}$$

where $K$ and $\mu$ are the elastic bulk and shear moduli, respectively, $\sigma'_{ij}$ is the deviatoric stress, and $\sigma_e = (3\sigma'_{ij}\sigma'_{ij}/2)^{1/2}$ is the effective stress. The effective plastic strain rate $\dot{\varepsilon}^p = (2\dot{\varepsilon}^p_{ij}\dot{\varepsilon}^p_{ij}/3)^{1/2}$ is obtained from (2.9) except that the tensile flow stress $\sigma_{ref} f(\varepsilon^p)$ in the denominator is replaced by the flow stress in (2.6) established from the Taylor dislocation model accounting for the strain gradient effect, i.e.,

$$\dot{\varepsilon}^p = \dot{\varepsilon}\left(\frac{\sigma_e}{\sigma_{flow}}\right)^m = \dot{\varepsilon}\left[\frac{\sigma_e}{\sigma_{ref}\sqrt{f^2(\varepsilon^p) + l\eta^p}}\right]^m, \tag{2.12}$$

where $\dot{\varepsilon} = (2\dot{\varepsilon}'_{ij}\dot{\varepsilon}'_{ij}/3)^{1/2}$. Equations (2.10)-(2.12) can be rearranged to give the stress rate in terms of the strain rate,

$$\dot{\sigma}_{ij} = K\dot{\varepsilon}_{kk}\delta_{ij} + 2\mu\left\{\dot{\varepsilon}'_{ij} - \frac{3\dot{\varepsilon}}{2\sigma_e}\left[\frac{\sigma_e}{\sigma_{ref}\sqrt{f^2(\varepsilon^p) + l\eta^p}}\right]^m \sigma'_{ij}\right\}. \tag{2.13}$$

This is the constitutive relation in CMSG, which involves the conventional stress and strain only, and no higher-order stress. The plastic strain gradient comes into play to enhance the incremental plastic modulus, similar to Acharya and Bassani (2000) and Acharya and Beaudoin (2000). CMSG also bears similarity with Ever et al.'s (2002) viscoplastic strain gradient plasticity theory because CMSG would have been viscoplastic if $\dot{\varepsilon}_0$ were not replaced by $\dot{\varepsilon}$ in (2.9). Furthermore, Huang et al. (2004) compared CMSG with the higher-order theory of mechanism-based strain gradient plasticity (Gao et al., 1999; Huang et al., 2000a, b) established from the same Taylor dislocation model (1934, 1938). The stress distributions predicted by the lower and higher-order theories are only different within a thin boundary layer whose thickness is around 10 *nm* (e.g., Shi et al., 2001; Huang et al., 2004).

## 2.4. EFFECTIVE PLASTIC STRAIN GRADIENT

The effective plastic strain gradient $\eta^p$ in CMSG (Huang et al., 2004) is the same as that in the higher-order theory of mechanism-based strain gradient plasticity theory (Gao et al., 1999), and is given by

$$\eta^p = \int \dot{\eta}^p dt, \quad \dot{\eta}^p = \sqrt{\frac{1}{4}\dot{\eta}^p_{ijk}\dot{\eta}^p_{ijk}}, \quad \dot{\eta}^p_{ijk} = \dot{\varepsilon}^p_{ik,j} + \dot{\varepsilon}^p_{jk,i} - \dot{\varepsilon}^p_{ij,k}, \tag{2.14}$$

where $\dot{\varepsilon}^p_{ij}$ is the tensor of plastic strain rate.

## 2.5. EQUILIBRIUM EQUATIONS AND BOUNDARY CONDITIONS

Since CMSG does not involve the higher-order stress, equilibrium equations remain the same as those in conventional continuum theories, i.e.,

$$\sigma_{ji,j} + f_i = 0, \tag{2.15}$$

where $f_i$ is the body force. The traction and displacement boundaries are also the same as their counterparts in conventional continuum theories, i.e.,

$$\sigma_{ji} n_j = t_i, \tag{2.16}$$

$$u_i = \bar{u}_i, \tag{2.17}$$

where $n_j$ is the unit normal, and $t_i$ and $\bar{u}_i$ are stress traction and displacement on the traction boundary and displacement boundary, respectively.

### 2.6. FINITE ELEMENT ANALYSIS FOR CMSG

Unlike the higher-order theory of mechanism-based strain gradient plasticity (Gao et al., 1999; Huang et al., 2000a, b), CMSG is a lower-order theory which does not involve the higher-order stress such that its governing equations are essentially the same as those in the classical plasticity. One can easily modify the existing finite element program to incorporate the plastic strain gradient effect. Specifically, we have implemented the constitutive relation (2.13) of CMSG in the ABAQUS finite element program via its USER-MATERIAL subroutine UMAT. Its only difference from classical plasticity is that the plastic strain gradient must be evaluated in UMAT. This is accomplished by the numerical differentiation within the element, i.e., to interpolate the plastic strain increment $\Delta\boldsymbol{\varepsilon}^p$ within each element via the values at Gaussian integration points in the isoparametric space, and then to determine the gradient of plastic strain increment via the differentiation of the shape function.

## 3. Fracture of the Bimaterial Interface between Elastic and Elastic-Plastic Solids

We use CMSG to investigate fracture of the bimaterial interface between sapphire and Nb as in Elssner et al.'s (1994) and Korn et al.'s (2002) experiments. Figure 1 shows a schematic diagram of the four-point bend specimen used in Elssner et al.'s (1994) and Korn et al.'s (2002) experiments. A single crystal sapphire layer is sandwiched between a single crystal and a polycrystalline Nb layer, which in turn are sandwiched by two alumina layers. There is a notch on the interface between the single crystal Nb and sapphire, and the notch length is 0.4 *mm*. The inner span of the four-point bend specimen is 9 *mm*, and the momentum arm is also 9 *mm* (Fig. 1). The specimen thickness (in the out-of-plane direction) is 2 *mm*.

TABLE 1. Elastic properties of Sapphire, Alumina, Single Crystal Nb and Polycrystalline Nb (Korn et al., 2002)

| | Young's modulus $E$ ($GPa$) | Poisson's ratio $\nu$ |
|---|---|---|
| Sapphire | 425 | 0.16 |
| Alumina | 390 | 0.27 |
| Single Crystal Nb | 145 | 0.36 |
| Polycrystalline Nb | 105 | 0.39 |

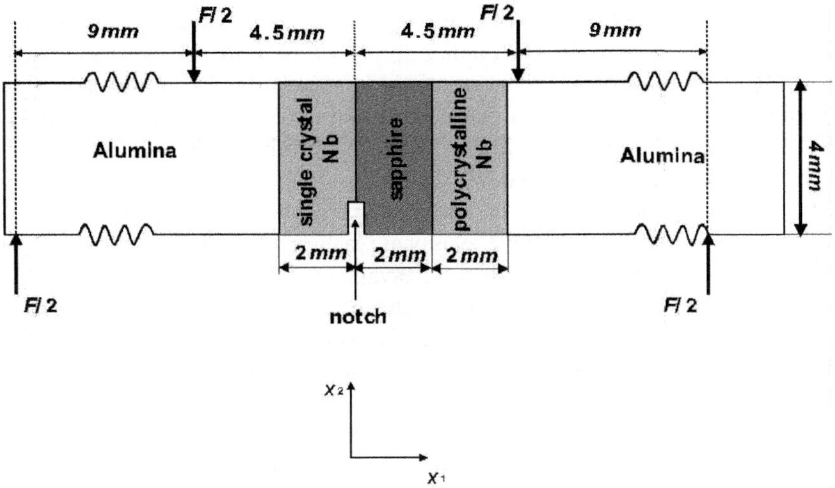


*Figure 1.* A schematic diagram of the four-point bend specimen used in Elssner et al.'s (1994) and Korn et al.'s (2002) experiments to study fracture of Nb/sapphire interface.

Sapphire and alumina are linear elastic. Their elastic properties, as well as those of single crystal and polycrystalline Nb, are given in Table 1 (Korn et al., 2002). The single crystal and polycrystalline Nb are both elastic-plastic with the power-law relation between stress $\sigma$ and plastic strain $\varepsilon^p$ in uniaxial tension,

$$\sigma = \sigma_{ref} f\left(\varepsilon^p\right) = \sigma_Y \left(\frac{E}{\sigma_Y}\right)^N \left(\varepsilon^p + \frac{\sigma_Y}{E}\right)^N, \qquad (3.1)$$

where $E$ is the elastic modulus, $\sigma_Y$ is the initial yield stress, $N$ is the plastic work hardening exponent, and $\sigma_{ref} = \sigma_Y \left(\frac{E}{\sigma_Y}\right)^N$. For single crystal Nb, we have taken the ratio of yield stress to Young's modulus as $\sigma_Y/E = 0.1\%$, which gives the yield stress $\sigma_Y = 145MPa$. The plastic work hardening exponent is $N = 0.05$ to represent easy glide in single crystal deformation. For polycrystalline Nb, the plastic work hardening exponent is $N = 0.24$, the yield stress is $\sigma_Y = 105MPa$ (http://www.rembar.com/niobium.htm) which also gives the ratio of yield stress to Young's modulus as $\sigma_Y/E = 0.1\%$. The rate sensitivity of Nb is $m = 20$, and Burgers vector is $b = 0.25nm$. Besides depending on the shear modulus $\mu$, reference stress $\sigma_{ref}$, and Burgers vector $b$, the intrinsic material length $l$ in (2.7) depends also on the coefficient $\alpha$ in the Taylor dislocation model, which may vary from 0.3 to 0.5.

We focus on the effect of the coefficient $\alpha$ in the Taylor dislocation model (therefore the effect of the intrinsic material length $l$) on the normal stress around the crack tip at the Nb/sapphire interface. Figure 2 shows the distribution of normal stress $\sigma_{11}$ along the single crystal Nb/sapphire interface ahead of the notch tip for the coefficient $\alpha = 0, 0.1, 0.2, 0.3,$ and 0.5. Here $\alpha = 0$ (dashed line) corresponds to classical plasticity without the strain gradient effect; $\sigma_{11}$ is the tensile stress normal to the interface and it is normalized by the yield stress $\sigma_Y = 145MPa$ of the single crystal Nb. The distance $r$ to the notch tip ranges from 0.1 $\mu m$ to 200 $\mu m$. The applied force is $F = 100N$ for the 4-point bend test shown in Fig. 1. It is observed that, at a finite distance away from the notch tip, the strain gradient effect vanishes such that CMSG and classical plasticity give essentially the same stress distribution. This finite distance depends on the coefficient $\alpha$ in the Taylor dislocation model, and is about 2 $\mu m$ for $\alpha = 0.5$ and 0.5 $\mu m$ for $\alpha = 0.1$. Within this finite distance to the notch tip the normal stress given by CMSG is elevated due to the strain gradient effect. For example, classical plasticity ($\alpha = 0$) only gives $\sigma_{11} = 3.7\sigma_Y$ at $r = 0.1$ $\mu m$, which is far short to trigger cleavage cracking. For $\alpha = 0.5$, CMSG gives $\sigma_{11} = 15\sigma_Y$ at the same distance $r$ to the notch tip. This stress is nearly 4 times that of classical plasticity, and definitely exceeds the stress needed for cleavage cracking (about $10\sigma_Y$). For $\alpha = 0.3$, CMSG gives $\sigma_{11} = 11.4$ $\sigma_Y$ at the same $r$, which is also high enough for cleavage cracking. However, this stress becomes $\sigma_{11} = 5$ $\sigma_Y$ for $\alpha = 0.1$ and 6.7 $\sigma_Y$ for $\alpha = 0.2$ at $r = 0.1$ $\mu m$, which are not high enough to trigger cleavage cracking. Since the coefficient $\alpha$ in the Taylor dislocation model usually ranges from 0.3 to 0.5, the above results suggest that the stress level at a distance of 0.1 $\mu m$ away from the notch tip at the Nb/sapphire interface has indeed reached that for cleavage cracks.

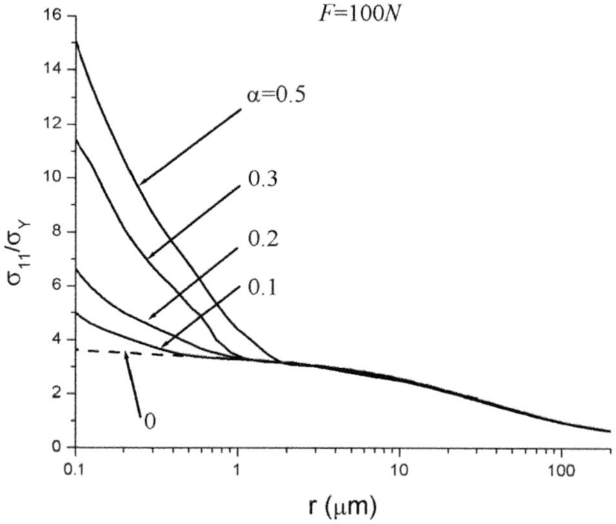


*Figure 2.* The distribution of normal stress $\sigma_{11}$ along the interface between the single crystal Nb and sapphire ahead of the notch tip for different values of the coefficient $\alpha$ in Taylor dislocation model, where $\alpha = 0$ corresponds to classical plasticity (without the strain gradient effect), and $\sigma_{11}$ is the tensile stress normal to the interface normalized by the yield stress $\sigma_Y$ of single crystal Nb. The elastic properties of sapphire, alumina, single crystal and polycrystalline Nb are given in Table 1. The ratio of yield stress to Young's modulus is 0.1% for both single crystal and polycrystalline Nb. The plastic work hardening exponent $N = 0.24$ and 0.05 for polycrystalline and single crystal Nb, respectively. The rate sensitivity exponent $m = 20$ and Burgers vector $b = 0.25$ $nm$. The applied force (see Fig. 1) is $F = 100N$.

## 4. Concluding Remarks

We have used the conventional theory of mechanism-based strain gradient plasticity (CMSG) (Huang et al., 2004) to investigate the fracture of Nb/sapphire bimaterial interface studied in Elssner et al.'s (1994) and Korn et al.'s (2002) experiments. It is shown that, at a distance of 0.1 $\mu m$ away from the notch tip on the Nb/sapphire interface, the normal stress is high enough to trigger cleavage cracking. This cleavage cracking in presence of plastic flow is consistent with the aforementioned experiments.

## Acknowledgements

Y.H. acknowledges the support from NSF (grant # CMS-0084980), ONR (grant # N00014-01-1-0205, program officer Dr. Y.D.S. Rajapakse), and ASCI Center for Simulation of Advanced Rockets at the University of Illinois supported by US department of Energy through the University of California under subcontract B523819. K.C.H. acknowledges the support from the Ministry of Education of China (#20020003023 and Key grant project #0306). The support from NSFC is also acknowledged.

## References

Acharya, A. and Bassani, J.L. (2000) Lattice incompatibility and a gradient theory of crystal plasticity, *Journal of the Mechanics and Physics of Solids* **48**, 1565-1595.

Acharya, A. and Beaudoin, A.J. (2000) Grain-size effect in viscoplastic polycrystals at moderate strains, *Journal of the Mechanics and Physics of Solids* **48**, 2213-2230.

Arsenlis, A. and Parks, D.M. (1999) Crystallographic aspects of geometrically-necessary and statistically-stored dislocation density, *Acta Materialia* **47**, 1597-1611.

Ashby, M.F. (1970) The deformation of plastically non-homogeneous alloys, *Philosophical Magazine* **21**, 399-424.

Bailey, J.E. and Hirsch, P.B. (1960) The dislocation distribution, flow stress, and stored energy in cold-worked polycrystalline silver, *Philosophical Magazine* **5**, 485-497.

Beltz, G.E., Rice, J.R., Shih, C.F. and Xia, L. (1996) A self-consistent method for cleavage in the presence of plastic flow, *Acta Materialia* **44**, 3943-3954.

Bishop, J.F.W. and Hill, R. (1951a) A theory of plastic distortion of a polycrystalline aggregate under combined stresses, *Philosophical Magazine* **42**, 414-427.

Bishop, J.F.W. and Hill, R. (1951b) A theoretical derivation of the plastic properties of a polycrystalline face-centered metal, *Philosophical Magazine* **42**, 1298-1307.

Cottrell, A.H. (1964) *The mechanical Properties of Materials*, J. Willey, New York, p. 277.

Elssner, G., Korn, D. and Rühle, M. (1994) The influence of interface impurities on fracture energy of UHV diffusion-bonded metal-ceramic bicrystals, *Scripta Metallurgica et Materialia* **31**, 1037-1042.

Evers, L.P., Parks, D.M., Brekelmans, W.A.M. and Geers, M.G.D. (2002) Crystal plasticity model with enhanced hardening by geometrically necessary dislocation accumulation, *Journal of the Mechanics and Physics of Solids* **50**, 2403-2424.

Gao, H., Huang, Y., Nix, W.D. and Hutchinson, J.W. (1999) Mechanism-based strain gradient plasticity - I. Theory, *Journal of the Mechanics and Physics of Solids* **47**, 1239-1263.

Huang, Y., Gao, H., Nix, W.D. and Hutchinson, J.W. (2000a) Mechanism-based strain gradient plasticity - II. Analysis, *Journal of the Mechanics and Physics of Solids* **48**, 99-128.

Huang, Y., Xue, Z., Gao, H., Nix, W.D. and Xia, Z.C. (2000b) A study of microindentation hardness tests by mechanism-based strain gradient plasticity, *Journal of Materials Research* **15**, 1786-1796.

Huang, Y., Qu, S., Hwang, K.C., Li, M. and Gao, H. (2004) A conventional theory of mechanism-based strain gradient plasticity, *International Journal of Plasticity* **20**, 753-782.

Hutchinson, J.W. (1997) Linking scales in mechanics, in Karihaloo, B.L., Mai, Y.W., Ripley, M.I. and Ritchie, R.O. (eds.) *Advances in Fracture Research*, Pergamon Press, Amsterdam, pp. 1-14.

Jiang, H., Huang, Y., Zhuang, Z. and Hwang, K.C. (2001) Fracture in mechanism-based strain gradient plasticity, *Journal of the Mechanics and Physics of Solids* **49**, 979-993

Kocks, U.F. (1970) The relation between polycrystal deformation and single crystal deformation, *Metallurgical and Materials Transactions* **1**, 1121-1144.

Kok, S., Beaudoin, A.J. and Tortorelli, D.A. (2002a) A polycrystal plasticity model based on the mechanical threshold, *International Journal of Plasticity* **18**, 715-741.

Kok, S., Beaudoin, A.J. and Tortorelli, D.A. (2002b) On the development of stage IV hardening using a model based on the mechanical threshold, *Acta Materialia* **50**, 1653-1667.

Kok, S., Beaudoin, A.J., Tortorelli, D.A. and Lebyodkin, M. (2002c) A finite element model for the Portevin-Le Chatelier effect based on polycrystal plasticity, *Modelling and Simulation in Materials Science and Engineering* **10**, 745-763.

Korn, D., Elssner, G., Cannon, R.M. and Rühle, M. (2002) Fracture properties of interfacially doped Nb-$Al_2O_3$ bicrystals: I, fracture characteristics, *Acta Materialia* **50**, 3881-3901.

Nix, W.D. and Gao, H. (1998) Indentation size effects in crystalline materials: A law for strain gradient plasticity, *Journal of the Mechanics and Physics of Solids* **46**, 411-425.

Nye, J.F. (1953) Some geometrical relations in dislocated crystals, *Acta Metallurgica et Materialia* **1**, 153-162.

Qu, S., Huang, Y., Jiang, H., Liu, C., Wu, P.D. and Hwang, K.C. (2004) Fracture analysis in the conventional theory of mechanism-based strain gradient (CMSG) plasticity, *Int. J. Fracture* (in press).

Shi, M., Huang, Y., Jiang, H., Hwang, K.C. and Li, M. (2001) The boundary-layer effect on the crack tip field in mechanism-based strain gradient plasticity, *International Journal of Fracture* **112**, 23-41.

Shi, M., Huang, Y. and Gao, H. (2004) The J-integral and geometrically necessary dislocations in nonuniform plastic deformation, *International Journal of Plasticity* **20**, 1739-1762.

Suo, Z., Shih, C.F. and Varias, A.G. (1993) A theory for cleavage cracking in the presence of plastic flow, *Acta Metallurgica et Materialia* **41**, 1551-1557.

Taylor, G.I. (1934) The mechanism of plastic deformation of crystals. Part I.–theoretical, *Proceedings of the Royal Society of London A* **145**, 362-387.

Taylor, G.I. (1938) Plastic strain in metals, *Journal of the Institute of Metals* **62**, 307-324.

Wei, Y. and Hutchinson, J.W. (1999) Models of interface separation accompanied by plastic dissipation at multiple scales, *International Journal of Fracture* **95**, 1-17.

Xue, Z., Huang, Y., Hwang, K.C. and Li, M. (2002) The influence of indenter tip radius on the micro-indentation hardness, *J. Eng. Mater. Tech.* **124**, 371-379.

# DEFORMATION AND DIFFUSION IN NANO-GRAINED METALS

WEI YANG, XIN-LING MA, HONGTAO WANG and WEI HONG
*Department of Engineering Mechanics, Tsinghua University,*
*Beijing 100084, china*

**Abstract**

Metals of nano-grain sizes deform by grains sliding pass each other, with little distortion occurring in the grain cores. Actual deformation minimizes the plastic dissipation and stored strain energy for representative steps of grain neighbor switching. A model of 9-grain cluster is proposed that engages both the Ashby-Verrall mechanism and the 30 degree rotation of closely linked grain pairs. A constitutive relation is derived that relates the creep strain rate linearly to the difference between the applied stress and a threshold stress. We also conduct investigation for the rapid stretching of nano-grained metals via molecular dynamics simulation. The low temperature nitriding process for surface-nanocrystallized iron is analyzed and the result indicates that the ultra-fine grain size and the gradient distribution hold the key for the unique concentration profile obtained in the experiment.

## 1. Introduction

When a grain shrinks below a critical size of dislocation depletion that ranges from 15 nm to 20 nm for various metals, the available experimental evidence suggests that dislocations seldom present within the grain. Nano-grained metals exhibit peculiar deformation behavior. Molecular dynamics simulation [1] indicated that copper of an average grain size of 5.2 nm deformed mainly by many small, independent slip events in the grain boundaries. Though dislocations were seldom observed within the grains, they did observe noticeable traces of stacking faults due to the passage of partial dislocations across the grains. Absence of interior dislocations does not cripple the capability of plastic deformation of nano-grained metals. Instead, material scientists had great expectation [2]on the super-ductility behavior of nano-grained materials at room temperature. Recently, Lu et al. [3]reported a stunning experiment of rolling nano-crystalline copper (about 30 nm grain size) at room temperature (0.23 of homologous temperature) to an elongation of 5100%. No texture seemed to form during rolling, and the deformed sample at 5100% elongation exhibited random orientations of the equiaxed nano-grains, with about the same grain size and slightly increased dislocation density as that of the as-deposited sample.

A small grain size renders unexpected features in the deformation of sub-micron metals. When the grain sizes fall within the range of about 10 nm, they can be regarded



*Q. P. Sun and P. Tong (eds.), IUTAM Symposium on Size Effects on Material and Structural Behavior at Micron- and Nano-Scales,* 77–84.

as crystallized grain cores glued together by viscous interlayers with atoms of short-range order. Nano-grain metals of such small grain sizes deform mainly by neighboring grains sliding pass each other with insignificant distortion in the grain shape. Ashby and Verrall [4]gave a vivid description on the deformation of four-grain cluster, where the inter-grain distance (measured from their mass centers) in the elongation direction increases, and that normal to the elongation decreases. Unfortunately, that process is incomplete since it cannot repeat itself. Beside the change in the inter-grain distance, rotation of the inter-grain link may be deemed necessary.

The present paper aims at exploring the mechanics framework under the above-mentioned scenario of nano-grains sliding pass each other. Section 2 describes the formulation on the kinematical and energetic aspects of a meso-scale cell assembling nano-grains, as well as its preliminary simulation. Section 3 describes a model of 9-grain cluster, with the outcome of the quantitative prediction on creep rate and creep threshold. The prediction agrees with the experimental result by Cai et al. [5] Section 4 describes the fast stretching of nano-grained metal through molecular dynamics simulation.

We also examine in Section 5 for the diffusion mechanism of low temperature nitriding on pure iron with SMAT that was accomplished recently at 300°C. The nitrogen concentration profile differs from that of the conventional crystalline iron. We simulate the nitriding process and the analysis indicates that the ultra-fine grain size and the gradient distribution hold the key for the unique concentration profile obtained in the experiment.

## 2. Theory

Consider an aggregate element that consists of $N$ grains. A set GB is defined as the pairs of grains that currently share the same grain boundaries. The set evolves as the grain switches its neighbors during the deformation. The grain boundary shared by the $\alpha$-th and the $\beta$-th grains is denoted by $\Gamma^{\alpha\beta}$. $\Gamma^{\alpha\beta}$ has the normal of $\mathbf{n}^{\alpha\beta}$ and the length of $l^{\alpha\beta}$. The perception of Ashby [6]on the geometrically necessary dislocations (GND) for a polycrystalline metal may be borrowed to simulate the integrity of interlayers separating the nano-grain cores. The discontinuity due to the relative gap/overlap in the GB normal and the relative rotation may be regarded as three "geometrically necessary super-dislocations:" two edge super-dislocations of equal amplitude but opposite signs located at both ends of $\Gamma^{\alpha\beta}$, and one edge super-dislocation at the center of $\Gamma^{\alpha\beta}$.

Geometrically necessary dislocations store elastic energy. One may calculate the strain energy from the interlayer distortion along $\Gamma^{\alpha\beta}$. The strain energy due to the gap/overlap motion is:

$$\Delta E_n = \frac{\mu}{\pi(1-\nu)} \sum_{\alpha,\beta\in GB} \left[(\boldsymbol{u}^{\alpha} - \boldsymbol{u}^{\beta}) \cdot \boldsymbol{n}^{\alpha\beta}\right]^2 \left( \ln\frac{l^{\alpha\beta}}{r_0} + \frac{1}{2} \right) \tag{1}$$

where $\mu$ denotes the shear modulus, $\nu$ denotes the Poisson's ratio, and $r_0$ the cut-off core of the dislocation. The strain energy due to the relative rotation can be calculated by the energy of a tilt grain boundary as:

$$\Delta E_r = \frac{\mu b}{4\pi(1-\nu)} \sum_{\alpha,\beta\in GB} l^{\alpha\beta} \left|\theta^\alpha - \theta^\beta\right| \ln \frac{1}{\left|\theta^\alpha - \theta^\beta\right|}. \tag{2}$$

Denote $\Delta g$ as the difference between the specific free energy of the short range ordered matter and that of the crystallized grain core. Above a ground energy state for the crystallized matters, extra energy stored in disordered interlayer is $E_n = \Delta g D \sum_{\alpha,\beta\in GB}^{3N} \delta^{\alpha\beta} l^{\alpha\beta}$, where $\delta^{\alpha\beta}$ is the thickness of $\Gamma^{\alpha\beta}$. The value of $\Delta g$ can be estimated as $\Delta g = \Gamma / r_{\text{triple}}$, where $\Gamma$ denotes the interface tension, and $r_{\text{triple}} = (\delta, 2\delta)$ is the radius of curvature between ordered and disordered phases near a triple junction.Grain boundary sliding dissipates plastic work. The rate of plastic dissipation per unit area can be written as

$$\dot{W}_{\text{slide}} = \frac{4}{\pi N D^2} \sum_{\alpha,\beta\in GB}^{3N} \tau_{\text{B}}(\theta^\alpha - \theta^\beta) \left[ \frac{\left|(\boldsymbol{v}^\alpha - \boldsymbol{v}^\beta)\times(\boldsymbol{x}^\alpha - \boldsymbol{x}^\beta)\right|}{D^\alpha + D^\beta} + \frac{D^\alpha \omega^\alpha + D^\beta \omega^\beta}{2} \right] l^{\alpha\beta} \tag{3}$$

The grain boundary sliding resistance $\tau_{\text{B}}(\theta^\alpha - \theta^\beta)$ depends on the current misalignment $\theta^\alpha - \theta^\beta$ between the $\alpha$-th and the $\beta$-th grains. Typically, the grain boundary sliding resistance decreases for higher angle grain boundary.

An energetically favorable deformation mode for the nano-grained metal should satisfy the following conditions: (1) the strain energy for the geometrically necessary dislocations, plus the accumulated plastic dissipation by grain boundary sliding, should be minimized; and (2) the minimized strain energy and plastic dissipation should be compensated by the work done under the prescribed loading against the strain $\varepsilon_{ij}$ of the aggregate.

Preliminary simulation is conducted according to the above scheme. The load is applied by incremental strain $\Delta\varepsilon$, followed by complete relaxation. The stress strain curves obtained are shown in Fig. 1 and they are parameterized by $\Delta\varepsilon$. Further study is needed to incorporate the effect of mass diffusion.

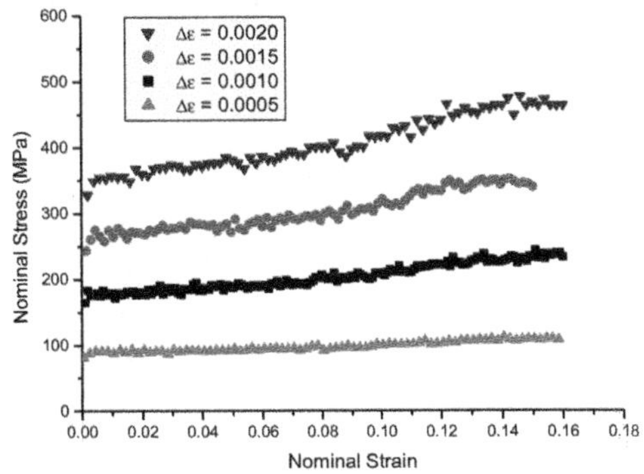


*Figure 1.* Simulated stress strain curve under strain incremental loading.

## 3. Model

The Coble creep mechanism was favored by Karch et al. [2]and by Lu et al. [3]in describing the strong dependence of the flow rate on grain size, the model nevertheless predicts elongated grains that contradicts the experimental finding by Lu et al. [3] For the room temperature superplasticity behavior of nano-grained metals below a homologous temperature less than 0.3, the grain boundary sliding itself should be the dominant mechanism. Moreover, the creep rate of nano-grained copper was found to be linear, rather than quadratic, with respect to the applied stress [3,5] Accordingly, neither the dislocation motion transverse the grains, nor the deformation rate in the grain interior should be the bottleneck of the deformation, though they might assist the deformation.

The 4-grain cluster model of Ashby and Verrall [4] with its concrete calculation and the perception of grains squeezing pass each other, should serve as the starting point for the mechanism model of nano-grained metals. Figure 2 illustrates an attempt to complete the Ashby-Verrall mechanism by adding a rotation of closely contacted grain pairs. The cell contains 9 regular hexagon grains and is termed the 9-grain cluster model. The diameter of the grain's circumferential circle is denoted as D. The cluster covers an area of $27\sqrt{3}D^2/8$. We assume the cluster has a uniform thickness D that is unchanged during the plane strain deformation. The mass conservation during grain boundary sliding, grain rotation, mass diffusion and ordering/disordering transition renders the total area of the cluster invariant.

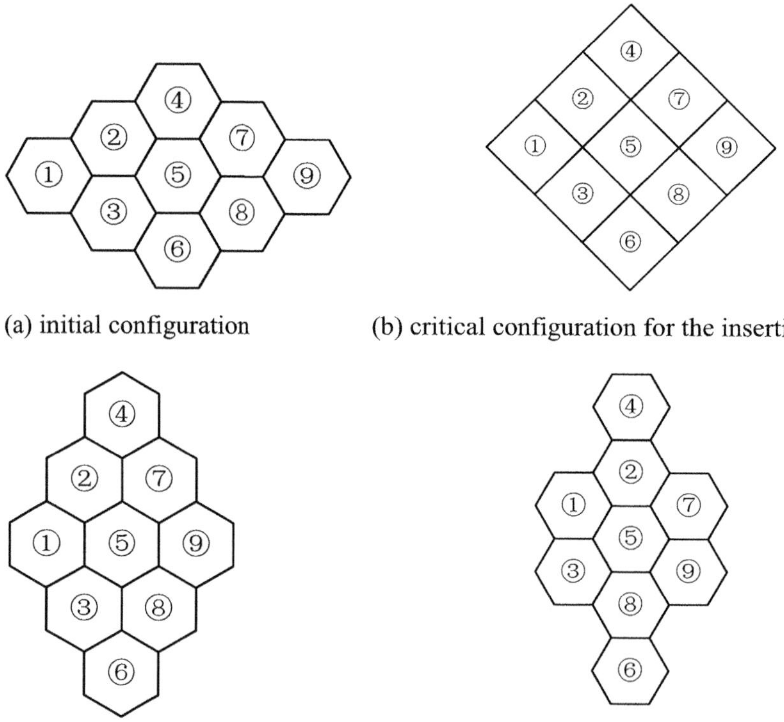


(a) initial configuration (b) critical configuration for the insertion process

(c) completion of insertion stage (d) rotation of 30 degrees

*Figure 2.* Model of 9-grain cluster.

Fully relaxed creep refers to a simple case where the loading rate $\dot{\varepsilon}$ is so slow that the deformation can be fully accommodated by mass diffusion, grain rotation and grain boundary sliding. Neither the cross-grain dislocation emissions nor the geometrically necessary dislocations along the grain boundaries are present in that case. The plastic dissipation within a representative grain is composed of four parts: (1) dissipation by mass diffusion from sources to the sinks; (2) dissipation by grain boundary sliding; (3) dissipation by fluctuation in grain boundary area; and (4) dissipation by fluctuated ordering/disordering transition. Neither fluctuation in the grain boundary area nor that in the volume of disordered region is recoverable. The energy is irreversibly dissipated in terms of heat. The last two contributions will provide a threshold stress for creep.

The macroscopic plastic power. Accordingly, the constitutive law for fully relaxed creeping behavior is:

$$\dot{\varepsilon} = C(\sigma - \sigma_{\mathrm{th}}) \tag{3}$$

where the creep rate C is given by

$$C = \frac{\sqrt{3}\Delta\varepsilon^2}{8D}\left[\frac{m^2 K_{\mathrm{B}} T D}{3\Omega\left\{nD_{\mathrm{V}} + D_{\mathrm{B}}\sum_{i=1}^{n}\frac{\delta}{l_i}\right\}} + \frac{\eta_{\mathrm{B}}}{\delta}s^2\right]^{-1} \tag{4}$$

and the threshold stress is:

$$\sigma_{\mathrm{th}} = \frac{1}{\Delta\varepsilon}\left[\left(\sqrt{2\sqrt{3}} - \sqrt{3}\right)\frac{4\Gamma}{3D} + \left(1 - \frac{\pi}{2\sqrt{3}}\right)\Delta g\right] \tag{5}$$

The linear creep law in (3) agrees in form with the experimental data by Cai et al. (1999) for nano-grained copper.

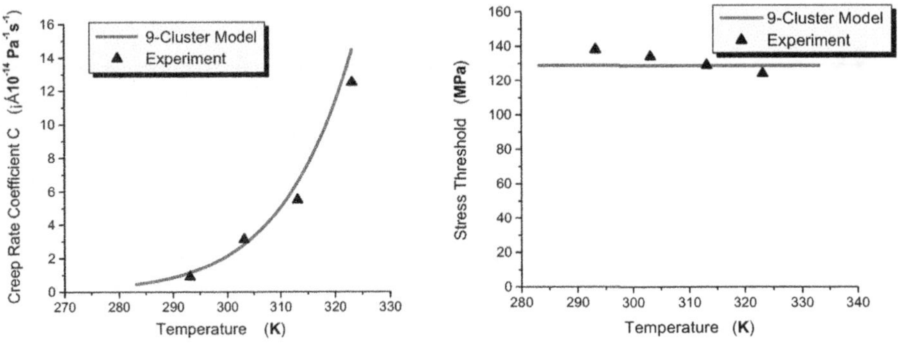


*Figure 3.* Creep rate (left) and threshold stress (right) for nano-grained copper.

To compare the prediction of the present model to the experiment of Cai et al. [5] the following material data for nano-crystalline copper are adopted. The elastic behavior of copper is characterized by E = 115Gpa and $\nu = 0.31$, the atomic volume of copper is $\Omega = 8.78\times10^{-30}\,\mathrm{m}^3$. The diffusion constants for copper are: $D_{0B} = 3\times10^{-9}\,\mathrm{m}^2/\mathrm{s}$, $\Delta Q_B = 0.72\mathrm{eV}$; $\delta = 0.5$ nm (see Cai et al. [5]; and $D_{0V} = 4.4\times10^{-6}\,\mathrm{m}^2/\mathrm{s}$, $\Delta Q_V = 1.98\mathrm{eV}$. The specific grain boundary energy for copper is 0.53J/m2. If an intermediate value of the triple junction radius is taken as $r_{triple} = 1.5\delta \approx 0.75\mathrm{nm}$, then $\Delta g = 7.95\times10^8\,\mathrm{J/m}^3$. The average grain size for the nanocrystalline copper tested by Cai et al. [5]and Lu et al. [3] is about 30 nm. The predicted result, as shown in Figure 3, agrees well with the experiment [5].

## 4. Fast Stretching of Nanocrystals

The method of molecular dynamic simulation is only capable of the case of fast stretching. The previous works concerning on this subject were not conducted to the regime of large deformation. Here we only present a simulation for copper under LJ potential. The profiles of the local crystalline order are plotted in Figure 4. The sites where local fcc order are colored as green, that of hcp as red, and those of amorphous state as blue. Two graphs in Figure 4 represent the different logarithmic strains in the longitudinal direction. As the amount of tension increases, one observes the bursts of defects (in the disordered form of stacking faults) within the grains, and the whitening areas gradually increase. Most stacking faults are arrested by the grain boundaries and rotate slightly toward the tensile direction. A different family of stacking faults may be observed during the later stages of tension, and some areas dissolve into sub-grains of smaller size.

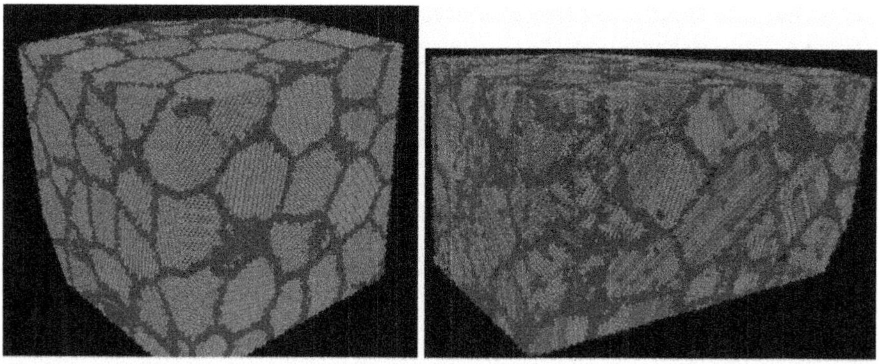

(a) 2% (b) 39%

*Figure 4.* Profiles of centro-symmetry parameter for stretching nanocrystalline Cu.

## 5. Diffusion in Nanocrystals

Recently Tong et al. [7]reported low-temperature (300° C) nitriding on the surface of pure Fe plate treated by SMAT. The essence of the treatment is to introduce a self-nanocrystalline surface layer with gradual transition to the conventional crystalline material. The concentration profile of nitrogen has a feature distinct from that of the traditional crystalline materials: a saturated layer of nitrogen concentration up to about 10 microns from the surface followed by rapid declination away from the surface layer. The ultra-fine grain size and the gradient distribution of the grain size along the depth hold the key for this unique nitriding distribution. A diffusion simulation that takes into account the grain size gradient reached good agreement with the experimental result, as shown in Figure 5.

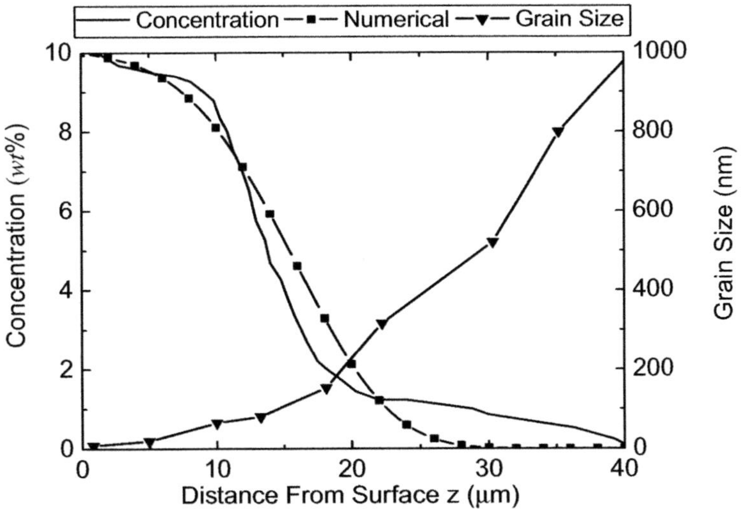


*Figure. 5.* Gradient characterization of pure Fe plate processed by SMAT [7] the solid curve with squares is the simulation of present work.

## 6. Conclusions

A micromechanics theory for the deformation of nano-grained metals is proposed. Preliminary numerical scheme is developed to simulate the case of incremental strain loading. The model of 9-grain cluster seems to capture the essence of the room temperature superplastic creep of nano-grained copper. Molecular dynamics simulation is carried out for fast stretching of nanocrystals to the extent of large deformation. The diffusion analysis under a variable grain size model predicts the low temperature nitriding curve.

**References**

1. Schiotz J., Di Tolla F. D., Jacobsen K. W. *Nature*, 391(1998): 561-563.
2. Karch J., Birringer R and Gleiter H. *Nature*, 330(1987): 556-558.
3. Lu L., Sui M. L. and Lu K., *Science*, 287(2000): 1463-1466.
4. Ashby M.F. and Verrall R.A. *Acta Metallurgica*, 21 (1973): 149-163.
5. Cai B., Kong Q.P., Lu L. and Lu K. *Scr. Metall.*, 41(1999): 755-759.
6. Ashby M.F. Philosophical Magzine, 21(1970): 399-424.
7. W. P. Tong, N. R. Tao, Z. B. Wang, J. Lu and K. Lu, Science, 2003; 299:686.

# MULTISCALE ANALYSIS OF STRENGTH ENHANCEMENT OF CELLULAR MATERIALS UNDER IMPACT LOADING

H. ZHAO and S. ABDENNADHER
*Laboratoire de Mécanique et Technologie,*
*ENS-Cachan/CNRS/University Paris 6*
*61, Avenue du président Wilson,*
*94235 Cachan cedex, France*

## Abstract

This paper presents a multiscale analysis of the macroscopic rate sensitivity of metallic cellular materials. It begins with a review of likely causes responsible for the strength enhancement of cellular materials, which include the rate sensitivity of the base material, the effects of the pressure of the air entrapped in the cell, the inertia effect and the shock enhancement. A testing method using 60 mm diameter Nylon Hopkinson pressure bars with suitable signal processing is used to investigate the rate sensitivity of various metallic cellular materials, i.e., 5056, 5052 aluminium honeycombs of different cell sizes and wall thickness, IFAM and Cymat aluminium foams. In order to identify the factor responsible for the strength enhancement of those materials, a multiscale analysis is performed on a model structure which is a square tube made of rate insensitive materials. At the macroscopic level, significant enhancement is experimentally observed under impact loading, whereas the crushing mode is nearly the same under both static and impact loading. At mesoscopic level, post mortem microhardess measurement provides a map of residual plastic strain everywhere in the cell wall. Numeric simulations and theoretical models give a satisfactory explanation of the role of the lateral inertia.

## 1. Introduction

Metallic based cellular materials (honeycomb, foam, hollow spheres) are promising structural materials which can be used in lightweight structure, impact energy absorption, acoustical wave attenuation, etc. A large number of experimental, numerical and analytical studies on the behaviour of cellular materials have been reported in the open literatures (Wierzbicki, 1983; Gibson & Ashby, 1988; Klintworth & Stronge, 1988; Reid & Peng, 1997; Deshpande & Fleck, 2000a; Banhart, 2001; Hanssen et al., 2002, Zhao, 2004).

Under quasi-static loading, many micromechanical analyses have been reported for honeycombs. Wierzbicki (1983) has developed an out-of-plane crushing model that gives an analytical prediction of the crushing pressure; Klintworth and Stronge (1988)



*Q. P. Sun and P. Tong (eds.), IUTAM Symposium on Size Effects on Material and Structural Behavior at Micron- and Nano-Scales,* 85–95.

have formulated the behaviour of in-plane crushing that takes account of localised deformation band effects. The behaviours of metallic foams have been studied by many authors. Under dynamic loading, the rate sensitivity of these materials is, however, not clear. In the case of metallic foams, Deshpande and Fleck (2000b) used a standard split Hopkinson pressure bar (SHPB) set-up (diameter 12.7 mm) with a polymeric output bar to test Alulight and Duocel foams. Results were obtained with considerable scatter partially because of small size of samples and no rate sensitivity was observed. Mukai et al. (1999) have investigated, using also a standard SHPB arrangement, Alporas foams obtained by direct foaming of aluminium melts with a blowing agent. Rate sensitivity was observed in their results. For aluminium honeycombs, Goldsmith and co-workers (1992) have reported some experimental work on out-of-plane crushing and on the ballistic perforation of honeycombs. They have fired a rigid projectile to a target made of honeycombs and have shown that mean crushing pressures sometimes increase up to 50% with respect to static results. Wu and Jiang (1997), Zhao and Gary (1998) have also found a significant enhancement of out-of-plane crushing, contrary to the generally admitted viewpoint of a negligible rate sensitivity of such materials.

This paper presents at first an experimental study on the behaviour of cellular materials under impact loading. Various factors that may be potentially responsible for the strength enhancement of metallic foam are discussed in Section 2. The testing method using large diameter Nylon pressure bars in SHPB set-up and experimental results on aluminium honeycombs, IFAM and Cymat foams are presented in Section 3. A multiscale analysis on a model-structure (square tubes) in Section 4 shows that the inertia effect is the main reason of measured macroscopic rate sensitivity of studied cellulaire materials.

## 2. Potential Causes for Rate Sensitivity

There exist four known causes for the rate sensitivity of the cellular materials: namely, the rate sensitivity of the base material, the pressure of the air entrapped in the cell, the inertia effect and the shock enhancement. All these potential sources may contribute to the macroscopic rate sensitivity in metallic cellular materials.

### 2.1. STRAIN RATE SENSITIVITY OF THE BASE MATERIAL

It has been shown in micromechanics analyses that the ratio of cellular material strength over that of the cell wall material depends on a power function of the relative density (Gibson and Ashby, 1988). From such an analysis, the rate sensitivity of the base material induces the macroscopic rate sensitivity of a cellular material. However, it is difficult to determine the local strain rate accurately because the strain is localised in a narrow band. As a consequence, the average nominal strain rate underestimates the actual local strain rate. An approximate relation between the strain rate of the cell wall materials and the nominal strain rate of the cellular material can be however evaluated. The ratio between the strain rate of the cell wall material and the nominal strain rate of the honeycomb ranges from 1 to 2 (zhao, 2004). Thus, the rate sensitivity of a metallic cellular material is comparable with that of the base metals.

## 2.2. PRESSURE INCREASE OF THE AIR ENTRAPPED IN THE CELL.

The strain rate sensitivity of a cellular material may be due also to the existence of a gas/fluid phase (Gibson and Ashby, 1988; Deshpande and Fleck, 2000a). Taking an open cell foam as an example, when the foam is loaded slowly, the gas filled in the foam can move our without resistance. Under impact, the gas entrapped in cells may have no time to move out. They will then take part in the whole resistance of cellular materials by the pressure increase due to the volume change. The quicker the loading, the more the air is entrapped.

The importance of this supplementary air pressure can be evaluated by the gas volume change (zhao, 2004). For a foam of 0.1 relative density filled with air (specific coefficient $\gamma = 1.4$), one can deduce that the amplitude of the contribution of air pressure is around 2 times initial pressure at 50% strain (only 0.1 times at 10% strain). The upper limit of air pressure is about 0.2 MPa. It can be neglected in most cases of metallic cellular materials.

## 2.3. MICRO-INERTIA EFFECT

Another possible factor that may cause strength enhancement is the micro-inertia effect. It was shown that the buckling of a column under impact compression occurs at a delayed time because of lateral inertia, so that the apparent critical buckling force is higher than the quasi-static one. Calladine and English (1984) explained in detail the role played by lateral inertia. They developed simple models to describe the buckling and post-buckling behaviour of a simple column under impact loading. Such a basic concept was used by a number of authors to explain observed strength increase under impact. Su et al. (1995) have given a classification of the rate sensitivity of different structures. Langseth et al. (1996) considered that the lateral inertia is a cause of the strength increase observed in steel and aluminium square tubes. In the case of cellular materials, it is observed for Balsa wood that the folding wave-length has become smaller (Vural and G. Ravichandran, 2003) under impact loading and it explains a higher crushing strength. Even for the case where there is no wavelength change, the enhancement is still possible in progressive folding because of inertia effects (Zhao and Abdennadher, 2004). Therefore, the micro-inertia effect is not negligible in the case of metallic cellular materials because of the high mass density of the metal base.

## 2.4. SHOCK WAVE EFFECT UNDER HIGH-VELOCITY IMPACT.

At high impact speed (>50m/s), Reid and Peng (1997) reported the strength enhancement by the formation of shock wave. If the constitutive relation is concave, a single shock wave front can be formed. However, such kind of concave constitutive function is unusual for solid materials in uniaxial stress condition. For cellular materials, the densification part of stress-strain curve is a concave function and the shock wave may be formed under usual one-dimensional stress conditions.

It is noted that such stress enhancement cannot be considered as a material characteristic. Such a shock enhancement can be numerically reproduced by using a rate insensitive model with the consideration of the densification and locking behaviour.

Besides, such shock enhancement occurs only at high impact velocities. For these reasons, the shock enhancement will not be considered in this study.

## 3. Testing using Large Scale Nylon Hopkinson Pressure Bars and Experimental Results

Above analyses show that the rate sensitivity of cellular materials comes from various sources. It is therefore difficult to predict the rate sensitivity by a simple theoretical model. Experimental studies are definitely the main investigating means. To perform tests at high strain rates, the most widely used method is the spilt Hopkinson pressure bar test.

### 3.1. STANDARD SHPB ANALYSIS

The SHPB (Split Hopkinson Pressure Bar), also called Kolsky apparatus, is a commonly used experimental technique to study constitutive laws of materials at high strain rates (Hopkinson, 1914; Kolsky, 1949) and the strain signals at the bar-specimen interface. From forces and velocities at both bar-specimen interfaces, the classical analysis assumes the axial uniformity of stress and strain fields in the specimen and an average stress-strain curve can be obtained (like those obtain from a quasi-static test) which lead to the so-called two-waves analysis. A summary of the development of the SHPB technique can be found in (Zhao and Gary, 1996, 1997).

### 3.2. WAVE DISPERSION CORRECTION FOR LARGE DIAMETER NYLON PRESSURE BARS

From an experimental point of view, impact tests on cellular materials using SHPB have two major difficulties. One is the large scatter due to the small ratio between the specimen size and the cell size. To overcome this difficulty, a large diameter pressure bar is necessary to host a larger specimen. Another is the weak signal due to the weak strength of cellular materials, which leads to a low signal-noise ratio. Here, soft nylon pressure bars are used to solve this problems. However, soft Nylon bars are viscoelastic materials and the wave dispersion increases greatly with the diameter of the bars. Kolsky's original SHPB analysis is based on the basic assumptions that the wave propagation in the bars can be described by a one-dimensional wave propagation theory. As the waves are not measured at bar-specimen interfaces to avoid their superposition, they have to be shifted from the position of the strain gages to the specimen faces, in terms of both time and distance. This shifting needs the knowledge of wave propagations along the bar.

Wave dispersions in a cylindrical bar have been extensively studied in the past decades. Pochhammer and Chree's harmonic wave propagation theory has been used in the data processing (Davies, 1948). The correction of wave dispersion due to a viscoelastic behaviour coupled with a geometrical effect is necessary for the large diameter Nylon bar. Such a correction is not equivalent to a simple additive correction of a nominal elastic geometrical dispersion and a one-dimensional viscous dispersion and attenuation. As the attenuation coefficient is also affected by the geometric effect, it has been proved (Zhao and Gary, 1995) that such an approach gives wrong results.

## 3.3. RATE SENSITIVITY OF CELLULAR MATERIALS – EXPERIMENTS

The 60 mm diameter Nylon Hopkinson pressure bar allows for an accurate measurement, because a reasonable large specimen can be used and the signal/noise ratio is improved by soft Nylon bars. It is applied to measure the rate sensitivity of various cellular materials.

### *3.3.1. Honeycombs*

The dynamic out-of-plan behaviours of 5056 and 5052 aluminium honeycombs were studied, which involved three different cell sizes and wall thicknesses. Figure 1a shows a typical progressive buckling pattern observed for all the tested honeycombs. Figure 1b presents typical mean stress (crushing force divided by cross section) – displacement curves. The bold part is under impact loading, which is followed by the quasi-static test. It shows that the enhancement under impact loading does exist.

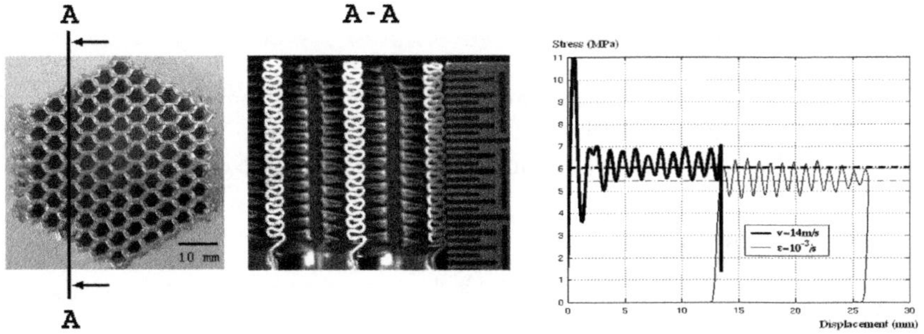


*Figure 1.* a) Progressive folding pattern; b) A typical mean stress- displacement curve during impact and subsequent quasi-static test

Such experimental results show that there is a strain rate enhancement of about 15%. It is not due to the air pressure because such increases are invariant with nominal strain. Inertia effect in the progressive folding process is supposed to be the main origin rather than the strain rate sensitivity of the base material, because the rate sensitivity of aluminium is much lower (Lindholm et al. 1971).

### *3.3.2. Rate sensitivity of IFAM AA6061 powder metallurgy foam*

IFAM AA6061 foam samples are obtained from Aluminium powder blowing process, using precursor materials made of pre-alloyed AA6061 powder and 0.5 wt.% titanium hydride powder as a blowing agent. The foaming was carried out in a batch furnace with indirect conductive heating and atmosphere circulation at 730°C. The final specimens are cylinders 45 mm in diameter and 60 mm long with closed outer skins. The density of the specimen is about 620 kg/m3. Figure 2a indicates that the main crushing mode is progressive folding. Figure 2b shows the flow stress at 10% of nominal strain with respect to the logarithmic value of the nominal strain rate. It shows an enhancement of about 15% for the mean flow stress. The observed rate sensitivity of IFAM foam is also supposed due to the micro-inertia effect, the same reason for the rate sensitivity in 5056 honeycombs.

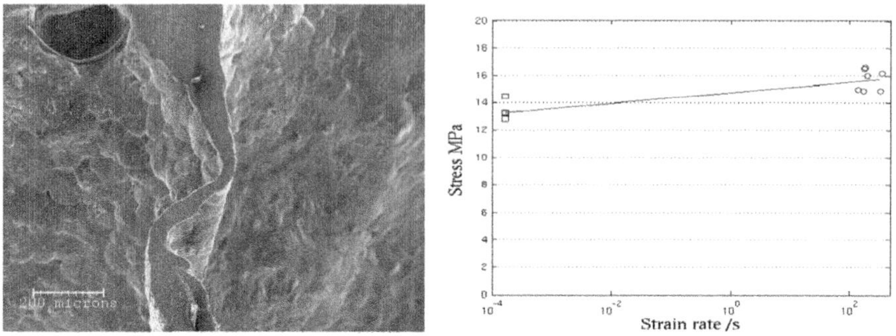


*Figure 2.* a) Progressive folding of cell wall of IFAM foam, b) Rate sensitivity of foams

### *3.3.3. Rate sensitivity of Cymat foam*

45 mm x 45 mm x 55 mm Cymat foam samples were cut from foam plates of about 200 mm thickness manufactured by the gas injection process. The average density is about 250 kg/m$^3$. Figure 3a shows the failure mode by multiple cracks of the cell walls of this brittle foam.

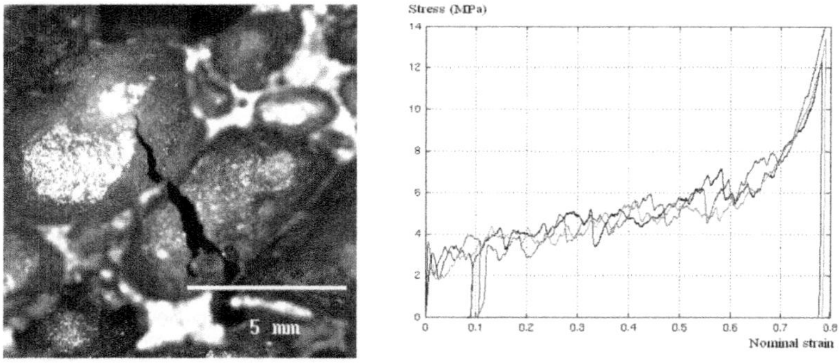


*Figure 3.* a) Fractures of Cymat foam, b) Typical mean stress- nominal strain curves

Tests under static and dynamic loading are performed. Typical mean stress-nominal strain curves are shown in Figure 3b. There is nearly no difference among these curves. The rate insensitivity of the Cymat foam gives an additional proof that the base material sensitivity of aluminium as well as the air pressure does not play a significant role. It indicates that the inertia effect involved in the successive folding is the main cause for the strength enhancement of aluminium Honeycombs and IFAM foams.

## 4. Multiscale Analysis of Inertia Effect in a Succesive Folding Process

In order to determine the factor responsible for strength enhancement under impact in a successive folding process, we studied a model-structure: commercially available brass square tube (after suitable partial annealing). The observed crushing mode of such tubes (right part of Figure 2) is also successive folding of walls. In addition, the behaviour of brass is known to be rate insensitive for a large range of strain rates. Experimental studies of brass square tubes are presented to confirm that strength enhancement under impact can occur without changes in the folding wavelength. Material characterization tests under static and dynamic loading up to 2500/s were performed to ensure that there is no rate sensitivity for brass in this range of strain rates.

### 4.1. EXPERIMENTAL STUDY OF THE CRUSHING STRENGTH OF BRASS SQUARE TUBES

#### *4.1.1. Experimental characterization of the rate sensitivity of brass.*

Tests under a comparable loading mode (static as well as dynamic) were performed to determine if the partially annealed base material (brass) is rate insensitive.Specimens were cut from heat treated brass tubes. Their dimensions were about 1.5mm thick, 2mm wide and 10mm long. Such specimens were tested with an MTS810 universal testing machine at 0.001mm/s. Dynamic tests were performed with a Split Hopkinson Pressure Bar or Kolsky apparatus. Figure 5 shows a comparison of the stress/strain response of annealed brass from quasi-static up to strain rates of 2500/s. There is no observable rate sensitivity.

#### *4.1.2. Strength enhancement of brass square tubes*

Static crushing tests were performed on heat-treated tubes with a universal testing machine. To avoid a global elastic buckling mode, the length of the tube specimen was chosen as 104 mm. The successive folding mode was observed. The load cell and displacement measurements of the test machine were used to record force-displacement signals. The force-displacement recordings were quite repeatable indicating that the heat-treatment applied is reproducible.

For dynamic loading, the direct impact configuration (Hauser, 1966) with a large scale SHPB system (diameter 80 mm, input bar of 6m and output bar 4m). A typical comparison between static and dynamic tests is shown in Figure 6. We can see that not only the initial peak load but also the successive peak load were measured.

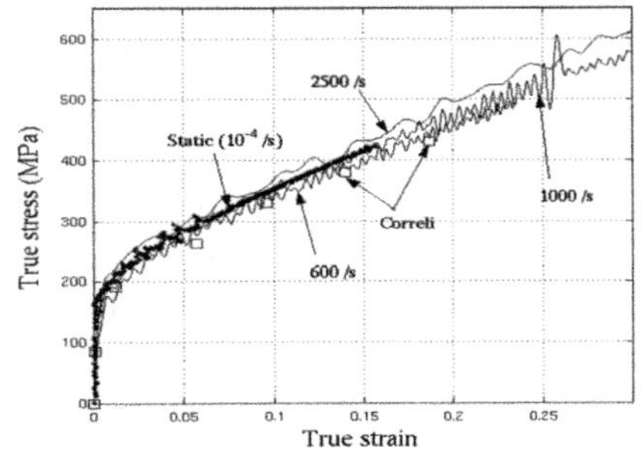


*Figure 5.* Rate sensitivity of base material

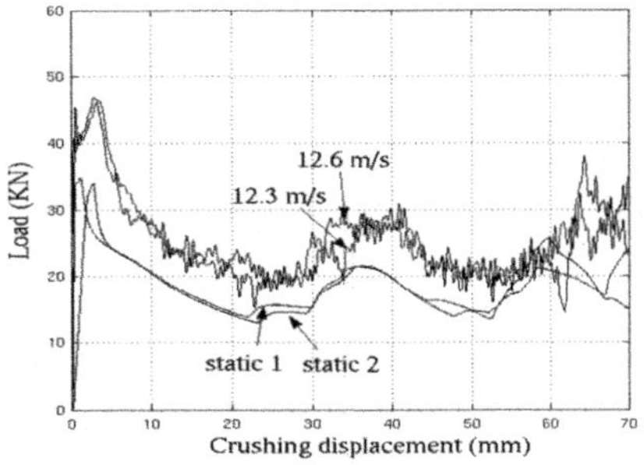


*Figure 6.* Strength enhancement of tubes

## 4.2. NUMERICAL AND THEORETICAL ANALYSES OF TUBE CRUSHING

In the crushing tests presented above, only force-displacement recordings were made. For static loading, some pictures were also taken of the tubes during tests. There are no simple methods to know what was happening in the tube, especially under impact loading. An alternative is to simulate the crushing of the tubes numerically. The explicit version of the commercial code LS-DYNA was used to perform the simulations under impact loading, and the implicit version was adopted for static loading. The velocity-time histories of both tube sides (measured velocities in both static and dynamic cases) were imposed by two rigid walls. Figure 7 provides a comparison between the force-time histories of the rigid wall under both static and dynamic loading. The simulated crushing tests also reproduce the strength enhancement under dynamic loading not only for the initial peak load but also for the successive peak load, which is another proof of the value of simulations.

### *4.2.1. Numerical folding events*

A careful examination of the sequence of events in successive crushing reveals that the folding cycles are composed of two stages. At the start, crush is obtained by bending in the middle of the flat plates (the two trapezoids around nodes B or B' in Figure 8) and there exist small areas around the four corner lines (the two adjacent triangles around node A in Figure 8) which remain vertical (no significant lateral displacement). In the simulated images of a deformed tube (left part of Figure 8), the corner line between the two triangles around node A is straight whereas the flat plate is already bent. Subsequently, the second stage begins with the buckling of the corner line areas as shown on the right of Figure 8.

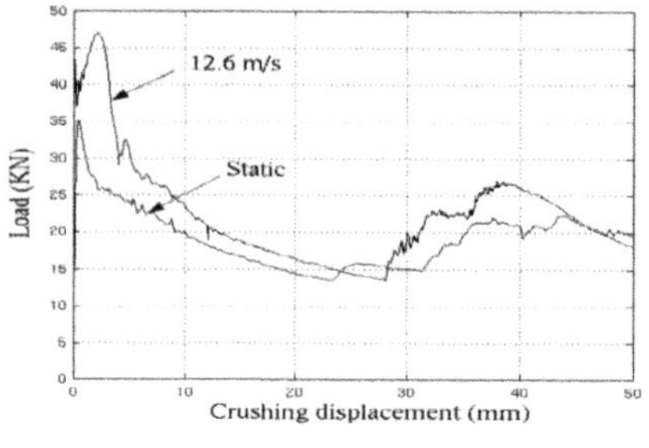


*Figure 7.* simulated enhancement

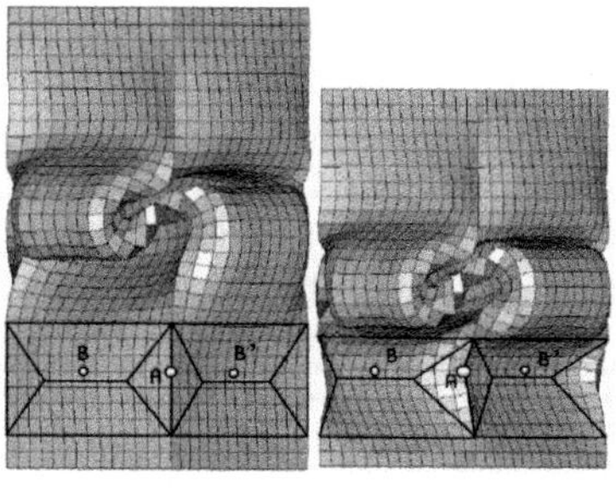


*Figure 8.* Folding events

The buckling of these edge zones (the two adjacent triangles around node A in Figure 8) corresponds to a decrease of the global crushing load. The simulation provides the displacement in two perpendicular lateral directions at node A at the corner with the corresponding global crushing force. It shows that the time when the corner node begins to move laterally correspond to the time when the global crushing force begins to decrease. It reveals that in the regions close to the edge, elements are more compressed (higher strain) under dynamic loading because of inertia, This is similar to what was found for a straight column. As the base material (brass) strain hardens, a higher strain is reached under dynamic loading before buckling leading to higher crushing forces.

#### *4.2.2. Theoretical analysis of an idealized box column model*

A kinematical box model can be extracted from Fig.8. it is possible to derive a theoretical analysis. Similar box model has been used by many authors (Abramowicz & Wierzbicki, 1989; Wierzbicki & Huang, 1991; Markiewicz et al., 1996; Meng & Al-Hassani, 1983). Table 1. gives a comparison between model estimates and experimental results. It offers a good explanation of the successive peak load.

Table 1. Comparison of theoretical and experimental enhancement

| | $F^{static}_{initial\ peak}$ | $F^{impact}_{initial\ peak}$ | $F^{static}_{successive\ peak}$ | $F^{impact}_{successive\ peak}$ |
|---|---|---|---|---|
| Model | 35 kN | 46 kN | 23 kN | 29 kN |
| Experiment | 35 kN | 45 kN | 22 kN | 29 kN |

### 4.3. POST-MORTEM MICRO-HARDNESS MEASUREMENT FOR VALIDATION

The above simulations and analyses suggest that strain hardening of the base material combined with inertia is the origin of the strength enhancement under dynamic loading. To prove that, micro-hardness tests on the crushed tubes at various crushing stages were performed. The micro-hardness of the material reveals its average strain hardening state. It therefore provides an experimental check of the local strain profile given by the simulation. Qualitative and quantitative matching was then sought between the simulations and the experiments. Figure 9 shows a typical micro-hardness measurement operation. Three crushing states under static loading were used for this matching. The first is the one just after the first peak load (9a left), the second is the one just after the second peak load, which is the successive peak load (middle). The last one corresponds to the case after full crushing (right). Figures 9b and 9c show the specimen preparation and an indentation mark from a micro-hardness measurement. It should be noted that the measurements presented in Figure 9d were made in the middle of the plate (half thickness position). The hardness near the plate surfaces is quite different because of bending effects. Such measurements prove that (i) just after the first initial peak load, the plastic strain is nearly uniform in a section. (ii) Afterwards, only the region near the corner line is loaded. These observations confirm the theory developed above. Furthermore, an excellent match between simulation and post mortem microhardness measurement was found (Figure 9e), validating the results given by the simulations.

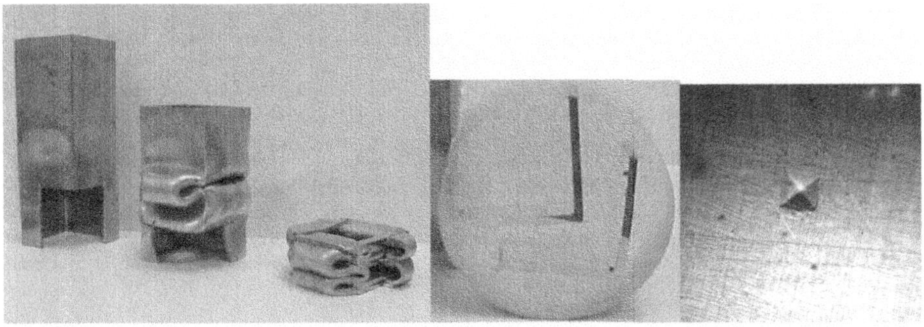

a) b) c)

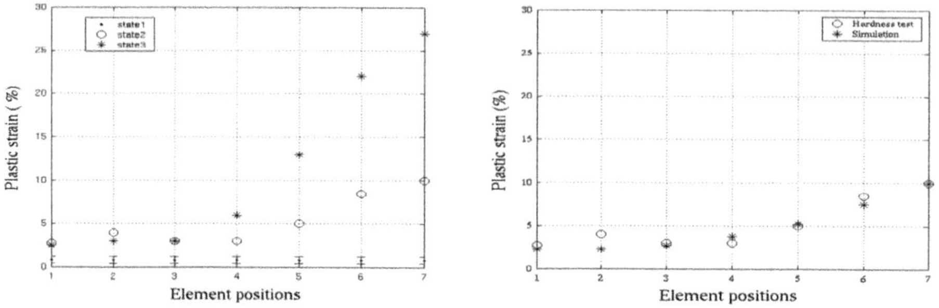


d) e)

*Figure 9.* a-e. Micro-hardness measurement compared with simulations

## REFERENCES

Abramowicz, W., Wierzbicki, T., 1989. Axial crushing of multicorner sheet metal columns. Journal of Applied Mechanics 56, 113-120.

Banhart, J., 2001. Manufacture, characterisation and application of cellular metals and metal foams. Progress in Material Science 46, 559-632.

Calladine, C.R., English, R.W., 1984. Strain-rate and inertia effects in the collapse of two types of energy-absorbing structure. International Journal of Mechanical Science 26 (11-12), 689-701.

Davies, R.M., 1948. A critical study of Hopkinson pressure bar. Philosophical Transaction of Royal Society A240, 375-457.

Deshpande, V.S., Fleck, N.A., 2000a. Isotropic constitutive models for metallic foams. Journal of Mechanics and Physics of Solids 48, 1253-1283.

Deshpande, V.S., Fleck, N.A., 2000b. High strain rate compression behaviour of aluminium alloy foams. International Journal of Impact Engineering 24, 277-298.

Gibson, L.J., Ashby, M.F, 1988. Cellular Solids. Pergamon Press.

Goldsmith, W., Sackman, J.L., 1992. An experimental study of energy absorption in impact on sandwich plates. International Journal of Impact Engineering 12 (2), 241-262.

Hauser F.E., 1966. Techniques for measuring stress-strain relations at high strain rates. Experimental Mechanics 6, 395-402.

Hanssen, A.G., Hopperstad, O.S., Langseth, M., Ilstad, H., 2002. Validation of constitutive models applicable to aluminium foams, International Journal of Mechanical Sciences 44(2), 359-406.

Hopkinson, B., 1914. A method of measuring the pressure in the deformation of high explosives or by the impact of bullets. Phil. Trans. Roy. Soc., A213, 437-452.

Klintworth, J.W., Stronge, W.J., 1988. Elasto-plastic yield limits and deformation laws for transversely crushed honeycombs. International Journal of Mechanical Science 30(3-4), 273-292.

Kolsky, H., 1949. An investigation of the mechanical properties of materials at very high rates of loading. Proceeding of Physical Society B62, 676-700.

Langseth, M., Hopperstad, O.S., 1996. Static and dynamic axial crushing of square thin-walled aluminium extrusions. International Journal of Impact Engineering 18(7-8), 949-968.

Lindholm, U.S., Bessey R.L., and Smith G.V., 1971. Effect of strain rate on yield stress, tensile strength and elongation of three aluminium alloys. Journal of Materials 6, 119-133.

Markiewicz, E., Ducrocq, P., Drazetic, P., Ravalard, Y., 1996. Calculation of the dynamic axial crushing response of complex prismatic sections. International Journal of Crashworthiness 1, 203-224.

Meng, Q., Al-Hassani, S.T.S., Soden, P.D., 1983. Axial crushing of square tubes. International Journal Mechanical Science 25, 747-773.

Mukai, T., Kanahashi, H., Miyoshi, T., Mabuchi, M., Nieh, T.G., Higashi, K., 1999. Experimental study of energy absorption in a closed cell aluminium foam under dynamic loading. Scripta Materialia 40(8), 921-927.

Reid, S.R., Peng, C., 1997. Dynamic uniaxial crushing of wood. International Journal of Impact Engineering 19(5-6), 531-570.

Su, X.Y., Yu, T.X., Reid, S.R., 1995. Inertia sensitive impact energy-absorbing structures part I : Effects of inertia and elasticity. International Journal of Impact Engineering 16(4), 651-672.

Vural, M., Ravichandran, G., 2003, Dynamic response and energy dissipation characteristics of balsa wood: experiment and analysis, International Journal of Solids and structures 40, 2147-2170.

Wierzbicki, T., Huang, J., 1991. Initiation of plastic folding mechanism in crushed box column. Thin-walled structures 13, 115-143.

Wierzbicki, T., 1983. Crushing analysis of metal honeycombs. International Journal of Impact Engineering 1(2), 157-174.

Wu, E., Jiang, W.S., 1997. Axial crush of metallic honeycombs. International Journal of Impact Engineering 19(5-6), 439-456.

Zhao, H., Gary, G., 1995. A three dimensional analytical solution of longitudinal wave propagation in an infinite linear viscoelastic cylindrical bar. Application to experimental techniques. Journal of Mechanics and Physics of Solids 43(8), 1335-1348.

Zhao, H., Gary, G., 1996. On the use of SHPB techniques to determine the dynamic behaviour of materials in the range of small strains. International Journal of Solids and structures 33(23), 3363-3375.

Zhao, H., Gary, G., 1997. A new method for the separation of waves. Application to the SHPB technique for an unlimited measuring duration. Journal of Mechanics and Physics of Solids 45, 1185-1202.

Zhao, H., Gary, G., 1998. Crushing behaviour of aluminium honeycombs under impact loading. International Journal of Impact Engineering 21(10), 827-836.

Zhao, H., 2004, Cellular materials under impact loading, IFTR-AMAS Edition, Warsaw, Poland. ISSN 1642-0578.

Zhao, H. and Abdennadher, S., 2004, On the strength enhancement under impact loading of square tubes made from rate insensitive metals, Int. J. Solid Struct., in press.

# PARTICLE SIZE EFFECTS IN NANOCOMPOSITES

A. ZAOUI*, V. MARCADON* and E. HERVÉ*#
* *Laboratoire de Mécanique des Solides, Ecole Polytechnique, CNRS 91128 Palaiseau cedex, France*
# *Université de Versailles Saint-Quentin, 78035 Versailles cedex, France*

## Abstract

Classical Eshelby's inclusion-based micromechanical models are known to be unable to take the particle size of matrix-inclusion composites into account and then to predict particle size effects, which are expected to be significant in nanocomposites. Nevertheless, improved micromechanical models, derived from the recent "morphologically representative pattern-based approach", which can take better into account the account the structural morphology of inhomogeneous materials, are shown to be able to predict at least relative size effects, *i.e.*, to predict an overall behavior depending on the particle size normalized by some morphological length scale associated with the particle spatial distribution. When, in addition, absolute physical lengths, such as the coil size of an amorphous polymer matrix or the thickness of an interphase layer of structurally modified polymer between particles and the surrounding matrix, are introduced into these models, they are also able to predict intrinsic size effects. Illustrations of such "relative" as well as "absolute" size effects in nanocomposites are reported in this paper.

## 1. Introduction

Nanoparticle-reinforced polymers are expected to exhibit specific mechanical properties when compared to classical composites for the same volume fraction, due to the shorter distance between the particles and the larger specific interfacial area. Two kinds of size effects can be considered. Indirect ones can come from percolation phenomena associated with the formation of connected aggregates of quasi-rigid particles which are responsible for a sharp overall stiffening of the composite: such aggregates can be formed at lower particle volume fractions when the particle size is smaller; when the particle volume fraction is fixed, the overall behavior can also depend on the relative size of the particles as compared with the mean distance between neighbor particles. More intrinsic size effects can also be expected: they could originate in a decrease of the molecular mobility of parts of the amorphous polymer matrix which are confined between neighbor particles. This could happen when the mean distance between the particles is of the order of (or even shorter than) the polymer coil size or the mean distance between cross-links. One can also speculate about the formation of a thin layer of polymer matrix with a modified structure (and then modified mechanical



*Q. P. Sun and P. Tong (eds.), IUTAM Symposium on Size Effects on Material and Structural Behavior at Micron- and Nano-Scales,* 97–106.

properties) at the interface with the particles: since the thickness of such layers would depend mainly on the polymer microstructure, whereas the particle size is fixed independently, this could result in overall properties depending on the particle size for a given polymer matrix.

Such relative or absolute size effects cannot be considered as proved experimentally in an unambiguous manner since the particle size or the mean distance between nearest neighbors can hardly be modified without changing at the same time the polymer microstructure, the particle-matrix interface, the spatial distribution of the particles, etc. Nevertheless, prospective models and simulations can be developed without delay for a better understanding of such potential effects in order to direct relevant further experimental investigations or to suggest material processing routes.

This paper reports on such a prospective contribution which aims at replying to the following questions: do we have to conclude necessarily from the known fact that classical micromechanical models derived from Eshelby's solution of the inclusion problem [1] are particle size-independent that a Cosserat-type, second-gradient or nonlocal theoretical framework must be adopted in order to predict nanoparticle size effects? Would improved micromechanical models, such as those derived from the so-called "morphologically representative pattern approach" [2, 3] (referred to as the "MRP-based approach" in the sequel), which take the composite morphology into account better than the classical Eshelby-type ones, not be able to express any particle size effect? In what follows, after a brief reminder of the ins and outs of the MRP-based approach (section 2), this method is first shown to be able to predict an indirect dependence of the overall behavior of matrix-inclusion composites on the particle size, scaled by the mean distance between nearest neighbour particles (section 3). If in addition internal length scales of the polymer matrix of nanocomposites, as suggested by Molecular Dynamics simulations, are introduced into similar micromechanical models, intrinsic size effects can also be predicted (section 4).

For the sake of simplicity, isotropic elasticity and spherical particles are considered only; extensions to viscoelasticity and to ellipsoidal particles could be performed easily but they would not throw additional light on the main questions addresser here.

## 2. The "Morphologically Representative Pattern-based Approach"

It can be stated easily that, since the Eshelby problem refers to a single inclusion embedded in an infinite matrix, the resultant stress and strain fields do not depend on the inclusion size; the same conclusion holds obviously for the inhomogeneous inclusion problem, and then for most of the classical bounds and estimates (Hashin-Shtrikman's bounds and estimates, the Mori-Tanaka model, the self-consistent scheme, etc.), which can be reduced to specific Eshelby-type problems. Similarly, the three-phase model [4] as well as Hashin's Composite Sphere Assemblage [5] (the "CSA" in the sequel), though they deal with composite spheres with two different radii, exhibit the same property since the ratio between these radii depends only on the volume fractions of the composite constituents.

On the other hand, it can be emphasized that all these classical models use a quite limited morphological information: Hashin-Shtrikman's bounds and estimates take only two-point correlation functions into account whereas the three-phase model and the CSA express the matrix continuity in a very simple way. Now, it can be noticed that size size effects in heterogeneous materials are frequently connected with the influence of the spatial distribution and of the volume fractions of the constituents. For instance, the formation of percolation paths consisting of adjacent particles, which may contribute to the construction of a quasi-rigid skeleton and to a sharp overall stiffening of the composite, can occur for a critical volume fraction depending on the particle size: for spherical particles with the same diameter $d$, a straight percolation path with the length $L$ needs a minimum critical particle volume $v_{cr}$ given by $v_{cr} = \pi L d^2 / 6$, which depends on $d$. It can then be expected that improved micromechanical approaches which would take better into account the composite morphology could exhibit some size sensitivity.

The MRP-based approach derives from a generalization of the basic idea of Hashin's CSA which, unlike classical treatments, consists in the consideration of finite composite elements instead of points: here, a given multiphase material is decomposed, after some morphological analysis, into $N$ patterns or "morphological phases" ($\gamma$) consisting of identical composite representative domains $D_{\gamma i}$. Unlike the classical "point" approach which needs a statistical description of the spatial distribution of all the points which belong to the same "mechanical phases", the "pattern" approach combines a deterministic description of small, but finite, well-chosen "composite patterns" and a statistical representation of the distribution of the pattern centers; in many cases, essential morphological features can be expressed in this way much more directly and easily than by using point correlation functions. For linear elasticity, the general Hashin-Shtrikman procedure [6] can be applied to this description with a polarization stress field which is no more uniform within the morphological phases ($\gamma$) but has identical values at homologous points of the domains of the same pattern. With help of adequate definitions of strain and stress averages over homologous points in each pattern [2] and use of the Green technique, the polarization stress field can be optimized and new Hashin-Shtrikman-type bounds can be obtained if the distribution of the pattern centers is isotropic or ellipsoidal (in the sense of Willis [7]). Like for the classical "point" Hashin-Shtrikman bounds, these bounds can be obtained from the solution of $N$ elementary inclusion-matrix problems where each domain $D_{\gamma i}$ is embedded in an infinite homogeneous matrix with extremum moduli. These problems can be solved numerically [3] in the general case. Similarly, new generalized (computer-aided) MRP-based Hashin-Shtrikman-type or self-consistent estimates can be obtained [8].

This treatment can be applied in particular to the CSA and isotropy or to the CCA and transverse isotropy, where composite spheres or cylinders with identical radii define the elementary patterns; the corresponding inclusion-matrix problems reduce to a single one since no length scale exists and an analytical solution, which improves significantly on classical Hashin's (and, of course, on Hashin-Shtrikman's) bounds, can be derived [9], whereas the three-phase model results as the associated self-consistent estimate. Several extensions can be derived from that. The most straightforward one consists in considering $n$-layered sphere (or cylinder) assemblages instead of Hashin's original CSA (or CCA), which leads to $n$-phase bounds or estimates [10, 11]. More significantly, several ($N$, say) families (instead of one set only) of similar patterns can be dealt with by the MRP approach. As illustrated in Figure 1, this results in $N$ inclusion-matrix problems whose solution yields bounds or estimates for the overall behavior.

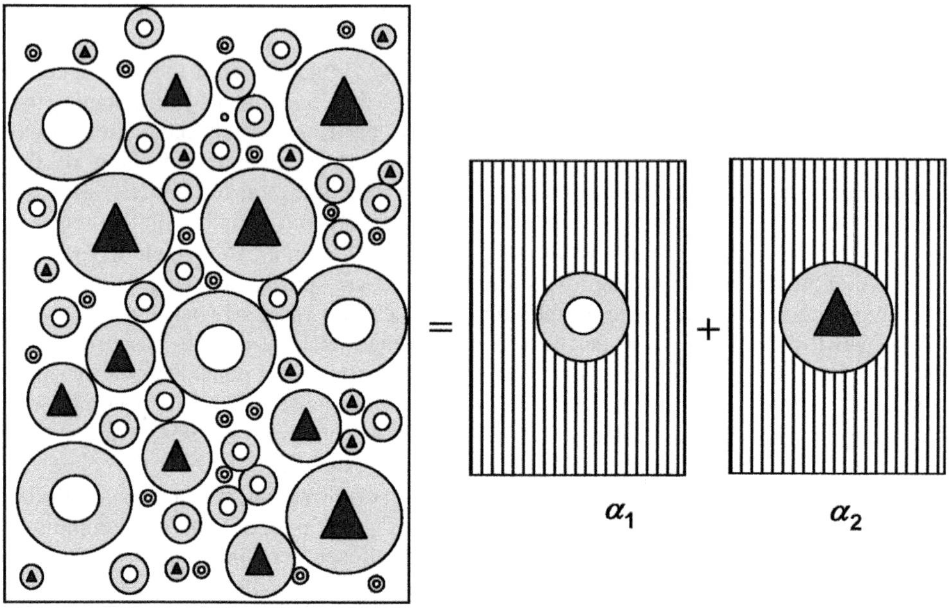


*Figure 1.* Generalized Hashin assemblage. Two families of different similar patterns modelled with help of two inclusion-matrix problems with relative frequency $\alpha_1$ and $\alpha_2$ (the infinite matrix is constituted with the wished reference medium)

A special case of such a "generalized Hashin assemblage" is obtained when the $N$ different pattern families are constituted with $n$-layered spheres. The associated self-consistent model [12] is used in the following for the description of various size effects.

## 3. "Relative" Size Effects

Indirect ("relative") size effects can be modelled by use of the $N$-pattern, $n$-sphere generalized self-consistent scheme: for instance, a porous material with a known void size distribution, discretized as $\alpha_i(c_i)$ ($i = 1$ to $N$) for $N$ values of the local porosity, can be described by $N$ patterns with $n = 2$; each pattern, with the porosity $c_i$ and the frequency $\alpha_i(c_i)$, is a composite sphere whose radii ratio is given by $c_i^{1/3}$. The resulting void size sensitivity of the overall behavior is only relative since it depends actually on $\alpha_i(c_i)$ and not on the absolute void sizes. Similar comments could be done, *e.g.*, for the use of the four-phase model ($N = 1$, $n = 3$) for the description of percolation effects in particulate composites with part of the polymer matrix confined between contiguous particles [13].

The generalized Hashin assemblage framework can also be adopted for a more direct discussion of spherical particle size effects in nanocomposites. As a matter of fact, the classical CSA is based on a geometrical fractal-like procedure according to which the whole space is filled up through the introduction of smaller and smaller similar composite spheres; consequently, it cannot deal rigorously with particles whose diameter $d$ is larger than some minimal value, $d_{min}$ say. Nevertheless, according to the MRP approach, a generalized Hashin assemblage can be defined for this case by stopping the above procedure when $d = d_{min}$ and by filling up the remaining space with the matrix material only. For an isotropic distribution of both phases, it can be shown [14] that the associated $N$-pattern, $n$-sphere generalized self-consistent scheme, with $N = 2$ and $n = 1$ and 2 respectively, is the one defined in Figure 2. Here, $c$ is the particle volume fraction and $c_1$, a partial particle volume fraction, has still to be defined.

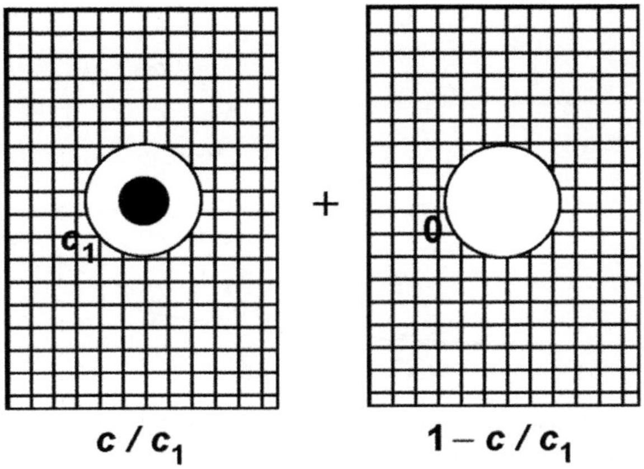


*Figure 2.* Generalized self-consistent scheme for $d \geq d_{min}$ (the infinite matrix is constituted with the effective medium)

To go further, let us first consider monodisperse particles with the same diameter $d$. The matrix shell thickness of the composite sphere can then be estimated as half the mean distance $\lambda$ between nearest neighbors in a random packing of spheres with the diameter $d$. This mean distance is a statistical variable which can be estimated and bounded [15]. Though a given distribution function $f(\lambda)$ could be dealt with easily, $\lambda$ is considered here, for the sake of simplicity, as a single parameter depending on the type of packing. Equivalently, the type of packing can be characterized by its maximum volume fraction $c_{max}$, correlated with $\lambda$, $c_1$ and $d$ by the relation

$$c_1 = \frac{c}{c_{max}} = \left(\frac{d}{\lambda}\right)^3 \qquad (1)$$

Although the correspondence is only an indirect one because isotropy is supposed here, reference can be made to cubic lattices for which the $c_{max}$ value is known to be equal to $\pi\sqrt{2}/6\,(\approx .74)$, $\pi\sqrt{3}/8\,(\approx .68)$, $\pi/6\,(\approx .52)$ for the fcc, bcc and sc (simple cubic) packing, respectively. Note that for the CSA, which cannot be obtained with single-sized spheres, $c_{max}$ would be equal to 1. The associated variation of the effective moduli with respect to the particle volume fraction $c$ is illustrated in Figure 3 for given bulk and shear moduli of the particles ($k_i$ and $\mu_i$) and the matrix ($k_m$ and $\mu_m$). It can be noticed that, for a given volume fraction c and a fixed value of $\lambda$, the stiffening efficiency

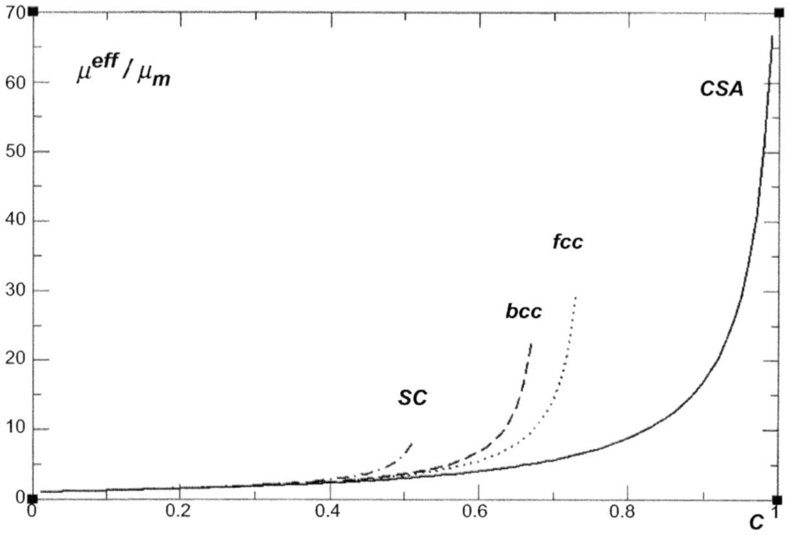


*Figure 3*. Particle size dependence of the effective shear modulus for fcc-like, bcc-like, cs-like and CSA packings ($\mu_i/\mu_m = k_i/k_m = 100$, $\nu_i = \nu_m = .125$)

of the particles (which is minimum for the CSA) is increasing with the particle size. This size effect is still a relative one since the particle diameter is always scaled by the packing distance $\lambda$.

A similar conclusion can be drawn when two sizes of particles are taken into account. For illustrative purpose, let us consider an equal number of large and small particles (diameters $D$ and $d$, respectively), located on the vertices of two inter-penetrating simple cubic lattices. This corresponds to a fcc close-packing of the associated tangent composite spheres. For a maximum matrix volume in the composite spheres, the partial volume fractions $c_D$ and $c_d$ of the particles are given by

$$c_D = \frac{c}{\alpha}\frac{1}{1+\beta^3}, \qquad c_d = \frac{c}{\alpha}\frac{\beta^3}{1+\beta^3}, \qquad \beta = d/D, \quad \alpha = \pi\sqrt{3}/16 \tag{2}$$

The isotropic analogue of this system can be modelled according to a $N$-pattern, $n$-sphere generalized self-consistent scheme, with $N = 3$ and $n = 2$, 2 and 1, respectively (Figure 4). The associated dependence of the effective moduli on the particle volume fraction $c$ is illustrated in Figure 5 for the same bulk and shear moduli of the particles ($k_i$ and $\mu_i$) and of the matrix ($k_m$ and $\mu_m$) as those considered in Figure 3 for monodisperse particles ($\beta = 1$). It can be noticed that the use of two particle sizes leads to an increase of the overall stiffening effect of the particles, which is larger when the size contrast is bigger. Nevertheless, this size effect is still a relative one since the overall properties of the composite do not depend on $d$ and $D$ separately but only on the ratio $\beta = d/D$.

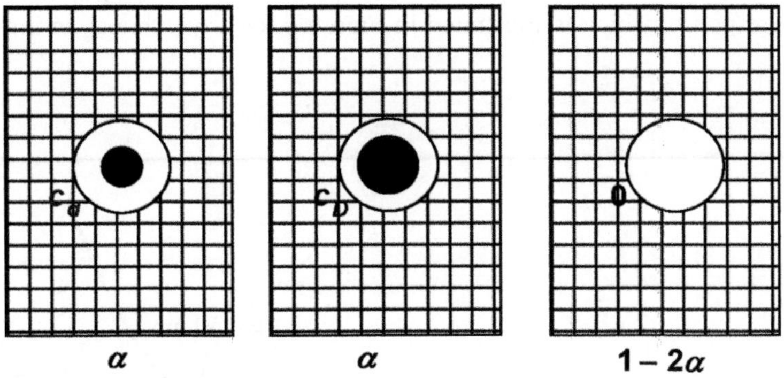


*Figure 4.* Generalized self-consistent scheme for two particle diameters $D$ and $d$ (the infinite matix is constituted with the effective medium)

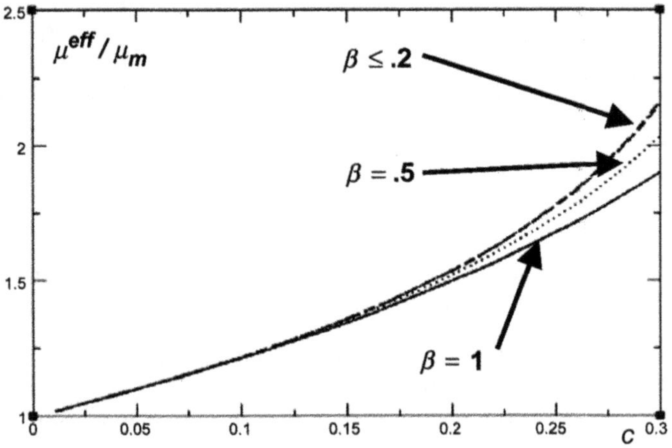


*Figure 5.* Particle size dependence of the effective shear modulus for two particle diameters $D$ and $d$, with $\beta = d / D$ ($\mu_i/\mu_m = k_i/k_m = 100$, $\nu_i = \nu_m = .125$)

## 4. "Intrinsic" Size Effects

The above methods can actually predict "intrinsic" size effects only if, in addition to morphological properties, absolute internal length scales are considered. Two illustrative examples of such effects are reported here:

– the first one is based on the assumption that, as soon as the mean distance between nearest neighbor single-sized spherical particles is decreased up to a critical value which is linked to the gyration radius of the statistical coils of the matrix polymer, that part of the polymer material which is confined between neighbor particles suddenly suffers a strong elastic stiffening. This results in a three-phase material with an elastic stiffness of the confined polymer material which is intermediate between the particles and the "free" polymer ones. Thus, two patterns must be used at least, one with a particle core and a confined polymer shell and another one for the free polymer only. The overall response, which is illustrated in Figure 6, clearly exhibits an intrinsic size effect.

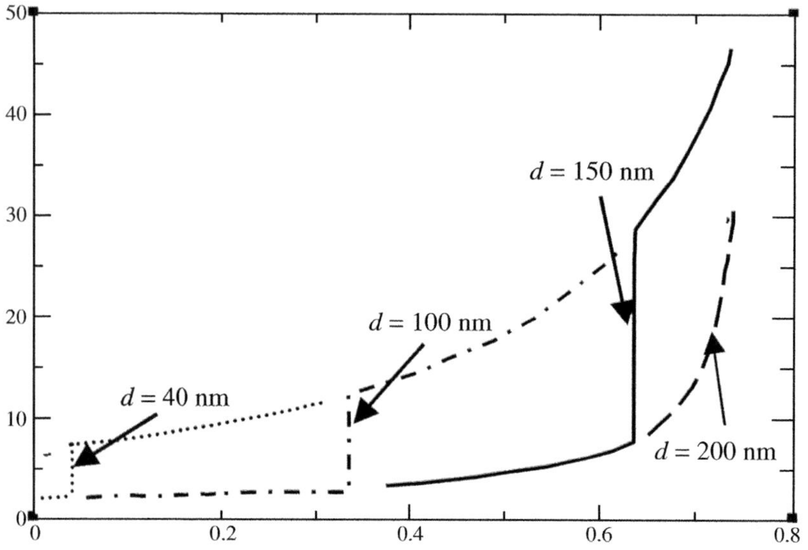


*Figure 6.* Predicted intrinsic size effect on the effective shear modulus (polymer gyration ratio R = 100nm; confined polymer moduli $\mu_{cp} = 10\mu_m$, $k_{cp} = 10k_m$, $\mu_i/\mu_m = k_i/k_m = 100$, $\nu_i = \nu_m = \nu_{c.p.} = .125$)

– another example can be developed from the fact that recent molecular dynamics simulations [16] indicate that the molecular structure of a thin polymer shell close to the interface with nanoparticles can be disturbed with respect to that of the bulk polymer: in this shell, the polymer mass density exhibits larger space fluctuations and the chain orientation is no more random (the chains are amost tangent to the interface). It can be inferred from these perturbations that the mechanical properties are also disturbed and different from the bulk properties. In addition, when the particle diameter is modified at fixed volume fraction, the thickness of the disturbed polymer shell is almost unchanged (see Figure 7, from [16]). According to a two-pattern model, as defined in Figure 8, an intrinsic size effect is predicted as soon as this thickness is not proportional to the particle diameter. This is illustrated in Figure 9 for silica particles and a virtual amorphous polymer matrix when the disturbed shell thickness is supposed to be fixed.

## 5. Conclusion

The above examples are an illustration of the fact that improved micromechanical models, developed within a fully classical framework, can predict at least relative size effects through an adequate description of the material morphology.

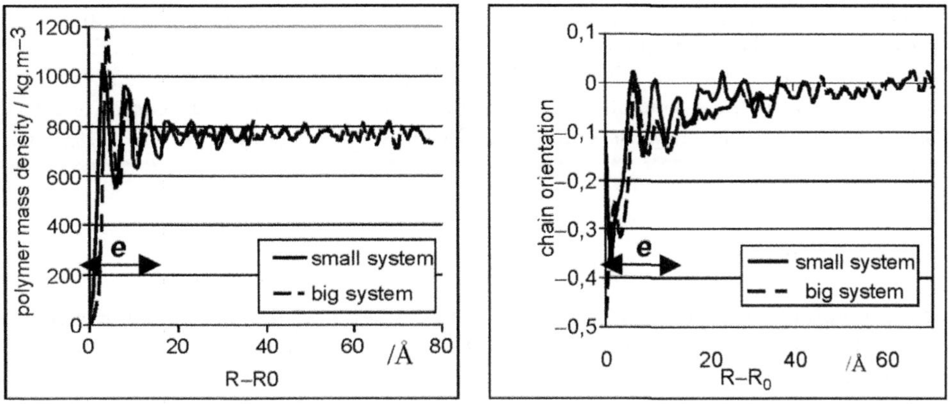


*Figure 7.* Radial variation of the polymer mass density and the chain orientation from the interface towards the bulk polymer as predicted in [16] by Molecular Dynamics simulations of two particle-polymer systems for the same volume fraction (4.44%) and particle radii R0 of 1.35nm ('small system') and 2.85nm ('big system')

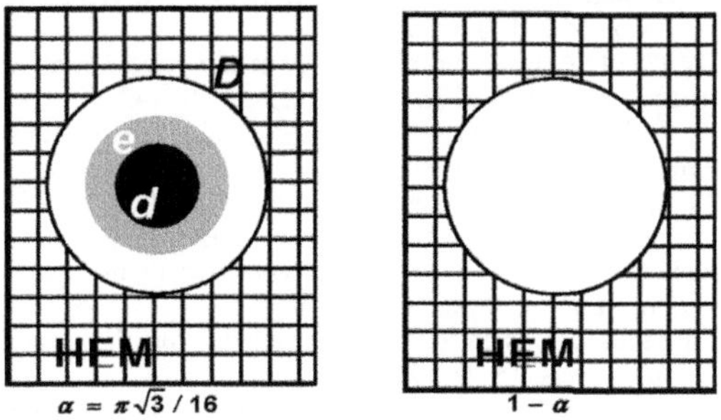


*Figure 8.* Two-pattern model used for predicting the overall properties of a particle-reinforced polymer with a disturbed polymer shell at the particle-matrix interface

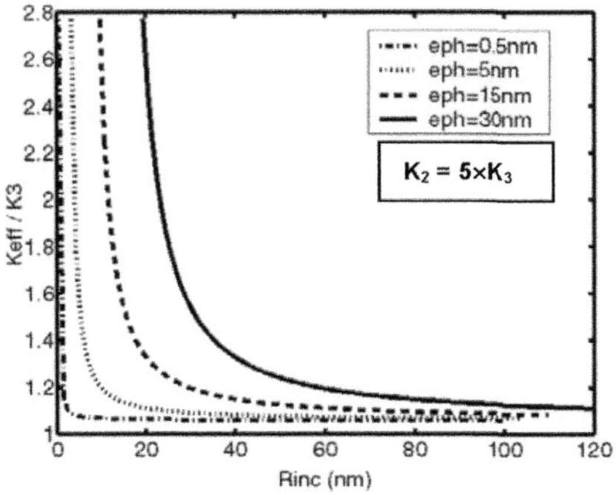


*Figure 9.* Intrinsic particle size effect on the effective bulk modulus of a particle-reinforced polymer with a disturbed polymer shell (thickness eph; bulk modulus $K_2$; $K_3 = k_m$) at the particle-matrix interface

When, in addition, physically-based internal scale lengths are introduced in the representation of the constitutive behavior of the constituents, intrinsic size effects can also be predicted. In view of a better understanding of such phenomena in nanoparticle-reinforced polymers, a cooperative experimental investigation of black carbon – epoxy systems is currently in progress within the CNRS Research Federation FM2SP (Paris, France).

## Acknowledgements

We are grateful to Pr D. Brown (LMOPS, Chambéry, France) for his efficient leadership in our collaboration in the field of Molecular Dynamics and also to Pr N. Alberola and Dr P. Mélé (LMOPS) for helpful discussions.

## References

[1] Eshelby, J.D. (1957) The determination of the elastic field of an ellipsoidal inclusion and related problems, *Proc. Roy. Soc. London*, **A 241**, 376-396.
[2] Stolz, C. and Zaoui, A. (1991) Analyse morphologique et approches variationnelles du comportement d'un milieu élastique hétérogène, *C. R. Acad. Sci. Paris*, **312**, série II, 143-150.
[3] Bornert, M., Stolz, C. and Zaoui, A. (1996) Morphologically representative pattern-based bounding in elasticity, *J. Mech. Phys. Solids*, **44**, 307-331.
[4] Christensen, R.M. (1979) *Mechanics of Composite Materials*, Wiley Interscience.
[5] Hashin, Z (1962) The elastic moduli of heterogeneous materials, *J. Appl. Mech.*, **29**, 143-150.
[6] Hashin, Z. and Shtrikman, S. (1963) A variational approach to the theory of the elastic behavior of multiphase materials, *J. Mech. Phys. Solids*, **11**, 127-140.
[7] Willis, J.R. (1977) Bounds and self-consistent estimates for the overall properties of anisotropic composites, *J. Mech. Phys. Solids*, **25**, 185-202.
[8] Bornert, M. (1996) A generalized pattern-based self-consistent scheme, *Computational Material Science*, **5**, 17-31.
[9] Stolz, C., Hervé, E. and Zaoui, A. (1993) On the behaviour of inhomogeneous elastic media: variational approach and morphological analysis, *Mat. Sci. Forum*, **123-125**, 245-256.
[10] Hervé, E. and Zaoui, A. (1993) n-layered inclusion-based micromechanical modelling, *Int. J. Engng Sci*, **31**, 1-10.
[11] Hervé, E. and Zaoui, A. (1995) Elastic behavior of multiply coated fibre-reinforced composites, *Int. J. Engng Sci.*, **33**, 1419-1433.
[12] Bornert, M. Hervé, E., Stolz, C. and Zaoui, A. (1994) Self-consistent approaches and strain heterogeneities in two-phase elastoplastic materials, *Appl. Mech. Rev.* **47**, Part 2, 66-76.
[13] Albérola, N.D. and Mélé, P. (1997) Composites particulaires: modélisation du comportement visco-élastique assortie du concept de percolation, *C. R. Acad. Sci. Paris*, **319**, série II, 1129-1134.
[14] Bornert, M. (1996) *Morphologie microstructrurale et comportement mécanique; caractérisations expérimentales, approches par bornes et estimations autocohérentes généralisées*. Ph. D. thesis, Ecole Nationale des Ponts et Chaussées, France
[15] Torquato, S. (1995) Mean Nearest-Neighbor Distance in Random Packings of D-Dimensional Hard Spheres, *Phys. Rev. Lett.*, **74**, 2156-2159.
[16] Marcadon, V., Brown, D., Hervé, E. and Zaoui, A. (2004) Modelling particle size effects on the behaviour of nanocomposites – Comparison between molecular dynamics and micromechanical approaches, *3rd Int. Conf. Computational Modeling and Simulation of Materials*, Sicily, Italy, June 2004 (to be published in *Advances in Science and Technology*, Ed. P. Vincenzini et al., Vol. 42 to 46, Techna Group Pub., Faenza, Italy)

# EFFECTS OF THE STRUCTURAL TOPOLOGY AND CONNECTION SIZE ON THE STRENGTH OF CIRCULAR HONEYCOMBS UNDER IN-PLANE COMPRESSION

T.X. YU [a] and D. KARAGIOZOVA [a,b]
[a] *Department of Mechanical Engineering, The Hong Kong University of Science & Technology, Clear Water Bay, Kowloon, Hong Kong*
[b] *Institute of Mechanics, Bulgarian Academy of Sciences, Acad. G. Bonchev Street, Block 4, Sofia 1113, Bulgaria*

## 1. Introduction

Foams made of ductile materials often have applications in energy absorbing devices as a core material for various structures. In order to improve the design of these structures it is important to understand the deformation mechanism of cellular materials with different topology when subjected to multi-axial loadings. Extensive studies on the mechanical properties of honeycombs and foam materials have been reported in the literature during the past decade considering different aspects [1-9]. Many studies are related to the limit (or yield) surface of 2D [2-4] or 3D [5-7] materials as a characteristic of the strength of a cellular material under biaxial or multi-axial loading. It has been established that the influence of elasticity and different kinds of imperfections notably affects the strength of the material under multi-axial loading [4,7] leading to its significant reduction. It was emphasised by Chen et al [4] that the switch of the deformation mode from cell wall stretching to cell wall bending is the major reason of the reduction of the hydrostatic strength of hexagonal honeycombs in comparison with the deviatoric strength.

In order to understand the deformation mechanism with no stretching, a honeycomb with a perfect periodic structure as hexagonally packed circular cells under in-plane compression is examined in the present study. Due to the fabrication procedures, the circular cells within a honeycomb material are usually connected by finite segments rather than just at points. The length of a connecting segment is not negligible and in some materials they are spread along arches with an average characteristic angle of $12^{\circ}$ [8]. The segment length can be related to the cell size if we assume that the 'gap', *d*,



*Q P. Sun and P. Tong (eds.), IUTAM Symposium on Size Effects on Material and Structural Behavior at Micron- and Nano-Scales,* 107–114.

between the cells (Fig. 1) remains constant regardless of the cell radius, $R$. In this case, scaling of a honeycomb material will lead to an increase of the angle of connecting segments when increasing the cell radius for $t/R$ = const ($t$ is the thickness of the cell wall) as shown in Fig. 1, where the scaling parameter $\eta = R/R_0$. The scaling ratio $\eta = 1$ is referred to the material employed in [8] for which $R_0 = 3.5$ mm, $t = 0.15$ mm and $\delta = 7^\circ$ ($d = 0.004R_0$).

For energy absorbing purposes, honeycombs are often made of lightweight materials, e.g. aluminium alloys which do not exhibit strain hardening properties. In that case, the initial high peak of the collapse force is followed by a structural softening, so that the force associated with the plateau region of the response can be significantly lower than the initial collapse force [10]. Therefore, a prediction based on the initial collapse stress of the material may overestimate the energy absorption capacity of the cellular material. A more realistic estimate of the energy absorption characteristics of honeycombs made of ductile materials and subjected to an in-plane compression can be obtained on the basis of the analysis of the post-collapse mechanism and the associated stresses, which can provide understanding of the evolution of the limit surface during post-collapse.

The objective of the present study is by analysing the post-collapse response of circular honeycomb material under an in-plane compression to estimate the plateau stresses depending on the direction of loading. The influence of the size of the connecting segment between two neighbouring cells is studied in order to examine the effect of the topology of the honeycomb material on the 'shape' of the limit surface and therefore, on the material strength depending on the direction of loading. A structural approach using the limit analysis and the concept of an equivalent structure is employed to describe the large plastic deformations during post-collapse [11].

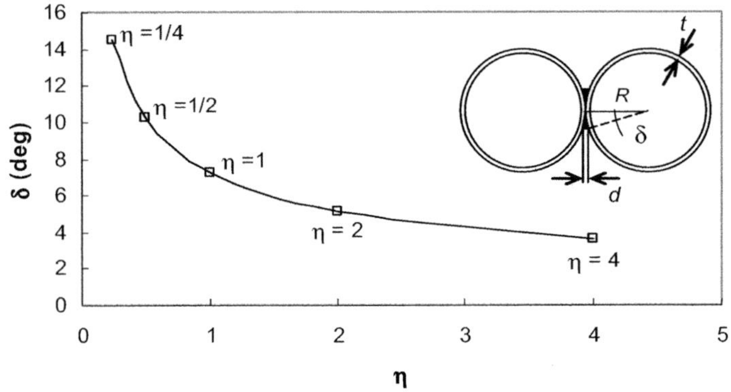


*Figure 1.* Variation of the angle of the connecting segment depending on the scaling parameter, $\eta$.

## 2. Mechanisms of Deformation

The deformation patterns in a honeycomb material produced by in-plane compression along different loading directions of compression and the characteristic features of the evolution of the deformed cells during the post-collapse can be identified from the experimental results [8] as well as from the numerical simulations [11]. These basic characteristics of deformation are used in the present analysis to construct representative blocks consisting of collapsible structures with different orientations, which can accommodate the collapse mechanisms depending on the direction of loading.

Three particular directions of in-plane compression are analysed, namely the uniaxial compressions in *X*- and *Y*-directions and an equi-biaxial in-plane compression. The corresponding configurations of the representative blocks are depicted in Fig. 2 assuming asymmetric deformation modes that develop at an advanced phase of post-collapse. It is seen that such a block, which consists of four rings, can represent different deformation modes. The uniaxial compression in the *Y*-direction causes a collapse of the four cells (Fig. 2(a)), while the compression in the *X*-direction results in squashing of the rings within a 'shear' zone [8]. Therefore, the collapsible representative block consists of two deformable and two undeformable rings (Fig. 2(b)). On the other hand, three deformable rings and one undeformable ring located at the centre of the representative block can accommodate a 'flower-like' deformation mechanism resulting from a biaxial compression (Fig. 2(c)).

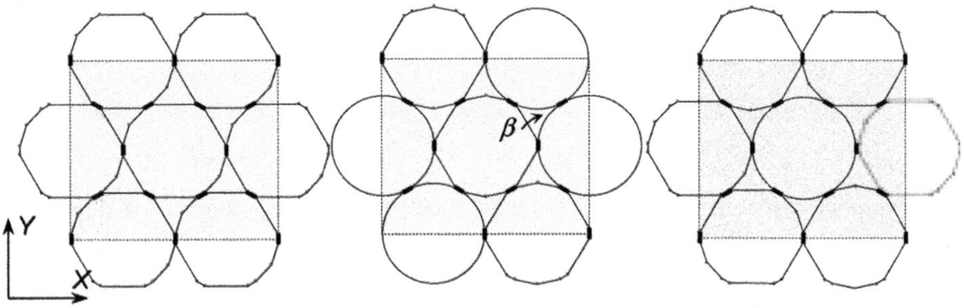


*Figure 2.* Deformation mechanisms of the representative block depending on the direction of loading. (a,b) Uniaxial compression in the *Y*- and *X*-direction, respectively; (c) Equi-biaxial compression.

## 3. Structural Approach

The post-collapse response of the various representative blocks can evolve when applying rotation, say $\beta$ (Fig. 2), which increases incrementally from zero, to certain arms [11]. Since the input energy to the system (i.e. the work done by the external forces) is equal to the plastic dissipation energy, *D*, due to the rotation at the plastic hinges, the following relationship is used to determine the variation of the collapse force per unit cell depth, $F(\beta)$, in a uniaxial direction of compression:

$$\Delta D(\beta) = F(\beta)\Delta U(\beta)\,, \tag{1}$$

where $\Delta U(\beta) = U(\beta_i) - U(\beta_{i-1})$, $i = 1,2,...$ is the variation of the displacement in the direction of loading.

The current side lengths of the representative block, $d_x(\beta)$ and $d_y(\beta)$, are then used to obtain the in-plane stresses

$$\sigma_x(\beta) = F_x(\beta)/d_y(\beta) \quad \text{and} \quad \sigma_y(\beta) = F_y(\beta)/d_x(\beta). \tag{2}$$

The equilibrium of the equivalent structure shown in Fig. 2(c) results in

$$D(\beta_i) - D(\beta_{i-1}) = f_x(\beta_i)\left[(U_x(\beta_i) - U_x(\beta_{i-1})) + \frac{d_x(\beta_i)}{d_y(\beta_i)}(U_y(\beta_i) - U_y(\beta_{i-1}))\right]. \tag{3}$$

The variations of the stresses $\sigma_x(\beta) = \sigma_y(\beta)$ during the evolution of the deformed shape of the representative block under an equi-biaxial compression are

$$\sigma_x(\beta_i) = \frac{D(\beta_i) - D(\beta_{i-1})}{d_y(\beta_i)\cos\alpha}\left[(U_x(\beta_i) - U_x(\beta_{i-1})) + \frac{d_x(\beta_i)}{d_y(\beta_i)}(U_y(\beta_i) - U_y(\beta_{i-1}))\right]^{-1}, \tag{4}$$

where $\alpha$ is the relative angle between the global and local coordinate systems [11] of the block in Fig. 2(c).

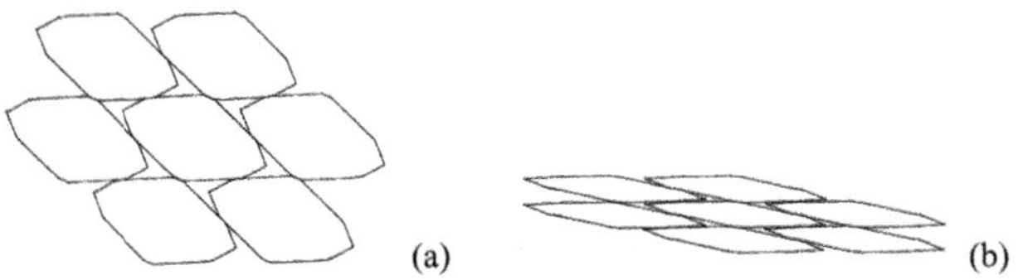

*Figure 3.* Evolution of the shape of the representative block under compression in the *Y*-direction; $\delta = 10°$. (a) $\gamma = 12.6$; (b) $\gamma = 21.5$.

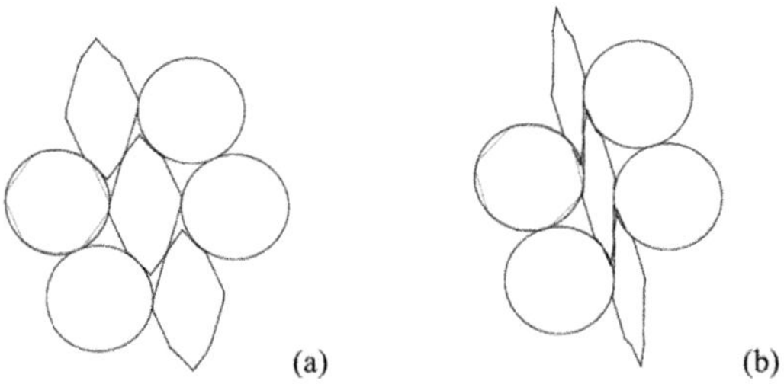

*Figure 4.* Evolution of the shape of the representative block under compression in the *X*-direction; $\delta = 10°$. (a) $\gamma = 19.9$; (b) $\gamma = 41.4$.

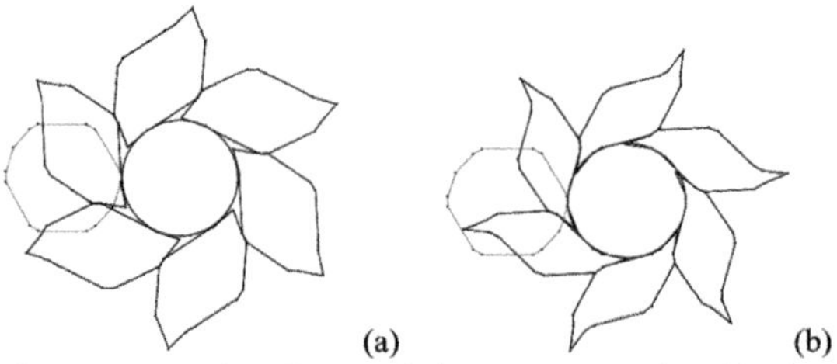

*Figure 5.* Evolution of the shape the shape of the representative block s under equi-biaxial compression; $\delta = 10°$. (a) $\gamma = 23.8$; (b) $\gamma = 36.6$.

Deformed shapes for different values of the energy dissipation ratio, $\gamma = D/M_0$, $M_0 = \sigma_0 t^2/4$ being the fully plastic bending moment of a cell wall per unit depth, are shown in Figs. 3-5 for connecting segments of 10°.

## 4. Relative Variation of the Deviatoric and Hydrostatic Stresses

In order to relate the response of the representative blocks to some general characteristics of cellular materials such as the relative density, the remote stresses associated with the different representative blocks are compared when dividing the actual stresses by $\sigma_0(\rho^*/\rho_s)^2$, where $\sigma_0$ and $\rho_s$ are the yield stress and the density of the wall material, and the relative density of hexagonally arranged circular honeycombs is

$$(\rho^*/\rho_s) = (\pi/\sqrt{3})(t/R). \tag{5}$$

The variations of the stresses associated with the three directions of loading with the energy dissipation during post-collapse process are given in Fig. 6 and they can be used to estimate the relative variation of the deviatoric and hydrostatic stresses during post-collapse so as to interpret the 'evolution' of the limit surface.

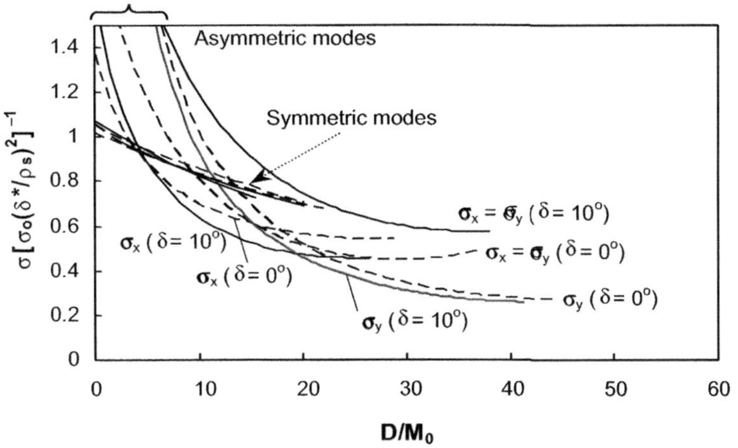


*Figure 6.* Variations of the stresses associated with the three directions of loading with the energy dissipation during post-collapse.

It is seen that the relative variation of the stresses for large values of the energy dissipation ratios $\gamma$ are in qualitative agreement with the experimental results for the plateau stresses of circular honeycombs obtained by Papka and Kyriakidis [8] despite their material (polycarbonate) exhibited non-linear strain-hardening. Likewise the experimental results, the smallest stresses correspond to a uniaxial compression in the *Y*-direction. An increase of stresses associated with the applied equi-biaxial loading ($\sigma_x = \sigma_y$) indicates a more rapid densification of the material when compressed bi-axially.

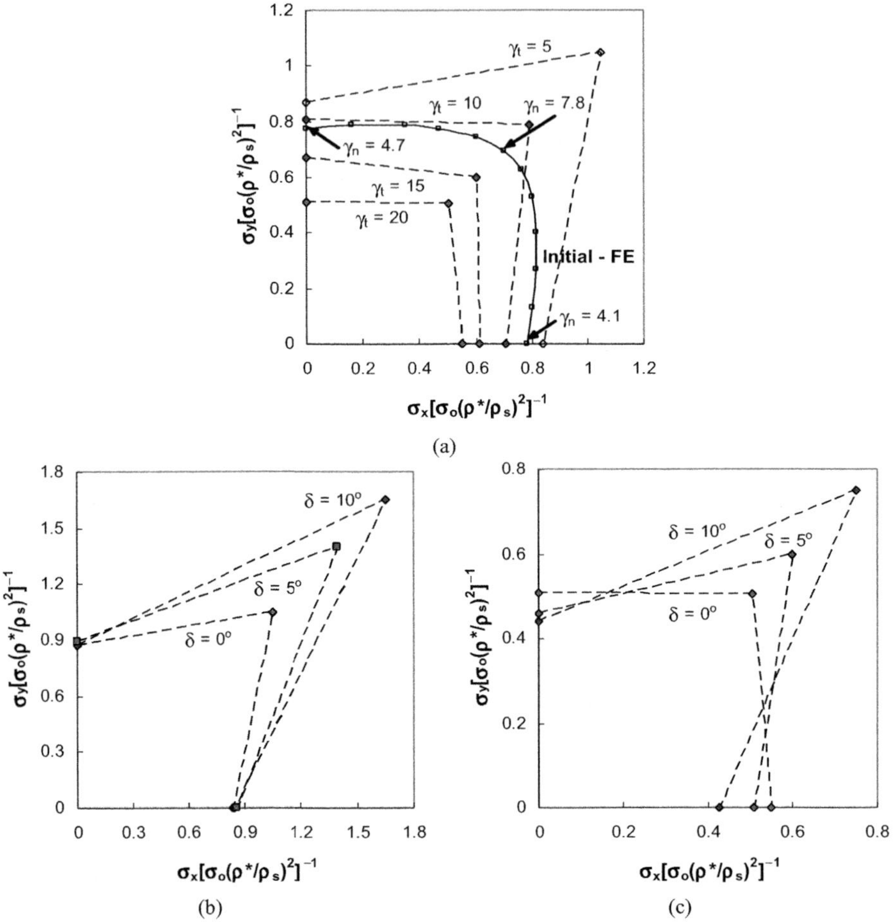


*Figure 7.* 'Limit' surfaces for different lengths of the connecting segments between cells. (a) Comparison between the numerical and theoretical predictions, $\delta = 0^{\circ}$; (b) $\gamma = 5$; (c) $\gamma = 20$.

A summary of the stresses associated with the three directions of loading and different energy dissipation during post-collapse process is shown in Fig. 7. The three directions of loading can be regarded as extreme points on the limit surface at constant plastic energy dissipation, $D$. Several combinations of the biaxial stresses, i.e. ($\sigma_x$, $\sigma_y = 0$), ($\sigma_x = 0$, $\sigma_y$) and ($\sigma_x = \sigma_y$) are shown in Fig. 7(a) for a point connection between the cells in comparison with the numerically obtained initial limit surface [11]. It is evident that the 'limit' surface evolves and the size varies with the dissipated energy during post-collapse but the stresses do not increase excessively under biaxial compression. Stresses $\sigma_x = \sigma_y$ smaller than the corresponding uniaxial values are observed within a certain range of the dissipated energy (Fig. 6). In Fig. 7(a), $\gamma_t$ denotes the theoretical ratio $D/M_0$ while $\gamma_n$ referrers to the numerical simulations.

The variations of the stresses $\sigma_x$ and $\sigma_y$ for a segment connection of $10^o$ are compared in Fig. 6 with those stresses in the case of point connection. It is evident that there is no proportional variation of the stresses during post-collapse. Figures 7(b,c) show the dependences of the stress combinations ($\sigma_x$, $\sigma_y = 0$), ($\sigma_x = 0$, $\sigma_y$) and ($\sigma_x = \sigma_y$) on the angles of connecting segments. It is seen that for small energy dissipation ratios the hydrostatic stresses increase notably with the increase of the angle of the connecting segments while the deviatoric stresses remain almost unchanged. An important observation is that for large values of $\gamma$ a kind of 'softening' of the structural response occurs with respect to the uniaxial compression when increasing the angle of the connecting segments. A 'hardening' of the response is still observed when an equi-biaxial compression is applied.

These results suggest that a honeycomb material with smaller cells will exhibit a larger strength under an equi-biaxial compression when assuming a constant 'gap' between the cells; that is, the relative values of the hydrostatic and deviatoric stresses become size-dependent.

The variation of the stresses associated with uniaxial or bi-axial compression is related to the particular mechanisms of deformation. It is seen from Figs. 3-5 that the post-collapse of any representative block is dominated by bending and shear. The natural curvature of the cells contributes to a response, which does not exhibit the high anisotropy characteristic of the hexagonal honeycombs. Consequently, not only the initial imperfections and other irregularities of the material [4] but also the shape of the cells even in a perfectly periodic structure can lead to the reduction of the hydrostatic stresses associated with the particular post-collapse behaviour of a cellular material.

It should be noted that the conclusions from this study related to the evolution of the limit surface during the post-collapse are strictly applicable to ductile materials with small strain-hardening, which is the characteristics of many metals. An increase of the strain-hardening will reduce the difference between the initial peak value and the plateau stress. However, the increase of the strain-hardening will raise the material strength in all directions of compression. Therefore, the conclusion with respect to the relative values of the collapse stresses associated with the uniaxial and equi-biaxial compressions will remain valid. This conclusion is supported by experiments reported in the literature. The experimentally observed collapse stresses of polycrystalline circular honeycombs [8] under equi-biaxial compression exhibit only a small increase in comparison with the collapse stresses associated with the uniaxial compression.

## 5. Conclusions

The effect of the structural topology of hexagonally packed circular honeycombs with perfectly periodic structure on the strength of the material under in-plane compression is studied for large deformations that occur during the post-collapse response. The analysis indicates that the stresses associated with an equi-biaxial compression are not excessively larger than the corresponding uniaxial stresses in contrast to the response of hexagonal honeycombs. The comparable values of the deviatoric and hydrostatic stresses are due to the particular deformation mechanism as the curvature of the cell walls invokes bending without stretching.

It is shown that the size of the connecting segments between the cells influences the 'shape' of the limit surface, which evolves during post-collapse. For equal plastic dissipation energy, the increase of the angle of the connecting segments leads to a structural 'softening' at an advanced stage of deformation, while the hydrostatic stress increases. In view of this behaviour, the finite length of the connecting segments acts as an non-scaling factor and it can be expected that smaller cells will exhibit a larger strength under an equi-biaxial compression when assuming a constant 'gap' between the cells, i.e. the relative values of the hydrostatic and deviatoric stresses become size-dependent.

**Acknowledgment**
The support from the Hong Kong Research Grants Council (RGC) in collaboration with the National Science Foundation of China under grant N_HKUST 601/01 is gratefully acknowledged.

## References

1. Gibson, L.J. and Ashby M.F.: *Cellular Solids: Structure and Properties*. 2nd edition, Cambridge University Press, 1997.
2. Klinthworth, J.W. and Stronge, W.J.: Elastoplastic yield limits and deformation laws for transversely crushed honeycombs. *Int. J. Mech. Sci.*, **30**(1988), 273-292.
3. Triantafyllidis, N. and Schraard, M.W.: Onset of failure in honeycomb under general in-plane loading. *J. Mech. Phys. Solids*, **46**(1998), 1089-1124.
4. Chen, C., Lu, T.J. and Fleck, N.A.: Effect of imperfections on the yielding of two-dimensional foams. *J Mech. Phys. Solids*, **47**(1999), 2235-2272.
5. Gibson, L.J., Ashby, M.F., Zhang, J. and Triantafillou, T.C.: Failure surfaces for cellular materials under multiaxial loads I. Modelling. *Int. J. Mech. Sci.*, **31**(1989), 635-663.
6. Deshpande, V.S. and Fleck, N.A.: Multi-axial yield behaviour of polymer foams. *Acta Materialia*, **49** (2001), 1859-1866.
7. Miller, R.E.: A continuum plasticity model for the constitutive and indentation behaviour of foamed metals. *Int. J. Mech. Sci.*, **42**(2000), 729-754.
8. Papka, S.D. and Kyriakides, S.: Biaxial crushing of honeycombs – Part I: Experiments. *Int. J. Solids Struct.*, **36**(1999), 4367-4996.
9. Yu, T.X. and Karagiozova, D.: Plastic collapse of honeycombs under in-plane biaxial compression, *Key Engineering Materials*, 2004 (to appear).
10. Shim, V.P-W. and Stronge, W.J.: Lateral crushing in tightly packed array of thin-walled metal tubes. *Int. J Mech Sci.*, **28** (1986), 709-728.
11. Karagiozova, D. and Yu, T.X.: Post collapse characteristics of ductile circular honeycombs under in-plane compression. (2004). Submitted to *Int. J. Mech. Sci.*

# TIME-DEPENDENT PLASTICITY IN THE SUBMICRON REGIME

A.H.W. NGAN[1], G. FENG[2], B. TANG[1], H. LI[1] and P.C. WO[1]
*1) Department of Mechanical Engineering,*
*The University of Hong Kong,*
*Pokfulam Road, Hong Kong, P.R. China*
*2) Now at Department of Materials Science and Engineering,*
*Stanford University, Stanford, CA 24305, USA*

## Abstract

Nanoindentation and nano-scratch tests are used to study the characteristics of time-dependent plasticity in a range of materials. The submicron size effects on indentation creep are studied on a range of materials including single-crystal $Ni_3Al$, polycrystalline pure Al and fused quartz samples at room temperature. The stress exponents are found to exhibit a very strong size effect varying from a small value approaching unity for the smallest plastic indent, to over 100 as the indent size gets larger, indicating a rapid change of the dominating mechanism. Nano-scratch tests reveal that the transition from elastic to plastic deformation during scratching is a stochastic process. The coefficient of friction also exhibits a strong size effect from a low value for elastic scratching to a higher value for plastic scratching.

## 1. Introduction

The pioneering experiments by Fleck et al. [1] and others about a decade ago have triggered a lot of interest in the solid mechanics and materials science communities to understand how materials strength varies with size in the submicron length scale. Apart from being an academic interest, the rapid development of new capabilities and design methodologies in fabricating electro-mechanical systems at the micron or submicron length scales also calls for an urgent need to understand better how materials behave mechanically in the submicron length scale, as the successful design and operation of these micron machines rely on satisfactory material performance at this length scale.

While static strength has been the focus in the past, other mechanical properties such as fatigue, fracture, impact strength, creep, wear, etc., are



*Q. P. Sun and P. Tong (eds.), IUTAM Symposium on Size Effects on Material and Structural Behavior at Micron- and Nano-Scales,* 115–124.

obviously equally important. However, much less has been known about how these properties vary with length scale in the submicron regime. As an example, fatigue behaviour in bulk metals are well-known to be governed by microstructural mechanisms such as persistent slip bands, but these are micron-scaled phenomena and so if the material size is in the submicron regime, the expectation is that the fatigue behaviour will be very different. The relevant experimental database, however, is still largely blank at the moment.

Over the last few years, the present authors have been focusing on time-dependent characteristics of plasticity in the submicron regime [2-7]. In our experiments, depth-sensing indentation, more commonly known as nanao-indentation, was used to probe the creep properties of a wide range of materials. In this paper, we review some recent discoveries in submicron scale time-dependent plasticity.

## 2. Incipient Plasticity and Creep in Nanoindentation

In a typical indentation experiment on an annealed crystalline sample, it is well-known that incipient plasticity will be generated when the load reaches a certain value. Before this critical load, the deformation is elastic as defined by the zero residual plasticity upon load removal, and above this critical load, a sudden displacement burst will be observed. When the load is held at a value below the critical value, no creep can be observed, which further confirms that the deformation is purely elastic. Figure 1 shows an example of this in aluminium, in which no change in tip displacement can be observed during the first loading cycle in the low-load regime [2]. However, recent research has shown that the critical load at which plasticity occurs is not a material constant but is a quantity that varies with, for example, the duration of load hold [3], the loading rate [8], and, most likely, the temperature at which the experiment is carried out. In fig. 1, a strain burst can be observed after a certain waiting time in the second load cycle at a higher load value. This waiting time will be shorter as the load increases [3], and an interesting observation from fig. 1 is that as long as the strain burst has occurred, creep begins simultaneously, as can be seen from the gradual but continuous increase in displacement after the strain burst in the second load cycle.

Experiments on a number of metals including In, Cu, Al, $Ni_3Al$ intermetallics, etc. reveal that the type of nanoindentation creep described above in fact has enormous rate when compared to ordinary creep of the same material in the bulk conditions [2,4]. Figure 2 shows the creep behaviour of $Ni_3Al$, a high-temperature intermetallic, under a small indentation load at room temperature [4]. The tip displacement $h$ continues to increase at an apparently constant rate of 0.02 nm/s even after a long holding time of 10 min, corresponding to an indentation creep rate $\dot{h}/h$ [9] of ~$10^{-5}$ $s^{-1}$, which is an enormous creep rate for $Ni_3Al$ at room temperature. Similar results were obtained for other materials studied [2,4-7]. The high dislocation density underneath the indenter, together with the proximity to free surface of the deformation zone which means a very short diffusion distance, are thought to be the reasons for the high speed of indentation creep in the submicron regime [7].

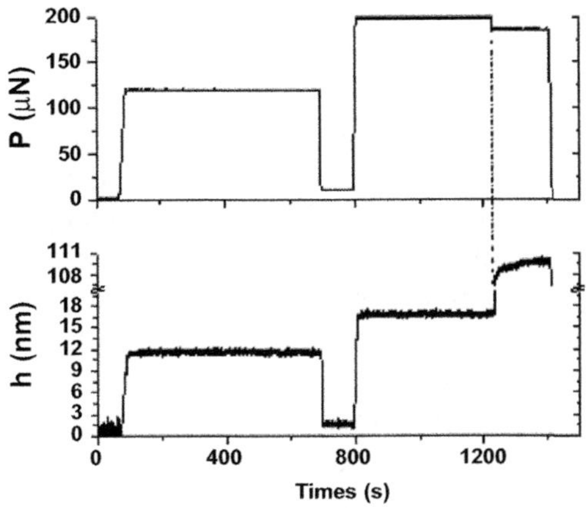


*Figure 1.* Load and displacement vs time plots during constant load indentation in aluminium at room temperature.

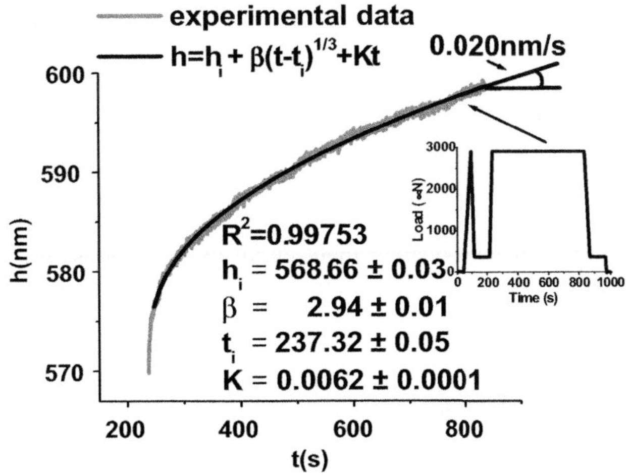


*Figure 2.* Displacement-time curve at constant indentation load of 3000 μN in Ni3Al(111). Berkovich indenter.

The influence of creep during nanoindentation is perhaps best illustrated by its effects on the measurement of hardness and elastic modulus using the well-known Oliver-Pharr approach [10]. In the Oliver-Pharr method of estimating the tip-sample contact area without the need of post-indentation imaging, the unloading behaviour of the sample is assumed to be purely elastic, so that Sneddon's classical solution in contact mechanics can be applied to measure the contact stiffness. In reality, however, creep may accompany the elastic recovery during unloading, even for high-melting materials such as $Ni_3Al$ at room temperature. This is likely to happen when (i) the unloading rate is low, (ii) the load hold before unloading is too short, or (iii) the peak load is too high [2,5]. In very soft materials, creep effects are almost unavoidable. The occurrence of creep during unloading will lead to an overestimation of the unloading stiffness if creep effects are mild, or may even cause the unloading curve to exhibit a negative slope if creep effects are strong. Fig. 3(a) shows an example of the latter situation in amorphous selenium at 311K [6]. Here, during unloading, creep is so significant that the indenter continues to sink into the sample even though the load on it is reducing, resulting in a negative apparent unloading stiffness. Figure 3(b) shows the elastic modulus calculated using the Oliver-Pharr method as a function of a factor measuring the significance of the creep. Because of the negative slope of the unloading curve, the elastic modulus estimated using the Oliver-Pharr method will be negative, and this is obviously unphysical.

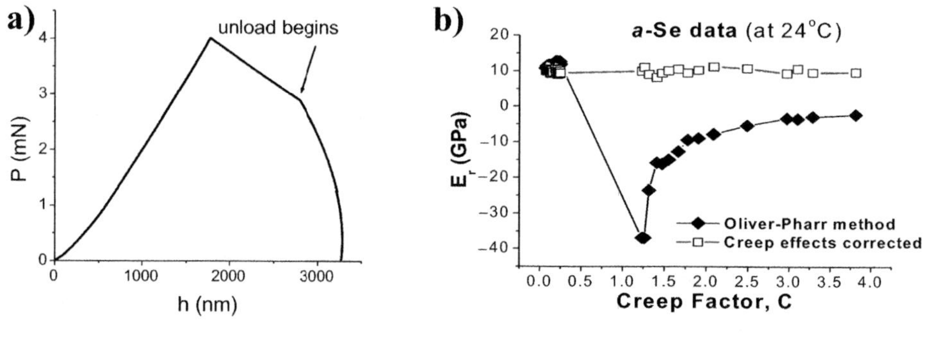


*Figure 3. (a)* A typical load-displacement curve in amorphous selenium at 311K. (b) Elastic modulus of amorphous selenium measured by the Oliver-Pharr method and after creep correction.

In a series of papers [4, 6, 11], the present authors have developed a rather simple method by which the creep effects can be corrected within the Oliver-Pharr framework. This involves obtaining a creep-free unloading stiffness $S$ via the following correction formula:

$$\frac{1}{S} = \frac{1}{S_u} + \frac{\dot{h}_h}{|\dot{P}|}, \tag{1}$$

where $S_u$ is the apparent stiffness at the onset of the unloading, $\dot{h}_h$ is the creep rate at the end of the load hold preceding the unloading, and $\dot{P}$ is the unloading rate. The corrected stiffness $S$ should now be purely elastic and is ready to be used in the Oliver-Pharr equations to calculate the elastic modulus and hardness. The correction formula in (1) was proven to be valid for the general case of power-law viscoelasticity within a Maxwell material model [11], and has been found to work satisfactorily in a range of materials, including common metals to very soft materials such as polymers, solder-joint metals [6], and bio-tissues [12]. In fig. 3(b) above, the modulus of selenium calculated after creep correction is found to be satisfactory at about 10GPa, no matter how significant the creep is. Here, the creep effect is characterized by a creep factor defined as $C = \dot{h}_h / S |\dot{P}|$.

## 3. Size Effect of Nanoindentation Creep

To investigate how nanoindentation creep varies with the indent size, a convenient parameter characterizing creep has to be used. The creep rate $\dot{\varepsilon}$ would not be a good parameter for this purpose because even in the bulk condition, it varies with the applied stress $\sigma$, usually through a power law

$$\dot{\varepsilon} \sim \sigma^n , \tag{2}$$

but in the nanoindentation situation, $\sigma$ itself is well-known to exhibit an indentation size effect [13] and so a separate size effect in creep cannot be conveniently detected if $\dot{\varepsilon}$ at different indent sizes are compared. However, the stress exponent $n$ in eqn. (2) is a parameter reflecting the mechanism for the creep, and this can be used to determine whether a change in the deformation mechanism exists as the indent size varies. The present authors [2,7] have performed a systematic study on this in a range of materials including In, $Ni_3Al$, Al and fused quartz. To measure $n$, the creep rate and the mean stress were calculated using the following scaling relations due to Bower et al. [9]:

$$\dot{\varepsilon} \sim \dot{h}/h \qquad \text{and} \qquad \sigma \sim P/h^2 , \tag{3}$$

where $h$ stands for tip displacement, $P$ the indentation load, and the dot refers to a time derivative. Figure 4(a) shows the log-log plot of the strain rate versus stress in the low-melting metal indium at room temperature under two constant indentation loads, one before the major strain burst in this material and one after, as shown in fig. 4(b) [2]. It can be seen that the creep curve at the smaller load in fig. 4(b) exhibits a steady-state stress exponent of about 1.5, while that at the higher load exhibits a stress exponent of about 6.2. This suggests a change in creep mechanism, from possibly a Herring-Nabarro-type mechanism to a dislocation climb mechanism, across the strain burst at ~120 μN. The strain burst is therefore believed to be associated with dislocation generation, but before the strain burst, the material can still deform at room temperature by a linear creep mechanism without involving dislocations. This is because room temperature is already ~ 0.7 of the absolute melting point of indium.

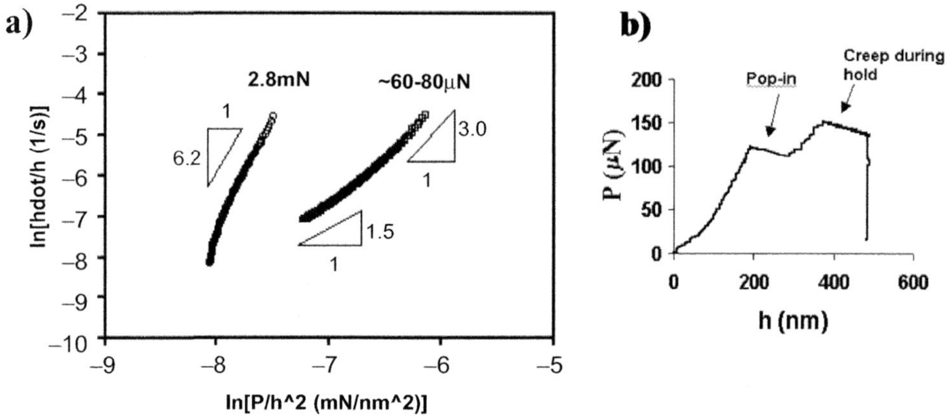


*Figure 4. (a)* Log $\dot{\varepsilon}$ vs log σ plots in indium at room temperature. (b) Load-displacement graph showing a major strain burst at ~120 μN in indium at room temperature.

In other materials for which room temperature is too low for diffusional creep to occur, no creep is detected before the strain burst, as shown earlier in fig. 1 for aluminium. In these materials, the stress exponent *n* measured after the strain burst still exhibits a significant size effect, as shown in Fig. 5. Here, the stress exponent in a number of higher melting materials exhibits a significant change from about unity at the smallest indentation load corresponding to sub-micron sized indents, to over 100 for high loads corresponding to micron sized indents [7]. These results suggest a rapid change of the creep mechanism, from linear diffusional flow ($n \sim 1$) for the smallest plastic indents, to athermal mechanisms with $n > 100$ for micron sized indents. Similar to the observation in indium in fig. 4(a), the occurrence of linear diffusion at the smallest indents is again thought to be due to the short diffusion path to free surface. In the two metals $Ni_3Al$ and Al in fig. 5, the linear mechanism for the small indent is thought to be pipeline diffusion, and the athermal mechanism for large indents is likely to be dislocation glide, which is well-known to be very insensitive to rate in these two metals. In the amorphous fused quartz, no dislocation activities can be present, and the athermal mechanism in large indents is thought to be propagation of micro-cracks underneath the indenter [7].

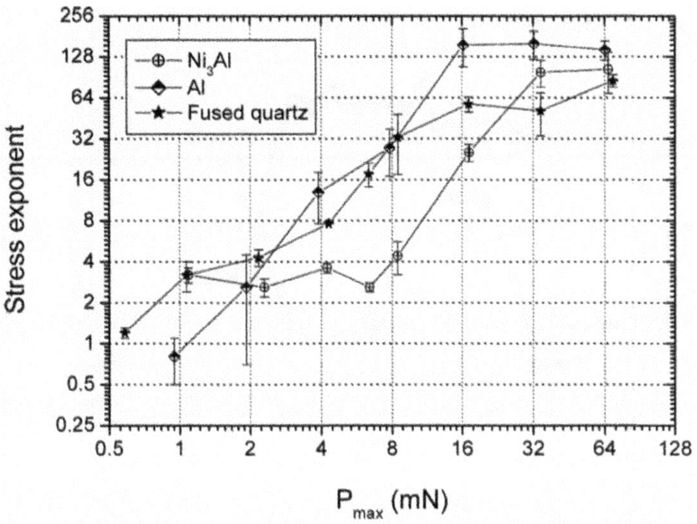


| Nominal Pmax (mN) | | 1 | 2 | 4 | 8 | 16 | 32 | 64 | 128 |
|---|---|---|---|---|---|---|---|---|---|
| Initial indent depth (nm) | $Ni_3Al$ | 76.0 | 119.0 | 172.9 | 263.4 | 400.6 | 591.4 | 883.8 | 1298.3 |
| | Al | 308.3 | 452.3 | 647.4 | 948.1 | 1476.7 | 2124.3 | 3137 | 4470.7 |
| | Fused quartz | 65.7 | 105.2 | 164.3 | 262.2 | 384.1 | 556.0 | 807.7 | 1130.0 |

*Figure 5.* Stress exponent vs holding load measured in constant-load experiments. The table shows the corresponding indent sizes at various holding loads.

## 4. Incipient Plasticity and Nano-scratch Behaviour

There is currently no widely accepted explanation for the delay in the occurrence of the strain burst as illustrated in fig. 1. One explanation that has been proposed may be described as the "invisible defect theory" [3, 14]. It has been suggested that some kind of stable dislocation structure or lock may exist underneath the indenter even though the contact may appear to be elastic. Upon load removal, this dislocation structure would be unzipped by attraction to the sample surface, hence it would be 'invisible' and the deformation would appear to be elastic. However, if the load is held, the dislocation structure may climb and expand by absorbing vacancies from the nearby free surface, and when it attains the critical unstable size, the dislocation structure would become a rapid dislocation source leading to an avalanche event. The waiting time for the initial structure to reach the critical size would thus be the observed waiting time for the strain burst to occur under constant load. Another explanation proposed is simply homogeneous dislocation nucleation inside the highly stressed zone underneath the indenter [8]. In this picture, the waiting time is simply due to the kinetics of the homogeneous nucleation.

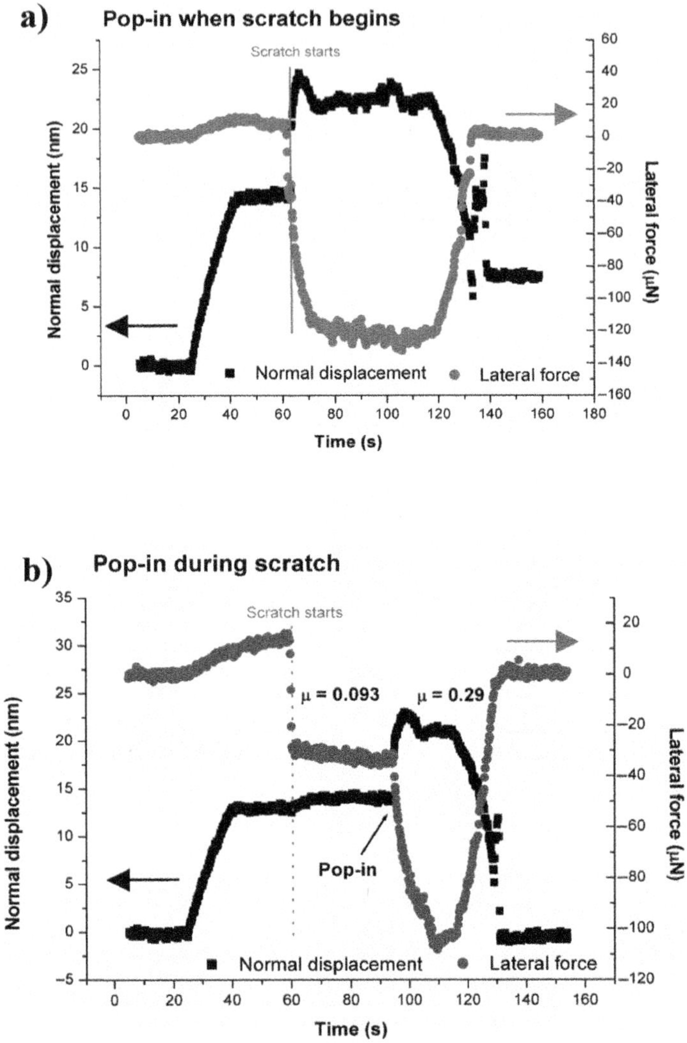


*Figure 6.* Normal displacement and lateral force in two types of sliding behaviour of an initially elastic contact in $Ni_3Al$. μ is the coefficient of friction.

An interesting experiment to illustrate that both the invisible defect theory and the homogeneous nucleation theory are rather insufficient to describe all the features of the delay in strain burst was recently carried out by Wo and Ngan [15]. Here, an initially elastic contact under uniaxial indentation load is subsequently slid by a lateral force after holding for some time under the uniaxial load, but before a strain burst has occurred. Three types of behaviour were observed in $Ni_3Al$ at room temperature. In the first type, which occurs when the uniaixial load is very small, the subsequent sliding behaviour is still elastic throughout. This is characterized by the continuity of the normal displacement of the tip at the start of the sliding. When the normal load is higher, however, a strain burst in the normal direction, characterized by a sudden discontinuity in the normal displacement, occurs during the subsequent sliding. In some cases, as illustrated in fig. 6(a), the strain burst occurs right at the beginning of the sliding, and in other cases, as shown in fig. 6(b), the strain burst occurs at some distance away from the start of the scratch. The situation in fig. 6(a) occurs at higher loads than that in fig. 6(b). In the situation in fig. 6(b), the location at which the strain burst occurs appears to be completely random. Since the position where the strain burst occurs is different from the position where the normal load is held, the invisible defect theory described above cannot explain the behaviour in fig. 6(b). A more likely explanation for this behaviour is that the indenter encounters some lattice inhomogeneities such as vacancy clusters or surface steps during sliding.

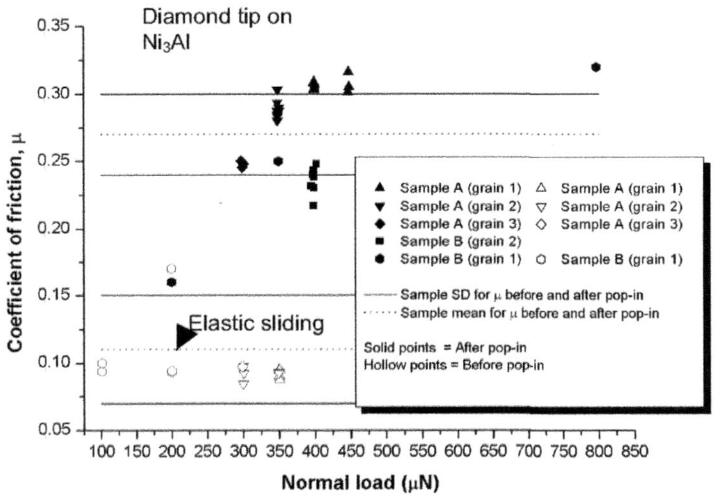


*Figure 7.* Coefficient of friction in different types of sliding behaviour in $Ni_3Al$.

Figure 7 shows the coefficient of friction, defined as the ratio of the lateral frictional force to the normal force, measured for the different types of sliding behaviour. It can be seen that for elastic sliding, the coefficient of friction between the diamond tip and the $Ni_3Al$ sample is rather low at about 0.09, but after a strain burst in the normal direction has occurred, the coefficient of friction increases to several times larger. The observation here has important implications on the design of sliding contacts in micro-machines. If the type of strain burst in fig. 6(b) occurs, the friction would be unpredictable, as the transition to plastic deformation may happen at any point during sliding. In design, the normal load range at which this type of random strain burst occurs should therefore be avoided.

## 5. Conclusions

In a wide range of materials significant creep occurs during submicron scaled mechanical contact. The creep effects are size dependent and can severely influence elastic modulus and hardness measurement using the nanoindentation technique. The occurrence of incipient plasticity also exhibits an interesting time-delay effect in a certain load range. Nano-scratch tests revealed that the transition from elastic to plastic deformation during scratching is a random process.

## Acknowledgements

This project was supported by two grants from the Research Grants Council of the Hong Kong Special Administrative Region, China (Project no. HKU 7062/01E and HKU 7083/02E).

## References

1. Fleck, N.A., Muller, G.M., Ashby, M.F. and Hutchinson, J.W.: Strain gradient plasticity: theory and experiment, *Acta Metall. Mater.* **42** (1994), 475-487.
2. Feng, G. and Ngan, A.H.W.: Creep and strain burst in indium and aluminium during nanoindentation, *Scripta Mater.* **45** (2001), 971-976.
3. Chiu, Y.L. and Ngan, A.H.W.: Time-dependent characteristics of incipient plasticity in nanoindentation of $Ni_3Al$ single crystal, *Acta Mater.* **50** (2002), 1599-1611.
4. Feng, G. and Ngan, A.H.W.: Effects of creep and thermal drift on modulus measurement using depth-sensing indentation, *J. Mater. Res.* **17** (2002), 660-668.
5. Ngan, A.H.W. and Tang, B.: Viscoelastic effects during unloading in depth-sensing indentation, *J. Mater. Res.* **17** (2002), 2604-2610.
6. Tang, B. and Ngan, A.H.W.: Accurate measurement of tip-sample contact size during nanoindentation of viscoelastic materials, *J. Mater. Res.* **18** (2003), 1141-1148.
7. Li, H. and Ngan, A H W.: Size effects of nanoindentation creep, *J. Mater. Res.* **19** (2004), 513-522.
8. Schuh, C.A. and Lund, A.C.: Application of nucleation theory to the rate and time dependence of incipient plasticity during nanoindentation, *J. Mater. Res.* (2004), in press.
9. Bower, A.F., Fleck, N.A., Needleman, A. and Ogbonna, N.: Indentation of a power-law creeping solid, *Proc. Roy. Soc. Lond. A,* **441** (1993), 97-124.
10. Oliver, W.C. and Pharr, G.M.: An improved technique for determining hardness and elastic modulus using load and displacement sensing indentation experiments, *J. Mater. Res.* **7** (1992), 1564-1583.
11. Ngan, A.H.W., Wang, H.T., Tang, B. and Sze, K.Y.: Correcting power-law viscoelastic effects in elastic modulus measurement using depth-sensing indentation, *Int. J. Solids Strut.* (2004), in press.
12. Kinney, J.H., Marshall, S.J. and Marshall, G.W.: The mechanical properties of human dentin: a critical review and re-evaluation of the dental literature, *Crit. Rev. Oral Biol. Med.* **14** (2003), 13-29.
13. Nix, W.D. and Gao, H.: Indentation size effects in crystalline materials: a law for strain gradient plasticity, *J. Mech. Phys. Solids* **46** (1998), 411-425.
14. Liang, H.Y., Woo, C.H., Huang, H.C., Ngan, A.H.W. and Yu, T.X.: Dislocation nucleation in the initial stage during nanoindentation, *Phil. Mag.* **83** (2003), 3609-3622.
15. Wo, P.C. and Ngan, A.H.W.: Incipient plasticity during nanoscratch in $Ni_3Al$, *Phil. Mag.* (2004), in press.

# ENERGY PENALTIES IN MATERIALS WITH MICROSTRUCTURES OF PHASE TRANSITIONS: MODIFIED GIBBS PHASE RULE

Y. HUO[1)] and I. MÜLLER[2)]
*1) Department of Mechanics and Engineering Sciences Fudan University, Shanghai, 200433, China, yzhuo@fudan.edu.cn*
*2) Technische Univ. Berlin, FB6, Sekr HF2, Thermodynamik, Strasse des 17. Juni, 135, D-10623 Berlin, Germany, im@thermodynamik.tu-berlin.de*

## Abstract

Complex microstructures are often formed in materials during phase transitions. It is important to understand the mechanism of the formation of such microstructures since they have a strong influence on the mechanical and physical properties of the materials. While the phase transition is governed by the difference of the chemical energies of the involved phases of the material, the formation of the microstructure is also affected by energy penalties resulted from the atomic distortions around the interfacial boundaries and constrains induced by the surrounding matrix. Based on a variational formulation, the energy terms penalizing the interfacial boundaries and the inhomogeneity are introduced and their effects on phase transition processes are discussed for diffusive and diffusionless phase transitions in alloys and solutions with fine lamellated microstructures. The results indicated that the Gibbs phase rule will take a different form and phase diagrams will be changed by the energy penalties. Moreover, jumps in the stresses, pressure or temperature will occur at the transition. The evolution of microstructures must be studied by considering the dynamics of the phase transition affected by the energy penalties as well.

## 1. Introduction

Materials can change their atomic/molecular structures drastically when the environmental conditions such as the pressure and the temperature have approached some certain critical values. This phenomenon is now called phase transition. The most well-known phase transitions to us are the water-vapor and water-ice transitions occurred at 100℃ and 0℃ on earth under the normal atmospheric pressure. So it is not surprising that they are among the first studied by scientists more than one hundred years ago. In the pioneer work of Van der Waals [1], a non-monotonic pressure-volume state function, the Van der Waals equation, was proposed to characterize the water-vapor phase transition. There are two descending braches representing the two phases, water and vapor, and one ascending branch separating them. This non-monotonic state function corresponds to a non-convex free energy function, which is now the starting point for the study of phase transitions.



*Q. P. Sun and P. Tong (eds.), IUTAM Symposium on Size Effects on Material and Structural Behavior at Micron- and Nano-Scales,* 125–132.

Gibbs [2] was one of the first researchers who established the general conditions for phase equilibrium, Gibbs phase rule, namely the Gibbs free energies (free enthalpies) of different phases should be equal for single component materials and the chemical potentials of each component at different phases should be equal for multi-component materials such as alloys and solutions. According to the Gibbs phase rule, the phase transition process should follow the convexification of the non-convex free energy.

However, the real transition process from one phase to the other should be more complicated than this simple picture due to the appearance of the phase mixture states in between. Capillary attraction of the liquid water in its vapor should raise the energy of the phase mixture was well known example of interface energies and studied extensively by Maxwell [3]. Two very different approaches were taken to treat the interface energy. In one approach as done by Laplace and Maxwell [3], the interface region is considered to be a singular surface without volume with an interface energy (surface tension) attached to it. In the other as done by Van der Waals [4] and later by Cahn and Hilliard [5], the interface region is taken as a region in which the density changes strongly and the interface energy is manifested by the spatial derivative of the density. The Laplace-Maxwell approach, even though suffers some physical defects is mathematically easier to work with and formed the bases for many later studies of phase transition processes in material sciences. The Van der Walls-Cahn-Hilliard approach evolves the variation of non-convex free energy with gradient interface energy term and is mathematically very difficult. Its solution in three dimensions is still an open problem. That is probably one of the reasons that the work of Van der Waals [4] was neglected until Cahn and Hilliard reestablished it in 1958 [5], see the Translator's Introduction by Rowlinson in [4] for more historical details.

Ericksen [6] studied the phase transition of a bar under tension in the context of nonlinear elasticity. Similar as the Van der Waals equation, he introduced a non-monotonic stress-strain relation that leads to a non-convex stored elastic energy. The minimization of such a free energy without considering the interface effect leads to the convexification. The best example of phase transition in elastic solids is the thermal elastic martensitic transformation in shape memory alloys. Observations have clearly shown that microstructures are formed during the phase transition and they cannot be predicted by only the non-convex free energy. One possible way to obtain such microstructures is to follow the Van der Waals-Cahn-Hilliard approach to introduce the strain gradient interface energy contribution. However, minimization of such a total energy in one dimension [7] allows only one single interface. Higher dimensional minimization is still too difficult. So, several works [8-10] are done in one-dimensional context to consider the effect of the boundary and the surrounding matrix on the phase transitions by adding an energy penalty for inhomogeneity. Fine microstructures with many interfaces can be indeed obtained by minimization. Local minimizers, bifurcation and metastability have been studied. Due to the mathematical complexity of such a variational approach, we tried in [11] to reestablish the Laplace-Maxwell approach with an energy penalty for inhomogeneity approximately obtained from the above variational formulation. Without much mathematical effort to calculate the minimizers, we obtained an analytical expression for the number of interfaces in phase mixtures and found that the stress jumps at the beginning of transition. In this work, we try to extend the method a little further to consider diffusive phase transitions of alloys and solutions for which the Gibbs phase rule of phase equilibrium is the equality of chemical potentials of each component in different phases. Emphasis is given to how the Gibbs phase rule will be changed by adding energy penalties for interface and inhomogeneity. In particularly, their effect on phase diagrams, which are the visualization of the Gibbs phase rule, will be discussed.

A brief review of the results in [8-11] will be given in the next section. Modified Gibbs phase rule will be studied in Section 3 followed by an example of phase diagrams in Section 4.

## 2. Energy Penalty for Interface and Inhomogeneity

Following [8-10], consider a nonlinearly elastic bar with a non-convex double-well energy density $f(u'(x))$. The total energy is

$$E = \int_0^L [f(u'(x)) + \frac{\varepsilon^2}{2} u''(x)^2 + \frac{\gamma^2}{2}(u(x) - u_h(x))^2]dx, \tag{1}$$

with boundary conditions

$$u(0) = u^0, \quad u(L) = D + u^0, \tag{2}$$

where $u(x)$ is the displacement of the bar, $u_h(x) = u^0 + dx$ is the displacement of a homogeneous bar subject to the same boundary conditions (2) with $d = D/L$, $\varepsilon$ and $\gamma$ are two small constants. Therefore, the strain gradient term penalizes interface and the last term penalizes inhomogeneity.

As $\varepsilon$ and $\gamma$ approach zero, the minimizing displacement of (1) should approach a piecewise linear function with its slopes in the left well of $f(u'(x))$, denoted as $u_x^-$, and in the right well as $u_x^+$, alternatively. Denote $L^{\pm}$ the total length of the intervals with slope $u_x^{\pm}$ and $N$ the number of points at which the slope has a change, the total energy (1) can be reduced under certain conditions [11] to the following simple form,

$$E = L^+ f(u_x^+) + L^- f(u_x^-) + \tau_1 N + \frac{\tau_2}{2} L\left(\frac{L^+}{N}\frac{L^-}{L}\right)^2 . \tag{3}$$

Where $\tau_{1,2}$ are two positive constants depending on $\varepsilon$, $\gamma$ and possibly $u_x^+ - u_x^-$. The mathematics required to search minimization of the above energy function (3) is much more simple than the energy functional (1) as shown in [11].

The first two terms in (3) are the common bulk energy of a material in phase mixture coming from the first integrand in (1), namely the non-convex free energy. The third one in (3) is the Laplace-Maxwell expression for the interface energy of singular surfaces resulted from the strain gradient term in (1). With only these three terms, it is evident that the minimization with respect to the number of interfaces $N$ is the smallest admissible value 1. Thus, it is the last term in (3), which is obtained from the last integrand in (1), that can produce microstructures. We shall use this idea to study phase transitions in binary alloys or solutions.

## 3. Modified Gibbs Phase Rule for Binary Alloys with Energy Penalties

Consider a volume $V$ as shown in figure 1a) with unit cross-section and height $H$ with temperature $T$, the material inside is a binary alloy/solution with $\overset{\circ}{m}_A$ mole atom $A$ and $\overset{\circ}{m}_B$ mole atom $B$. The alloys/solutions can be in two phases denoted as $\alpha$ and $\beta$. In order to have a quasi one-dimensional approximation the phases are assumed to be parallel lamellas as shown in the figure 1a). Similar shaped microstructures have been indeed observed in nature as shown in figure 1b) for perlite.

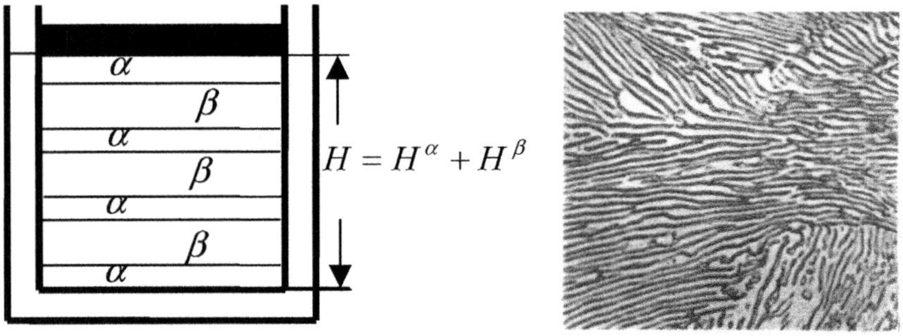


a) 1d model b) microstructures of perlite

*Figure 1.* Phase transitions of binary alloys/solutions with microstructures of parallel lamellas.

Following the idea in the last section, the total free energy of the system should be

$$E = \int_0^H \left\{ f(\rho_A, \rho_B, T) + \frac{\varepsilon^2}{2} \left[ \rho_A'(x)^2 + \rho_B'(x)^2 \right] \right\} dx + \int_0^H \frac{\gamma^2}{2} \left[ \left( m_A(x) - m_A^h(x) \right)^2 + \left( m_B(x) - m_B^h(x) \right)^2 \right] dx \quad (4)$$

Where $f(\rho_A, \rho_B, T)$ is a double well potential (the chemical free energy per unit volume) of the mole densities of the atoms $A$ and $B$, $(\rho_A, \rho_B)$. The term with the small constant $\varepsilon$ is the interface energy penalty taken as proportional to the square of the mole density gradients. The last term with the small constant $\gamma$ penalizes the inhomogeneity in which

$$m_{A,B}(x) := \int_0^x \rho_{A,B}(\xi) d\xi \text{ and } m_{A,B}^h(x) = \overset{\circ}{m}_{A,B} \frac{x}{H} \quad (5)$$

are the mole numbers of the atoms $A$ ($B$) in the interval [0,x] and the corresponding values for a homogenous body without microstructures, respectively.
Following the method in the last section, we can reduce the above energy functional to

an energy function of the mole densities in the two phase-wells $(\rho_A^{\alpha,\beta},\ \rho_B^{\alpha,\beta})$, the number of interfaces $N$ and total length of all intervals with $(\rho_A^{\alpha,\beta},\ \rho_B^{\alpha,\beta})$, $H^{\alpha,\beta}$. Namely, we should have

$$E = H^{\alpha} f(\rho_A^{\alpha},\rho_B^{\alpha},T) + H^{\beta} f(\rho_A^{\beta},\rho_B^{\beta},T) + \tau_1 N + \frac{\tau_2}{2} H \left[\frac{H^{\alpha}}{N}\frac{H^{\beta}}{H}\right]^2 . \tag{6}$$

It is more convenient to introduce the mole fractions $X^{\alpha,\beta} = \rho_A^{\alpha,\beta} / \rho^{\alpha,\beta}$ and the specific volumes per unit mole $v^{\alpha,\beta} := 1/\rho^{\alpha,\beta}$ as the natural variables of the specific free energy function density per unit mole and rewrite the total free energies for the α and β phases as

$$F^{\alpha,\beta} := H^{\alpha,\beta} f(\rho_A^{\alpha,\beta},\ \rho_B^{\alpha,\beta},T) = m^{\alpha,\beta} f(v^{\alpha,\beta},\ X^{\alpha,\beta},T). \tag{7}$$

For simplicity, we have used the same notation $f$ for the specific free energy per unit volume as a function of the mole densities and for the specific free energy per unit mole as a function of the specific volume and the mole fraction. Introducing the mole phase fraction $z := m^{\alpha}/m$ and the number of interfaces per mole $n := N/m$ with $m = m_A^{\circ} + m_B^{\circ}$ the total mole of the atoms $A$ and $B$, the total energy of the body (6) can be reduced to

$$\frac{E}{m} = zf(v^{\alpha},X^{\alpha},T) + (1-z)f(v^{\beta},X^{\beta},T) + \tau_1 n + \frac{\tau_2}{2} v \left[\frac{zv^{\alpha}(1-z)v^{\beta}}{nv}\right]^2 . \tag{8}$$

Note that the cross section of the volume is taken as unit. Under the constrains,

$$v = \frac{V}{m} = zv^{\alpha} + (1-z)v^{\beta} \text{ and } X = \frac{m_A^{\circ}}{m} = zX^{\alpha} + (1-z)X^{\beta}, \tag{9}$$

the minimization of (8) leads to the following equilibrium conditions

$$n/v = \left(\frac{\tau_2}{\tau_1}\right)^{1/3} \left(\frac{zv^{\alpha}}{v}\frac{(1-z)v^{\beta}}{v}\right)^{2/3}, \tag{10}$$

$$p_{\alpha} := -\frac{\partial f}{\partial v^{\alpha}} = p + \sqrt[3]{\tau_1^2 \tau_2}\,\frac{[(1-z)v^{\beta}/v]^{2/3}}{(zv^{\alpha}/v)^{1/3}}(1-\frac{zv^{\alpha}}{2v}), \tag{11}$$

$$p_{\beta} := -\frac{\partial f}{\partial v^{\beta}} = p + \sqrt[3]{\tau_1^2 \tau_2}\,\frac{(zv^{\alpha}/v)^{2/3}}{[(1-z)v^{\beta}/v]^{1/3}}(1-\frac{(1-z)v^{\beta}}{2v}), \tag{12}$$

$$\mu_A^{\alpha}(T,p_{\alpha},X^{\alpha}) = \mu_A^{\beta}(T,p_{\beta},X^{\beta}),\ \mu_B^{\alpha}(T,p_{\alpha},X^{\alpha}) = \mu_B^{\beta}(T,p_{\beta},X^{\beta}). \tag{13}$$

Where $\mu_{A,B}^{\alpha,\beta}$ are the chemical potentials of the atoms $A$ ($B$) in the corresponding phases, respectively. From (10), the microstructures, i.e. the number of interfaces per unit volume can be calculated. We refer to [11] for more discussions about this equation and its effect on energy barriers in phase transitions. It is clear from (11) and (12) that the energy penalties for interfaces and inhomogeneity have raised the pressures in the α and β phases, particularly at the beginning of the nucleation, namely, $p_\alpha$ ($p_\beta$) approaches infinite at $z = 0$ (1). From (10-12), we may write the force balance at the sidewall where the average pressure is $p$ as

$$pH = p^\alpha H^\alpha + p^\beta H^\beta - \left[ \tau_1 N + \frac{\tau_2}{2} H \left( \frac{H^\alpha}{N} \frac{H^\beta}{H} \right)^2 \right]. \tag{14}$$

This shows clearly that the penalizing energies act as a contracting force on the phase mixtures similar to the effect of surface tensions for convex surfaces. This seems to indicate that the energy penalty for the inhomogeneity introduced as the last term in (4) and (6) is a correction of the simplified assumption that the microstructures are parallel lamellas.

(13) will reduce to the commonly known Gibbs phase rule when $p_\alpha = p_\beta = p$ in case there are no energy penalties. However, the energy penalties introduced will raise the pressures inside the lamellas. Thus, now the chemical potentials of the atoms $A$ or $B$ have to be evaluated differently in different phases. So, we may call (13) the *modified Gibbs phase rule*.

## 4. Effect of Energy Penalties on Phase Transitions of Alloys/Solutions

In order to demonstrate how the energy penalties introduced above will affect the phase transition processes of binary alloys and solutions, we try to construct phase diagrams from the modified Gibbs phase rule (13).

Most known materials have their chemical potentials measured experimentally as data banks. For simplicity and for demonstration purposes, we prefer to use analytical expressions. Known examples are idea mixtures of idea gases and incompressible fluids. Thus, we choose a mixture of propane ($C_3H_8$) and butane ($C_4H_{10}$) and construct the gas-liquid phase transition diagram. As shown in the figure 2, the dashed lines are boundaries of phase regions without energy penalties and the solid lines are those with energy penalties. It is obvious that the energy penalties have a rather strong influence on the phase diagram.

Take a mixture of butane and propane with its mole fraction $X$ shown by the dotted line in Fig. 2, it is in the liquid phase for high pressure. As the pressure drops to cross the upper dotted line, the mixture would start to transform from liquid to gas if there is no energy penalties and the transition is governed alone by the chemical energies. However, the energy penalties for the interface and the inhomogeneity makes the nucleation of the new phases impossible at this pressure, so the mixture will be kept in the liquid phase until its pressure drops to cross the upper solid line. Now the chemical energy difference balanced the energy penalty of nucleation. Then, the gas phase can appear. Such a process of phase transition is well known for nucleation with convex surfaces, in particular with spherical surfaces [12]. For nucleation with lamellas, the introduce of the energy penalty for both the interface and the inhomogeneity resulted in a similar behavior, namely, both the phase transitions liquid-to-gas and gas-to-liquid become to be more difficult to occur.

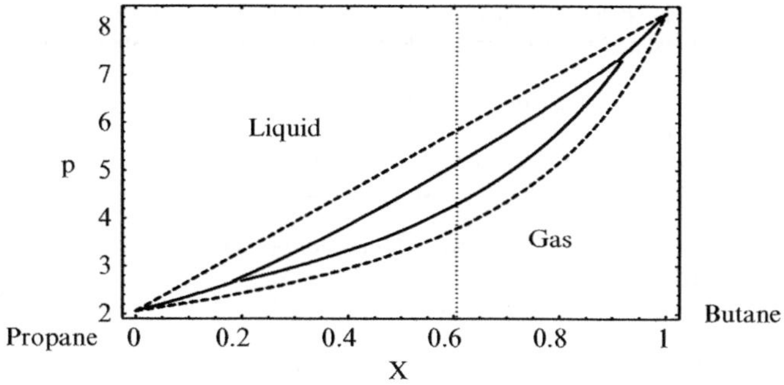


*Figure 2.* The effect of energy penalties on the phase diagrams of Propane-Butane mixture.

## 5. Discussions

One-dimensional model of phase transitions suffers some physical defects [7]. Adding energy penalty for inhomogeneity as in [8-11] does produce some nontrivial results for the microstructures observed during phase transitions, as also shown above. However, the physical meaning of such an energy penalty remains a mystery. One possible way to produce such a term was proposed in [10] by considering two coupled bar acting in parallel; one nonlinear elastic bar with martensitic variants, while the other linear elastic bar always in the austenitic phase. In some sense, this is an artificial one-dimensional simplification for the martensitic variants surrounded by the austenitic matrix.

The force balance (14) on the sidewall of a container with phase mixtures may give some help for a better understanding. As we know if the liquid phase is nucleated from the gas as a spherical droplet, the pressure inside the droplet is raised due to the surface tension [12]. This is generally true for nuclei with convex surfaces but not for plane surfaces as necessarily assumed in a one-dimensional model. However, introducing the energy penalty for inhomogeneity into the model does the same effect, namely raise the pressure inside the nuclei, thin lamellas now (14) seems to indicate that such an energy penalty might be due to the sidewall constrain, i.e. the constrains on the boundaries and/or surrounding matrix.

## Acknowledgement

This work was supported by Alexander von Humboldt Foundation of Germany and The National Science Foundation of China (10372023).

## References


1. Van der Waals, J.D., On the continuity of gaseous and liquid states, North Holland, 1988.
2. Gibbs, J.W., Collected Works, Longman, New York, 1928, Vol. 1.
3. Maxwell, J.C., Capillary Action, in Encyclopaedia Britannica, $9^{th}$ ed., 1876.
4. Van der Waals, J.D., The thermodynamic theory of capillary under the hypothesis of a continuous variation of density, Translated from Dutch by J.S. Rowlinson, Journal of Statistical Physics, 20 (1979), 197-244.
5. Cahn, J.W. and J.E. Hilliard, Free energy of a nonuniform system. I. Interfacial free energy, The Journal of Chemical Physics, 28 (1958), 258-267.
6. Eericksen, J.L., Equilibrium of bars, J. Elasticity, 5(1975), 191-202.
7. Carr, J., M.E. Gurtin and M. Slemrod, Structured phase transition on a finite interval, Arch. Rat. Mech. Anal., 86 (1984), 317-351.
8. Müller, S., Singular perturbations as a selection criterion for periodic minimizing sequences, Calculus of Variations, 1 (1993), 169-204.
9. Truskinovsky, L. and G. Zanzotto, Ericksen's bar revisited: Energy wiggles, J. Mech. Phys. Solids, 8 (1996), 1371-1408.
10. Vainchtein, A., T. Healey, and P. Rosakis, Bifurcation and metastability in a new one-dimensional model for martensitic phase transitions, Computer Methods in Applied Mechanics and Engineering, 170 (1999), 407-421.
11. Huo, Y. and I. Müller, Interfacial and inhomogeneity penalties in phase transitions, Continuum Mechanics and Thermodynamics, 15 (2003), 395-407.
12. Huo, Y. and I. Müller, Nucleation of droplet in a binary mixture, Meccanica, 38 (2003), 493-504.

# THE CAHN-HILLIARD PHASE-FIELD MODEL FOR TOPOLOGY OPTIMIZATION OF SOLIDS

M.Y. WANG and S. ZHOU
*Department of Automation & Computer-Aided Engineering*
*The Chinese University of Hong Kong*
*Shatin, NT, Hong Kong*

## Abstract

In this paper we present a method to optimize the topology of a solid structure based on the theory of phase transition with diffuse interface. A well-known physical model describing phase separation, the Cahn-Hilliard model is modified to reflect the objective of the design optimization. The topology optimization is formulated with a phase parameter that varies continuously at any position within the specified design domain. The relaxation of the parameter is driven by local minimization of the free energy of the system, including the design objective, the bulk energy, and the interface energy. The bulk energy plays a role to draw the design variable to its two distinct material phases (solid and void), while the interface energy smoothes the structure boundary to its minimum perimeter. With properties of mass conservation and energy dissipation of the model, complex morphological and topological material transitions such as coalescence and break-up of the boundary can be naturally captured. A thermodynamic equation for the phase parameter is obtained for the creation, evolution, and dissolution of controlled phase interfaces. We use a multigrid method to solve the resulting fourth-order Cahn-Hilliard equation. The capabilities of the method are demonstrated with three examples in 2-D for the mean-compliance minimization of structures.

## 1. The Problem of Solids Optimization

Let us consider the minimum compliance optimization problem of a statically loaded linear elastic structure under a single loading case (Bendsoe and Kikuchi 1988; Rozvany 1989; Haber et al. 1996). Let $\Omega \subseteq R^d$ $(d = 2$ or $3)$ be an open and bounded set occupied by the linear isotropic elastic structure. The boundary of $\Omega$ consists of three parts: $\Gamma = \partial\Omega = \Gamma_0 \cup \Gamma_1 \cup \Gamma_2$, with Dirichlet boundary conditions on $\Gamma_1$ and Neumann boundary conditions on $\Gamma_2$. It is assumed that the boundary segment $\Gamma_0$ is



*Q. P. Sun and P. Tong (eds.), IUTAM Symposium on Size Effects on Material and Structural Behavior at Micron- and Nano-Scales,* 133–141.

traction free. The displacement field $u$ in $\Omega$ is the unique solution of the linear elastic system

$$\begin{aligned} -\operatorname{div}\sigma(u) &= f && \text{in } \chi_\Omega(x)=1 \\ u &= u_0 && \text{on } \Gamma_1 \\ \sigma(u)\cdot n &= h && \text{on } \Gamma_2 \end{aligned} \tag{1}$$

with the strain tensor $\varepsilon$ and the stress tensor $\sigma$ at any point $x \in \Omega$, $E$ the elasticity tensor, $u_0$ the prescribed displacement on $\Gamma_1$, $f$ the applied body force known for all possible configurations of $\Omega$, $h$ the boundary traction force applied on $\Gamma_2$ such as an external pressure load exerted by a fluid, and $n$ the outward normal to the boundary.

The topology optimization problem of minimizing mean compliance is defined as:

$$\begin{aligned} &\inf_{\rho}\; l(u) = \int_\Omega f^T u d\Omega + \int_{\Gamma_2} h^T u dS \\ &\text{Subject to } (u,\rho) \in S(\Omega) \end{aligned} \tag{2}$$

The admissible space $S(\Omega)$ of pair $(u,\rho)$ for the problem is defined as $S(\Omega) = \{(u,\rho) \in U \times H\}$, with $u$ and $\rho$ satisfying

$$a(u,\rho) := \int_\Omega E(\rho)\varepsilon(u) : \varepsilon(v) d\Omega = \int_\Omega f^T v d\Omega + \int_{\Gamma_2} h^T v d\Gamma, \quad \forall v \in U \tag{3}$$

with ':' representing the second order tensor operator. This is the weak form of the equilibrium equation of the elastic system, where the set of kinematically admissible displacements $U$ is specified as

$$U := \left\{ v \in W^{1,2}(\Omega);\, v = 0 \text{ on } \Gamma_1 \right\} \tag{4}$$

and the space of feasible designs $H$ is

$$H = \left\{ \rho \in L^\infty(\Omega);\; 0 \le \rho \le 1 \text{ a.e. in } \Omega \right\} \text{and } \int_\Omega \rho(x) d\Omega \le V \tag{5}$$

with a limit on the amount of material in terms of the maximum admissible volume $V$ of the design.

A dominant approach to the topology optimization is to treat the problem as a material distribution over a fixed reference domain (Bendsoe and Sigmund 2003). This approach offers many beneficial features particularly in terms of computational efficiency. This is evident from the successful applications of the methods in industry. However, the approach has some intrinsic features that are less desirable. The chief one is the dependency of the design variables of material density on the finite element grid. This dependency yields a number of drawbacks, including a lack of an unambiguous mapping from the model to design, under-resolved finite element solutions for the physical response, and spurious numerical behavior caused by the coupling with finite element models. These fundamental issues are often argued in the literature and other methods such as level-set and phase-field methods have been developed more recently (Wang *et al.* 2003; Wang and Zhou 2004).

In this paper we present a phase-field based method to address the topology optimization problem. We adapt the Cahn-Hilliard model from mechanics and material sciences, treating the optimization problem as the stability of a system containing instable components and reaching a stable equilibrium.

## 2. The Phase Field Model

Phase filed methods are a particular class of diffuse-interface models often used in the study of critical phenomena and fluid dynamics. The idea is to introduce a phase-field parameter that describes the state of the system continuously. A free energy can be defined when the system is not in equilibrium (Cahn and Hilliard 1958; Cahn 1961; Allen, Cahn and Hilliard 1971), such that:

$$\Psi = \int_{\Omega}\left(\psi(\rho)+\frac{\gamma}{2}|\nabla\rho|^2\right)d\Omega \tag{6}$$

where $\Omega \subset R^d$, $d = 2, 3$ is a bounded domain with sufficiently smooth boundary, $0 \leq \rho \leq 1$ represents the concentration of one of the two components of a binary composite. $\gamma > 0$ is a measure of non-locality and introduces the interface thickness. $\psi(\cdot)$ is called the bulk energy or the homogeneous free energy and is a double well function who take its minimum value at $\rho = 0$ and $\rho = 1$. The term $\frac{\gamma}{2}|\nabla\rho|^2$ accounts for the interface energy or the gradient energy. Isothermal equilibrium of the system is characterized by minimizing the total free energy (6) (Elliott 1989; Garcke and Rumpe 2001).

Cahn and Hilliard generalized the problem with introduction of the generalized chemical potential $w$, so that it is the functional derivative of the energy (6)

$$w = \psi'(u) - \gamma\Delta u \tag{7}$$

Assuming that the system is isothermal and according to mass flux law (Degroot and Mazur 1962), Fick's diffusion law and mass balance law, the kinetics of phase separation can be modeled using non-equilibrium thermodynamics (Cahn and Hilliard 1958; Cahn 1961; Allen, Cahn and Hilliard 1971) as

$$\begin{aligned} u_t &= \nabla(M(u)\nabla w) \\ w &= \psi'(u) - \gamma\Delta u \end{aligned} \tag{8}$$

with boundary condition:

$$\frac{\partial c}{\partial n} = \frac{\partial w}{\partial n} = 0 \tag{9}$$

where $M > 0$ is called mobility. $n$ is the outer normal direction on the boundary. This yields a fourth-order, nonlinear parabolic diffusion equation describing the phase separation dynamics (Kim 2002).

## 3. The Cahn-Hilliard Model for Topology Optimization of Structures

Our method for topology optimization of a solid structure is based on the approach of Cahn and Hilliard for spinodal decomposition in binary alloys (Cahn and Hilliard 1958; Cahn 1961; Allen, Cahn and Hilliard 1971). In our model, the components of the alloys are replaced with the solid material phase and the void phase. In the initial state, corresponding a temperature higher than the critical temperature in alloy solidification process, the system is in a mixed uniform state, i.e., $\rho \in [0,1]$. As time goes on, such as in the cooling process to the critical temperature, phase separation drives the materials into homogeneous solid ( $\rho = 1$ ) or void ( $\rho = 0$ ) regions separated by regularized boundaries quickly. It is because the bulk energy is taken to be a double-welled potential such that $\Psi(1) = \psi(0) = 0$ and is non-zero only in the transition region where $0 < \rho < 1$. The thickness of the intermediate density region is proportional to $\gamma^2$. At certain moment, the bulk energy reaches its minimal point, but the interface energy is still very high for the long length interface. Further on, with the disappearing of small homogeneous regions, the length of interface is shorten and the system separates into two layers finally. This is illustrated in Figure 1.

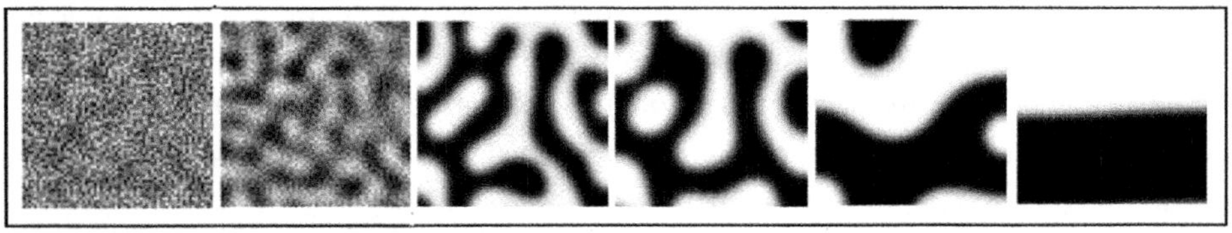

*Figure 1.* The evolution of a Cahn-Hilliard system

However, when considering elastic effect in the evolution process, the system will have more complicated behavior. The elastic interactions arising from a difference of lattice spacing between two coherent phases can have a strong influence on the phase separation behavior of the components. And if the elastic moduli are different in the two phases, the elastic interactions may accelerate, slow down or even stop the phase separation process (Fratzl, Penrose and Lebowitz 1999; Jou 1997). Garcke (2003) proposed that the evolution of the system is governed by diffusion equations for the concentrations of the components and by a quasi-static equilibrium for the mechanical part. The resulting system of equations is elliptic-parabolic and can be understood as a generalization of the Cahn-Hilliard equation. The elasticity terms enter the diffusion equation only linearly. He also given a derivation of the system and prove an existence and uniqueness result for it.

Following the above analysis of Cahn-Hilliard systems, we introduce an elastic energy term to the total free energy in Eq. (6) as,

$$\inf_{\rho} J(u) = \int_{\Omega} f^T u dx + \int_{\Gamma_2} h^T u dx + \int_{\Omega} \left[ \psi(\rho) + \frac{\gamma^2}{2} |\nabla \rho|^2 \right] dx \tag{10}$$

with $u$ and $\rho$ satisfying Eq. (3). The bulk energy is defined as

$$\psi(\rho) = \frac{1}{4} \rho^2 (\rho - 1)^2 \tag{11}$$

So the modified Cahn-Hilliard equation is given as

$$\begin{aligned} u_t &= \nabla(M(\rho)\nabla w) \\ w &= \psi'(\rho) - \gamma^2 \Delta\rho + F'(\rho) \end{aligned} \tag{12}$$

with the boundary condition

$$\frac{\partial c}{\partial n} = \frac{\partial w}{\partial n} = 0 \tag{13}$$

where

$$M(\rho) = \rho(1-\rho) \tag{14}$$

The mobility is a thermodynamically reasonable choice as suggested in (Kim 2002). This mobility significantly lowers the long-range diffusion across bulk regions, and is particularly appropriate when study flows with immiscible components (Kim 2002). In our problem of optimization of structural topology, checkerboard patterns are strictly forbidden which means different phases are also immiscible. Furthermore, it is reasonable to depress long-range diffusion by such mobility, in order to increase efficiency of the optimization process.

## 4. Multigrid Algorithm for Numerical Solutions

The Cahn-Hilliard equation is a nonlinear fourth-order PDE in space, and it is very difficult to solve by an explicit method. Here we use the multigrid method introduced in detail in (Kim 2002). It is robust and stable even with the elastic energy introduced to the system.

Following the principles of error smoothing and coarse grid correction of multigrid methods (Trottenberg 2001), we discretize Eq. (12) according to Crank-Nicholson algorithm and expand the nonlinear item by Taylor series. For numerical stability of the algorithm, the time step is set no more than half of the grid width. Therefore, this condition limits the convergence rate, and it usually takes a large number of iterations for convergence.

## 5. Numerical Examples

Here we illustrate the phase-field model for the mean compliance optimization problems that have been widely studied in the relevant literature (cf. Bendsoe and Sigmund 2003; Rozvany 1989). For clarity in presentation, the examples are in 2D under plane stress condition. The mesh used is of $96 \times 48$ quadrilateral elements in FEM analysis of the whole structure or a half of the structure due to the geometric symmetry. The volume ratio is specified to be $0.5$.

### 5.1 MBB EXAMPLE

The first example is known as MBB beams related to a problem of designing a floor panel of a passenger airplane in Germany. The floor panel is loaded with a unit concentrated vertical force $P = 1N$ at the center of the top edge. It is has a fixed support and a simple support at its bottom corners respectively. The design domain has a length to height ratio of 4:1 (Figure 2). The evolution of the optimization is shown in Figure 3, from the initial state to the final solution, where $m$ is the iteration number.

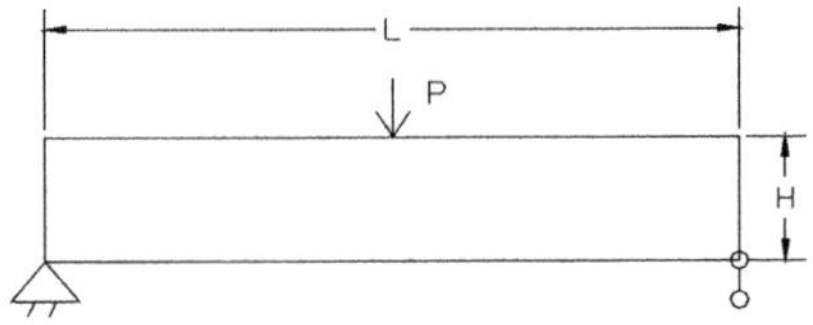


*Figure 2.* The MBB structure with fixed-simple supports

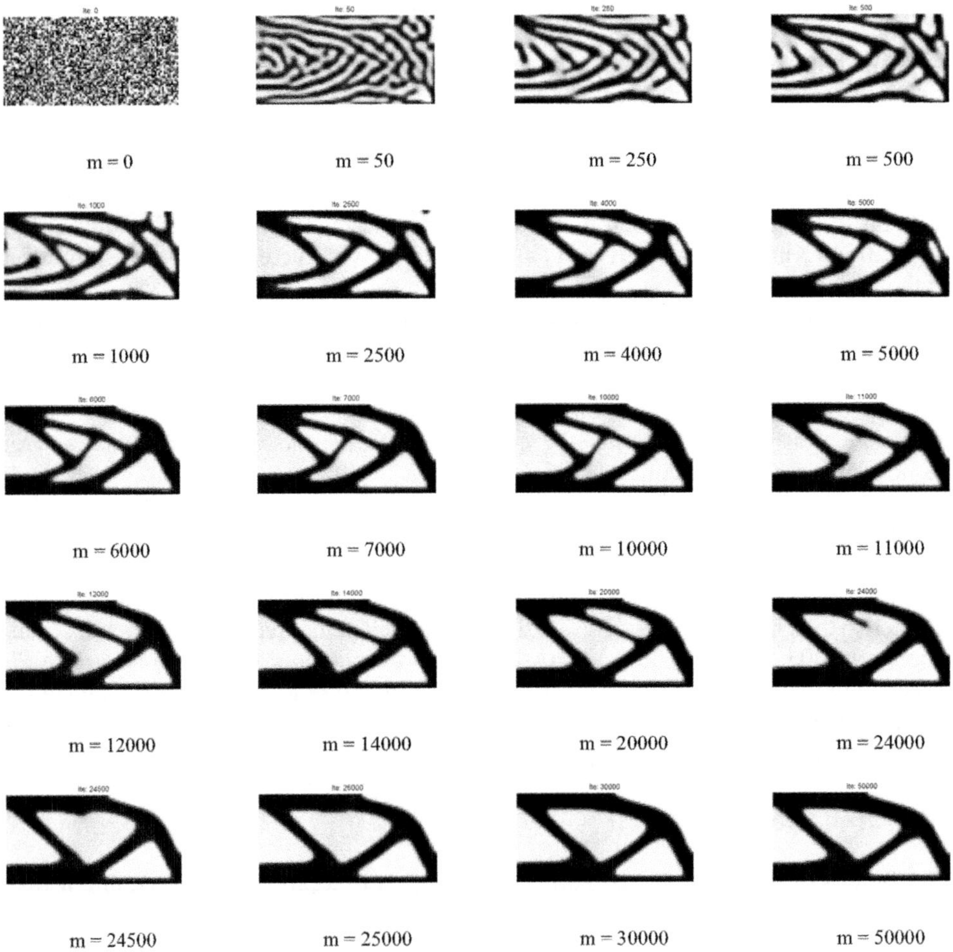


*Figure 3.* The optimization process for the MBB beam

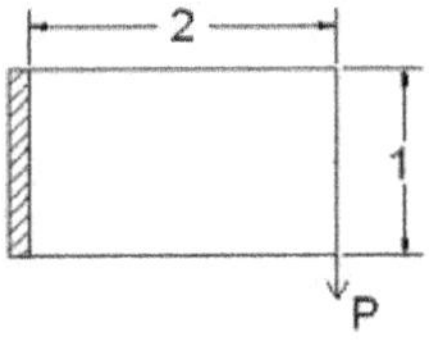


*Figure 4.* The Cantilever beam structure

## 5.2 CANTILEVER BEAM EXAMPLE

The second example is a cantilever beam with a concentrated vertical force $P = 10N$ at the bottom of its free vertical edge. The design domain has a length to height ratio of 2:1 (Figure 4). The evolution of the phase transition process is shown in Figure 5, from the initial state to the final solution.

## 5.3 BRIDGE STRUCTURE EXAMPLE

A bridge-type structure is considered last. A rectangular design domain of $L$ long and $H$ high with a ratio of $L:H = 2:1$ is loaded vertically at its bottom with multiple

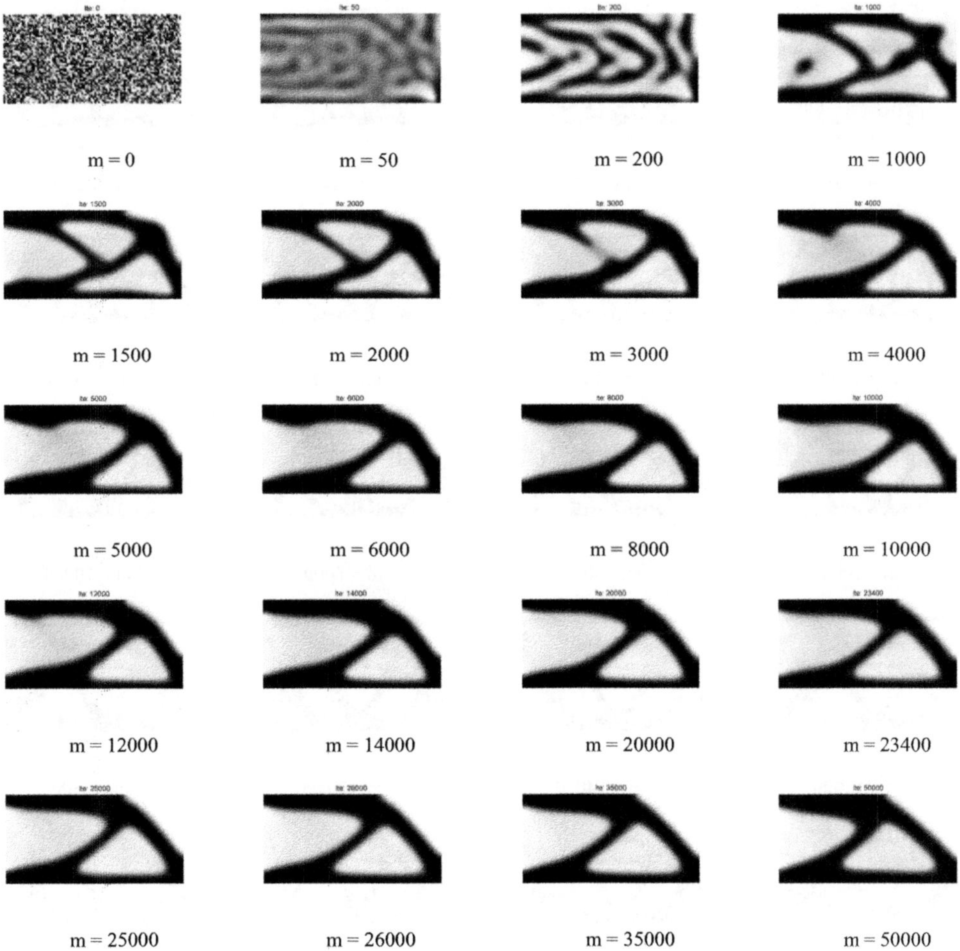


*Figure 5.* The evolution process for the cantilever beam

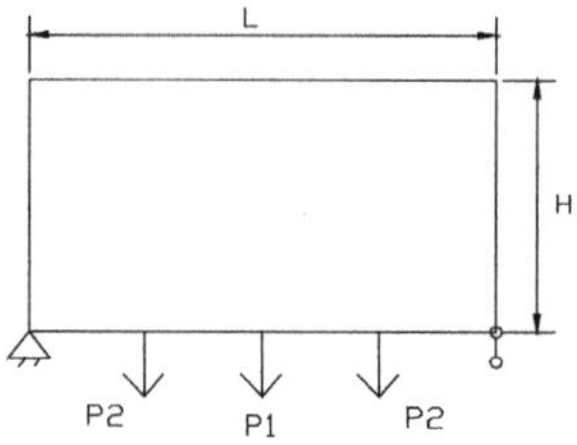


*Figure 6.* A bridge type structure with multiple loads

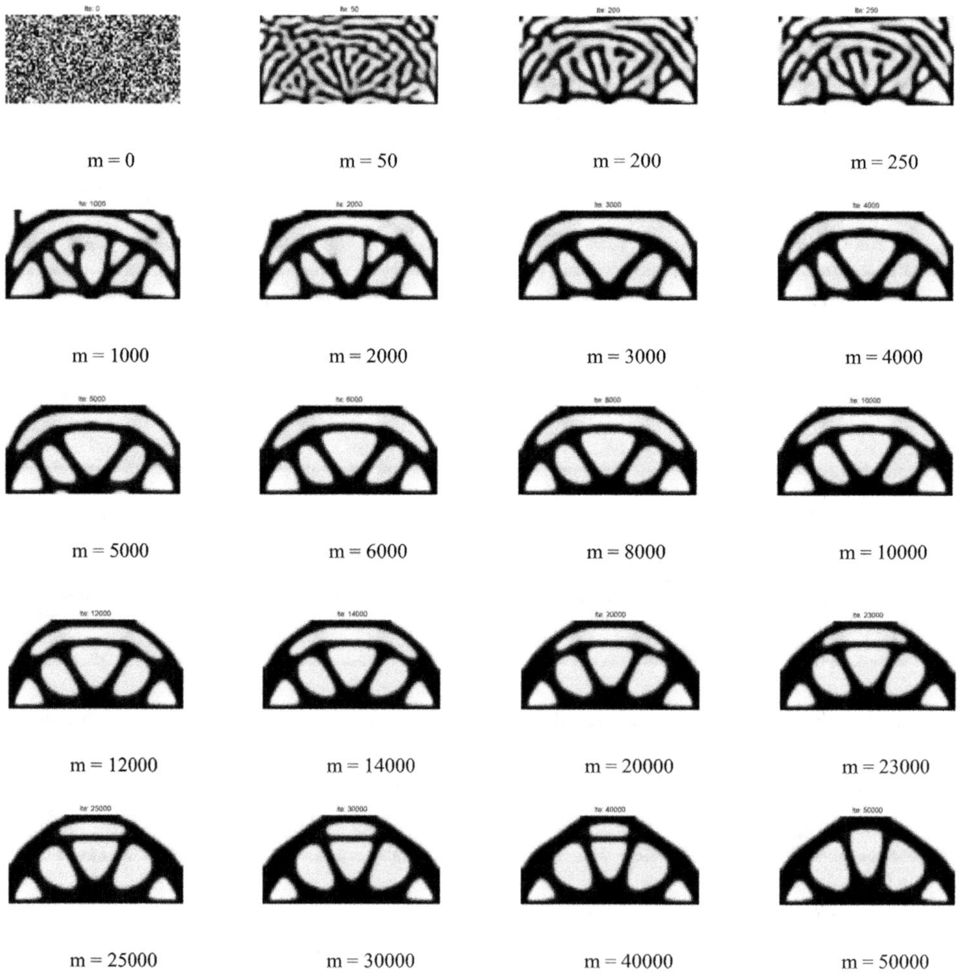


*Figure 7.* The evolution process for the bridge-type structure

loads $P_1 = 40N$ and $P_2 = 20N$ as shown in Figure 6. The left bottom corner of the beam is fixed, while it is simply supported at the right bottom corner. Figure 7 shows the solution process, while Figure 8 shows the changes of the different energy terms during the phase transition process.

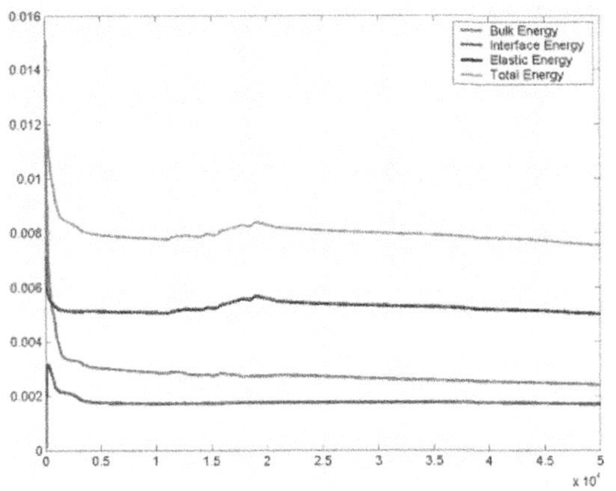


*Figure 8.* Changes of the energy terms during iteration

## References


Allen S. and Cahn J. (1979) A microscopic theory for antiphase boundary motion and its application to antiphase domain coarsening, *Acta Metallurgica* **27**, 1085-1095.

Bendsoe, M. P. and Sigmund, O. (2003) *Topology Optimization: Theory, Methods and Applications*. Berlin: Springer.

Bendsoe, M. P. and Kikuchi, N. (1988) Generating optimal topologies in structural design using a homo - genisation method. *Computer Methods in Applied Mechanics and Engineering* 71, 197-224.

Cahn J. (1961) On spinodal decomposition, *Acta metallurgica* **9**, 795-801.

Cahn J. and Hilliard J. E. (1958) Free energy of a nonuniform system. I. Interfacial free energy, *Journal of Chemical Physics* **28(1)**, 258-267.

Cahn J. and Hilliard J. (1971) Spinodal decomposition: A reprise, *Acta Metallugica* **19**, 151-161.

Cahn J. and Taylor J. (1994) Surface motion by surface diffusion, *Acta Metallurgica* **42**, 1045-1063.

Degroot S. R. and Mazur M. (1962) *Non-equilibrium Thermodynamics*, North-Holland Pub., Amsterdam.

Elliott C. M. (1989) The Cahn-Hilliard model for the kinetics of phase separation, *International Series of Numerical Mathematics* **88**, 35-72.

Eyre D. (1993) Systems of Cahn-Hilliard equations, *SIAM J. Appl. Math.* **53(6)**, 1686-1712.

Fratzl P. Oliver, P. and Lebowitz J. (1999) Modeling of phase separation in alloys with coherent elastic misfit, *Journal of Statistical Physics* **95**, 1429-1503.

Garcke H. (2003) On Cahn-Hilliard systems with elasticity. Preprint.

Jou H. Leo P. and Lowengrub J. (1997) Microstructural evolution in inhomogeneous elastic media, *Journal of Computational Physics* **131,** 109-148.

Kim J. (2002) Modeling and simulation of multi-component, multi-phase fluid flows, Ph.D dissertation, University of Florida.

Rozvany, G. (1989) *Structural Design via Optimality Criteria*. Dordrecht: Kluwer.

Trottenberg, U., Oosterlee C.W. and Schuller A. (2001) *Multigrid*, Academic Press, San Diego.

Wang, M. Y., Wang, X., and Guo, D. (2003) A level set method for structural topology optimization, *Computer Methods in Applied Mechanics and Engineering* **192(1-2)**, 227-246.

Wang, M. Y. and Zhou, S. W. (2004) "Phase field: a variational method for structural topology optimization," *CMES: Computer Modeling in Engineering & Sciences* **6(6)**, 547-566.

# AN EXPERIMENTAL INVESTIGATION OF THE RELATIONSHIP BETWEEN MOLECULAR STRUCTURE AND LENGTH SCALES IN INELASTIC DEFORMATION OF AN AMORPHOUS THERMOPLASTIC

J. AGDE TJERNLUND, E. K. GAMSTEDT and P. GUDMUNDSON
*KTH Solid Mechanics, Royal Institute of Technology*
*Osquars backe 1, 100 44 Stockholm, Sweden*

Experimental observations have indicated that the presence of strain gradients has an influence on the inelastic behaviour of polymers as well as in other materials such as ceramics and metals. A brief review is presented on experimental results showing strain-gradient effects in both thermoplastic and thermoset polymers. The present study has experimentally quantified length scale effects in inelastic deformations of the polymer material polystyrene with respect to the molecular length. The experimental technique that has been used is nano-indentation to various depths with a Berkovich indenter. The hardness has been calculated with the method by Oliver and Pharr, and also by direct measurements of the indented area from atomic force microscopy. The experiments showed that the length scale effects in inelastic deformations exist in polystyrene at ambient conditions. The direct method gave a smaller hardness than the Oliver-Pharr method. It was also showed that the length scale parameter according to Nix and Gao increases with increasing molecular weight. For high molecular weights above a critical value of entanglement, there was no pertinent increase in the length scale parameter. The length scale parameter for strain-gradient plasticity has a size of around 0.1 μm for polystyrene.

## 1. Introduction

Strain gradient plasticity has been studied extensively during the last decade for metallic materials. Most of these studies have dealt with modeling of strain-gradient plastic deformation. Few studies have been concerned with non-crystalline materials, such as polymers, even though strain gradient effects have been observed in this class of materials. Analysis of inelastic deformation of polymers is also becoming more relevant, due to miniaturization of e.g. microelectronic components, microelectro-mechanical systems (MEMS), coatings, development of composite matrix systems etc. The dependency on time, temperature, hydrostatic pressure and strain rate is usually higher for polymers than for metallic materials, which makes the analysis more difficult. The first step would be to quantify the extent of the strain-gradient effect in various polymers, and to investigate the underlying mechanisms responsible for such an effect.



*Q. P. Sun and P. Tong (eds.), IUTAM Symposium on Size Effects on Material and Structural Behavior at Micron- and Nano-Scales,* 143–150.

## 2. Strain-Gradient Effects in Polymers

There are indications that structural dimensions do have an influence on the plastic behaviour of polymers. Briscoe et al. [1] have done indentation testing for a selection of organic polymers: polymethylmethacrylate , polystyrene, ultra-high molecular weight polyethylene. Krupi ka [2] studied indentations in the polymers 2.5-dimethyl-2.5-hexanediol divinyl ether and hexanediol divinyl ether. They all showed that the hardness evaluated from small indents is larger than that resulting from a deeper one, which is indicative of a higher plastic yield stress in the presence of larger strain gradients. Chong and Lam [3-5], who have performed indentation tests in epoxy materials and polycarbonate, have also demonstrated this length scale effect. Hochstetter et al. [6] have done nanoindentations in the amorphous polymers polystyrene, polycarbonate and diethylene glycol bis(ally l carbonate). They have also calculated the evolution of the average stress with the average strain beneath the tip by using the theoretical background developed for elasto-perfectly plastic materials. By comparing the results they concluded that the approach does not work well and that the strain gradient can not be neglected.

Gamstedt et al. [7] have observed that the interfacial shear strength measured close to a fibre end in a single-fibre composite test was higher than the matrix shear strength macroscopically measured with an Iosipescu test for unsaturated polyester. A possible explanation is that the deviation results from a size effect. Since the carbon fibre diameter was small, the plastic deformation in the unsaturated polyester occurred in a small volume with large strain gradients. Fiedler and Schulte [8] measured the maximum values of interfacial shear stress in a composite material with epoxy matrix and glass fibres. These values were higher than the shear yield stress of the pure matrix.

Zhandarov et al. [9] has investigated the interfacial bond strength, toughness and load transfer between polymeric matrices (polyamide 6, polyamide 6.6, maleic anhydride grafted polypropylene and vinylester resin) and glass fibres having different diameters, by making single-fibre pull-out tests. The interfacial bond strength was characterized by an ultimate interfacial shear strength whose values were nearly independent of the fiber diameter. An effective shear-lag parameter was also measured, characterizing the load transfer efficiency between the fibre and the matrix. The value of decreased considerably with the fibre radius. The measured results for were used to calculate the radius of matrix material surrounding the fibre that was significantly affected by the presence of the fibre. The ratio of this radius to the fibre radius increases for decreasing fibre diameter.

## 3. Effects of Molecular Length

The length scale effects in inelastic deformati ons of a glassy thermoplastic, in this case polystyrene, have been experimentally quantified by micro-indentation to various depths. It has been examined if the molecular length affects the length scale in plastic deformation.

## 3.1. EXPERIMENTAL PROCEDURES

The examined materials were all linear atactic polystyrene, which is an amorphous thermoplastic. Because of the bulky vinyl group, polystyrene is not easily oriented even in extrusion [10]. Press moulding, which was used in the present investigation, would therefore most likely assure a random coil arrangement, with no preferential orientation of the molecular chains. In its undeformed state, polystyrene can be considered to be an isotropic material. All tests were done at room temperature, when the polymer is in its glassy state.

Six different polystyrene standards with various molecular weights and very narrow molecular weight distributions are examined. In addition, a polystyrene material with larger molecular weight distribution was used. From the molecular weight, the molecular contour length and the end-to-end distance, *R*, can be estimated [11]. In Table 1 the molecular weights, the polydispersity indices, the molecular contour length and the end-to-end distance *R* are given for the materials.

TABLE 1. The molecular weights, polydispersity indices, molecular contour lengths and the end-to-end distances for the used materials.

| $M_p$ [g/mol] | $M_w/M_n$ | Molecular length [nm] | $R$ [nm] |
|---|---|---|---|
| 4 920 | 1.03 | 15 | 4.7 |
| 10 050 | 1.02 | 30 | 6.8 |
| 19 880 | 1.02 | 59 | 9.5 |
| 52 220 | 1.02 | 150 | 15 |
| 96 000 | 1.03 | 280 | 21 |
| 189 300 | 1.02 | 560 | 29 |
| 56 300 | 1.25 | | |

After moulding, the specimens were cooled and fractured in liquid nitrogen to induce a more brittle fracture rather than bringing about gross molecular orientation in large plastically deformed zones, although some chain scission may occur close to the brittle fracture surface. The polymer close to the surface should then have similar properties as the polymer in the bulk. Details of the experimental procedures can be found in [12].

Several indentations to the depths 0.25, 0.5, 0.75, 1, 1.5 and 2 μm were made. The material is thought to be elastic-plastic and the following analysis is time independent, i.e. viscoelastic deformation is not considered. Polymers are generally viscoelastic, but in the glassy state under relatively high loading rates, the viscous contribution can be neglected. The hardness is defined as the mean pressure the material will support under load in the indented area

$$H = \frac{P_{max}}{A}, \tag{1}$$

where $A$ is the projected area of contact at peak load, $P_{max}$. The indenter geometry can be described by an area function $F(h)$ which relates the cross-sectional area of the indenter to the normal distance from its tip, $h$. This function can be derived from the load-displacement data [13].

An approach to capture length scale effects is to apply one of the models that already have been developed for metallic materials. A model proposed by Nix and Gao [14] is used here. It is employed as a phenomenological model since there is no direct coupling with the physical mechanisms in polymer yielding to plastic deformation in metals by dislocation movements. The hardness $H$ for a given depth of indentation $h_{max}$ is expressed as

$$\frac{H}{H_0} = \sqrt{\frac{h^*}{h_{max}} + 1}, \qquad (2)$$

where $H_0$ is the hardness in the limit of infinite depth and $h^*$ is a length scale parameter. The parameters $H_0$ and $h^*$ can then be identified from experimental data of $H$ and $h_{max}$.

## 3.2. RESULTS AND DISCUSSION

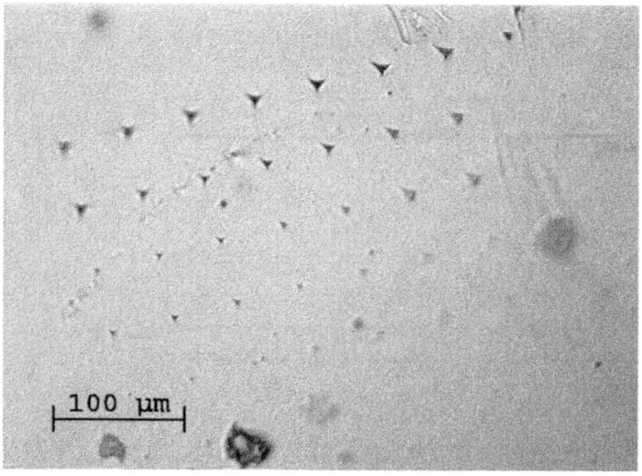


*Figure 1a.* Optical micrograph of a series of indentations on polystyrene with a molecular weight of 52 220 g/mol.

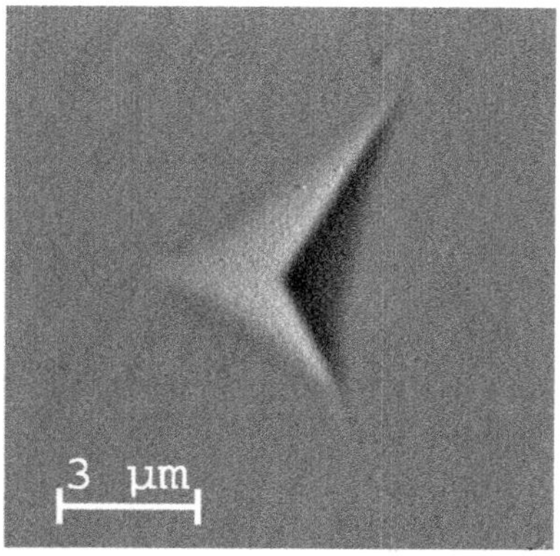


*Figure 1b.* Image from atomic force microscopy of an indentation on polystyrene with molar mass 4 920 g/mol where maximum load was 5.3 mN.

Micrographs of typical indentations are found in Fig. 1. These pictures were taken several months after the indentations were made. It can be observed that the indentations have not relaxed back to a flat surface. If it had, prominent viscoelastic (time dependent) deformation would have occurred. The hardness values were plotted against the maximum indentati on depths for each sample, as shown in Fig. 2 for low molecular weight. The data from the polystyrene with molecular weight 96 000 g/mol, 189 300 g/mol and the conventional polystyrene are very scattered and a length scale effect is not detectable. The variability in the data from samples with high molecular weight is caused by the surface roughness, which formed when breaking these samples. The uneven fracture surface indicates ductile fracture despite cooling the samples with liquid nitrogen. The degree of plastic deformation varies over the surface, which also gives rise to a variation in the hardness measurements. For the low-molecular materials in Fig. 2, there is a clear length-scale tendency, and henceforth we will concentrate on these data. From the atomic force microscope images for the polystyrene sample with molecular weight 4 920 g/mol the area of the triangle was measured which gave the same effect as the Oliver-Pharr method, i.e. decreasing hardness with increasing indentation depths. Eq (2) was used to make a least-squares adjustment to the hardness-depth data with respect to the asymptotic hardness in the limit of infinite depth, $H_0$, and the length scale parameter, $h^*$. The estimated parameters for the different materials are listed in Table 2. The asymptotic hardness in the limit of infinite depth, $H_0$, turned out to be relatively independent of molecular weight.

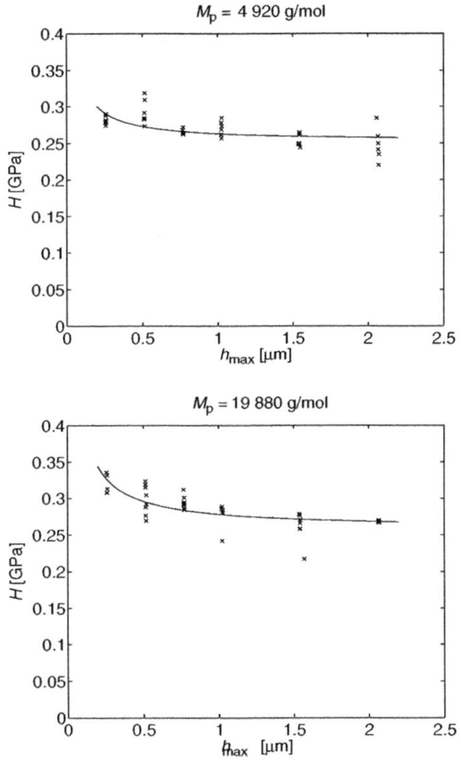


*Figure 2.* The hardness plotted against the maximum indentation depth for the samples with low-molecular weights with least-squares fits of the function in Eq (2).

TABLE 2. The molecular length, the end-to-end distance, $R$, the limit hardness, $H_0$, and the length scale parameter, $h^*$, for tested polystyrene materials.

| $M_p$ [g/mol] | Molecular length [μm] | $R$ [μm] | $H_0$ [GPa] | $h^*$ [μm] | $h^*/R$ |
|---|---|---|---|---|---|
| 4 920 | 0.015 | 0.0047 | 0.25 | 0.082 | 17 |
| 10 050 | 0.030 | 0.0068 | 0.25 | 0.10 | 15 |
| 19 880 | 0.059 | 0.0095 | 0.26 | 0.15 | 16 |
| 52 220 | 0.15 | 0.015 | 0.26 | 0.17 | 11 |

The length-scale relation in Eq (2) was then plotted with these values of $H_0$ and $h^*$ in the same figures as the measured data in Fig. 2. A comparison between the strain-gradient length scale parameter, the molecular contour length and the end-to-end distance can be made in Table 2. The length scale parameter is almost one order larger than the molecular length for small molecular weights but for larger molecular weights they are of the same order, but the end-to-end distance is quite small in comparison to the length scale parameter. The molecular lengths of extended molecules seem to correlate with the length scale of the strain-gradient effect, although longer structures, probably due to molecular entanglement should contribute. The ratio $h^*/R$ is almost constant for small indentation depths.

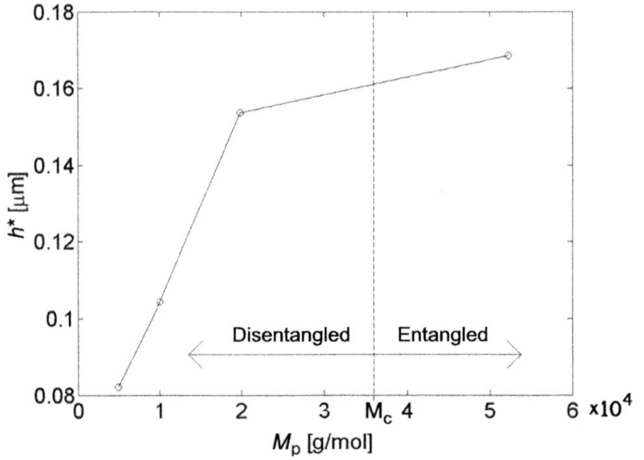


*Figure 3.* The length scale parameter, $h^*$, for polystyrene plotted against the respective molecular weight.

The length scale parameter, $h^*$, for the different molecular weights are plotted in Fig. 3. This plot shows that the length scale parameter increases with increasing molecular weight, i.e., with increasing lengths of the molecules. Between 19 880 and 52 220 g/mol there is a sudden decrease in the slope of the curve. One explanation to the sudden decrease in the slope of the curve in Fig. 3 might be associated with the degree of entanglement. The zero-shear-rate viscosity in thermoplastic polymer melts shows an abrupt increase in the molecular weight dependence at a certain critical molecular weight value $M_c$ [15]. The critical molecular weight value $M_c$ may be associated with the minimum chain length needed to generate chain entanglements. The low molecular-weight melts with a molecular weight lower than the critical $M_c$ are characterized by few or no chain entanglements. High molecular-weight melts are characterized by substantial amount of chain entanglements, which cause a restriction on the chain movements and imply a stronger molecular-weight dependence of the shear viscosity. The same entanglements are assumed to be present in the glassy state if the entanglements are retained after solidification and therefore the critical molecular weight might influence the strain gradient length scale. The critical value $M_c$ has been found to be 36 000 g/mol for polystyrene [15], which lies in between 19 880 and 52 220 g/mol, where the decrease in slope takes place, as seen in Fig. 3.

At the present stage one can only speculate about possible mechanisms that gives this length scale dependence in glassy thermoplastic polymers. Shear band formation is a mechanism for plastic deformation in polymers. There are some scales that can be identified from this mechanism. It is expected that microstructural dimensions like the thicknesses of and distances between shear bands could be of importance. From transmission electron microscopy of polypropylene samples which have been plastically deformed by scratching, Tang and Martin [16] detected shear band thicknesses of about 1 μm which is considerably higher than the contour length of the polymer molecules. Bowden [17] reports that slow-cooled polystyrene shows well-defined shear bands with a thickness of about 1 μm in uniaxial tension. After yielding, the shear bands were observed to multiply in number but not in thickness. By a low-frequency Raman scattering study of the nanostructure of plastically deformed thermoplastic glasses (linear polycarbonate and polymethylmethacrylate) induced by simple shear, Mermet et al. [18, 19] showed a significant heterogeneity in the deformed amorphous structure. The deformation is interpreted as coming from oriented macromolecular strands in the less cohesive spaces between more cohesive regions. The effect of macroscopic plastic deformation on these more cohesive regions is negligible in the glassy state. It is noteworthy that the size of these cohesive regions is considerable larger than the end-to-end distance of the molecules. The non-continuous nanostructure is believed to form for a wide range of materials at the transition from a liquid to a glassy solid [20].

DeMaggio et al. [21] has found a length scale in polystyrene by studying the thermal expansion of 7-300 nm thick Si-supported polystyrene films. The thermal expansion decreased if the film thickness was sufficiently small. For larger film thicknesses the thermal expansion approaches the thermoelastic bulk prediction. The film thickness dependence was modelled by assuming that a constrained layer close to the film-substrate interface has a zero thermal expansion coefficient. Another length scale effect has been observed by Du et al. [22] who measured yield stresses in polycarbonate and polystyrene thin films. They obtained values that were about twice the bulk values.

## 4. Conclusions

Indentation tests have been performed on polystyrene samples with different monodisperse molecular weights. It has been shown that length scale effects in inelastic deformations with large strain gradients exist in glassy polystyrene. The length scale parameter for strain-gradient plastic deformation was found to increase with increasing molecular weight. For a molecular weight between 19 880 g/mol and 52 220 g/mol the increase of the Nix-Gao length scale parameter with molecular weight is decreasing, which could be due to the fact that the molecular weight exceeds a critical value associated with the minimum chain length needed to generate chain entanglement. For molecular weights below the critical molecular weight the quotient between length scale parameter and the average length of a molecule between its endpoints is constant. The length scale parameter has a size around 0.1 μm for polystyrene.

## 5. Acknowledgement

The authors are grateful to Dr. Zhi-Hui Xu for performing the indentation tests and for fruitful discussions.

## References


1. Briscoe, B.J., Fiori, L. and Pelillo, E.: Nano-indentation of polymeric surfaces, *J. Phys. D: Appl. Phys.* **31** (1998), 2395-2405.
2. Krupi ka, A., Haraldsson, T., Johansson, M. and Hult, A. In-situ structure build-up and structure-property studies on UV-crosslinked vinyl ether-maleate resins, in A. Krupi ka Ph.D. thesis, *Use and Interpretation of Scratch Tests on Organic Coatings*, Department of Polymer Technology, Royal Institute of Technology, Sweden, 2002, pp. 53-69.
3. Lam, D.C.C. and Chong, A.C.M.: Indentation model and strain gradient plasticity law for glassy polymers, *J. Mater. Res.* **14** (1999), 3784-3788.
4. Chong, A.C.M. and Lam, D.C.C.: Strain gradient plasticity effect in indentation hardness of polymers, *J. Mater. Res.* **14** (1999), 4103-4110.
5. Lam, D.C.C. and Chong, A.C.M.: Characterization and modelling of specific strain gradient modulus of epoxy, *J. Mater. Res.* **16** (2001), 558-563.
6. Hochstetter, G., Jimenez, A., Cano, J.P. and Felder, E.: An attempt to determine the true stress-strain curves of amorphous polymers by nanoindentation, *Trib. Int.* **36** (2003), 973-985.
7. Gamstedt, E.K., Skrifvars, M., Jacobsen, T.K. and Pyrz, R.: Synthesis of unsaturated polyesters for improved interfacial strength in carbon fibre composites, *Compos. Part A.* **33** (2002), 1239-1259.
8. Fiedler, B. and Schulte, K.: Photo-elastic analysis of fibre-reinforced model composite materials, *Compos. Sci. Technol.* **57** (1997), 859-867.
9. Zhandarov, S., Pisanova, E., Mäder, E. and Nairn, J.A.: Investigation of load transfer between the fiber and the matrix in pull-out tests with fibers having different diameters, *J. Adhes Sci Technol.* **15** (2001), 205-222.
10. Evans, A.M., Kellar, E.J.C., Knowles, J., Galiotis, C., Carriere, C.J. and Andrews, E.H.: The structure and morphology of syndiotactic polystyrene injection molded coupons, *Polym. Eng. Sci.* **37** (1997), 153-165.
11. Flory, P.J.: *Statistical Mechanics of Chain Molecules*, John Wiley and Sons, New York, 1969.
12. Agde Tjernlund, J., Gamstedt, E. K. and Xu, Z.-H.: Influence of molecular weight on strain-gradient yielding in polystyrene, *Polym. Eng. Sci.*, in press, 2004.
13. Oliver, W.C. and Pharr, G.M.: An improved technique for determining hardness and elastic modulus using load and displacement sensing indentation experiments, *J. Mater. Res.* **7** (1992), 1564-1583.
14. Nix, W.D. and Gao, H.: Indentation size effects in crystalline materials: a law for strain gradient plasticity, *J. Mech. Phys. Solids* **46** (1998), 411-425.
15. Mark, J.E., Eisenberg, A., Graessley, W.W., Mandelkern, L. and Koenig, J.L.: *Physical Properties of Polymers*, Ch. 3, American Chemical Society, Washington DC, 1984.
16. Tang, H.X. and Martin, D.C.: Near-surface deformation under scratches in polypropylene blends, *J. Mater. Sci.* **38** (2003), 803-815.
17. Bowden, R.: The yield behavior of glassy polymers, in R. N. Haward (ed), *The physics of glassy polymers*, Applied Science, London, 1973, pp. 279-339.
18. Mermet, A., Duval, E., Etienne, S. and G'Sell, C.: Effect of a plastic deformation on the nanostructure of polycarbonate: study by low-frequency Raman scattering, *Polymer* **37** (1996), 615-623.
19. Mermet, A., Duval, E., Etienne, S. and G'Sell, C.: Low frequency Raman scattering study of the nanostructure of plastically deformed polymer glasses, *J. Non-Cryst. Solids* **196** (1996), 227-232.
20. Elliott, S.R.: Medium-range structural order in covalent amorphous solids, *Nature* **354** (1991), 445-452.
21. DeMaggio, G.B., Frieze, W.E., Gidley, D.W., Zhu, M., Hristov, H.A., Yee, A.F.: Interface and surface effects on the glass transition in thin polystyrene films, *Phys. Rev. Lett.* **78** (1997), 1524-1527.
22. Du, B., Tsui, O.K.C., Zhang, Q. and He, T.: Study of elastic modulus and yield strength of polymer thin films using atomic force microscopy, *Langmuir* **17** (2001), 3286-3291.

# MODELING OF AUSTENITE/MARTENSITE LAMINATES WITH INTERFACIAL ENERGY EFFECT

H. PETRYK, S. STUPKIEWICZ and G. MACIEJEWSKI
*Institute of Fundamental Technological Research,
Polish Academy of Sciences Świętokrzyska 21, 00–049 Warsaw,
Poland*

**Abstract**

Stress-induced formation and growth of internally twinned martensite plates within the austenite matrix in crystals of shape memory alloys is investigated. The recently developed model of evolving laminates is extended by taking into account the interfacial energy at three different scales. The interfacial energy density due to elastic strains at a twinned martensite boundary is calculated by the finite element method. Analytic relationships between characteristic dimensions of the rank-two laminate are derived by minimizing the size-dependent contribution of the interfaces to the total free energy. Specific examples have been calculated for the cubic-to-orthorhombic ($\beta_1$ to $\gamma_1'$) stress-induced transformation in a CuAlNi shape memory alloy.

## 1. Introduction

This paper deals with micro-mechanical modeling of stress-induced martensitic transformations in single crystals of shape memory alloys (SMA). The transformation is assumed to proceed by the formation and growth of parallel martensitic plates within the austenite matrix, as commonly observed in experiments. In turn, in many transformations the martensite plates are internally twinned, having the form of fine scale mixtures of two martensite variants (cf. Bhattacharya, 2003). A model of evolution of stress-induced microstructures of that type has recently been developed (Stupkiewicz and Petryk, 2002) by combining the micro-macro transition for sequential laminates with a local thermodynamic criterion of phase transformation. The phase transformation criterion used is based on a threshold value for the thermodynamic driving force at the phase transformation front, including in this way the intrinsic



*Q. P. Sun and P. Tong (eds.), IUTAM Symposium on Size Effects on Material and Structural Behavior at Micron- and Nano-Scales,* 151–162.

dissipation due to phase transition. Plasticity-like macroscopic constitutive rate equations have been derived in which the transformed volume fraction is incrementally related to the overall strain or stress. The model is developed within the framework of small strain elasticity with eigenstrains and takes into account different anisotropic elastic properties of austenite and martensite. This leads to the redistribution of internal stresses during transformation that can affect the evolution of the microstructure.

In the present paper, in order to study size effects the model outlined above is extended by taking into account the interfacial energy at three different scales: on the twin boundaries within a martensite plate, on the interfaces between austenite and twinned martensite plates, and on the crystal boundary. Analytic relationships between characteristic dimensions at the different levels of the rank-two laminate follow from minimization of the size-dependent contribution of the interfaces to the total free energy, in analogy to two-scale analyses (Khachaturyan, 1983; Roytburd, 1998). In this work, the interfacial energy density due to elastic strains at the interface between homogeneous austenite and internally-twinned martensite is calculated by the finite element method. The interfacial energy effect on the thermodynamic condition for phase transition is examined. Specific examples have been calculated for the cubic-to-orthorhombic ($\beta_1$ to $\gamma_1'$) stress-induced transformation in a CuAlNi shape memory alloy.

## 2. Modelling of evolution of laminated microstructures

We adopt the model that has been described in full detail in (Stupkiewicz and Petryk, 2002), therefore only a very brief account is given here. We study isothermal, stress-induced formation of parallel, internally twinned martensite plates of a total volume fraction $\eta$ in a single crystal of SMA. A representative volume element is considered (Fig. 1)

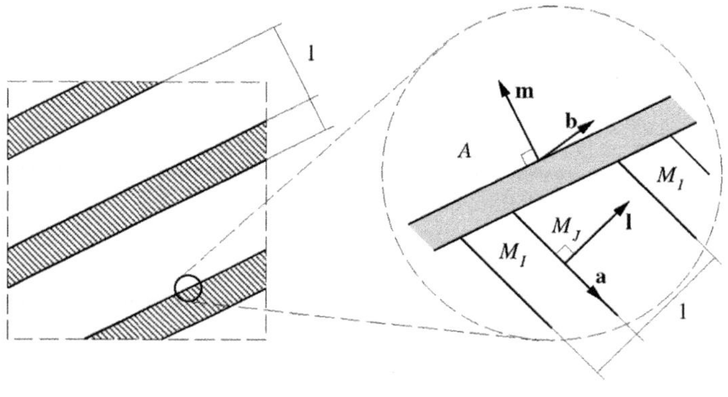


*Figure 1.* Schematic view of a representative volume element (rank-two laminate) with parallel, internally twinned martensite plates ($A$ – austenite, $M_I$ and $M_J$ – martensite variants $I$ and $J$)

with arrangement of martensite plates and twins taken to be periodic for simplicity. The basic rule for evolution of the laminated microstructure is obtained from the local thermodynamic criterion of martensite-to-austenite transition and transformed to the form

$$\dot{\eta} \geq 0, \quad f - f_c \leq 0, \quad (f - f_c)\,\dot{\eta} = 0, \tag{1}$$

where $f$ is the corresponding thermodynamic driving force, and $f_c \geq 0$ is a threshold value which is state-dependent in general. For reverse transformation, $\eta$ is to be replaced by $(1 - \eta)$. In the small strain setting, the bulk contribution to the overall Helmholtz free energy per unit representative volume of the austenite/martensite laminate reads

$$\bar{\phi}_v = \frac{1}{2}\,\bar{\boldsymbol{\sigma}} \cdot \tilde{\mathbf{M}}\;\bar{\boldsymbol{\sigma}} + \phi_0 + \eta\Delta^{am}\phi_0 + \frac{1}{2}\eta\,(1-\eta)\,\varepsilon^t\,\mathbf{Q}\,\varepsilon^t. \tag{2}$$

where $\bar{\sigma}$ is the macrosopic (average) stress, $\tilde{\mathbf{M}}$ is the effective elastic compliance tensor for the laminate, $\varphi_0$ and $\varphi_0 + \Delta^{am}\varphi_0$ is the Helmholtz free energy density at zero stress in a given temperature for austenite and martensite, respectively, and $\varepsilon^t = \lambda\,\varepsilon_I^t + (1 - \lambda)\varepsilon_J^t$ is the effective transformation strain of a strain of a martensite plate composed of two twin-related martensite variants $I$ (of volume fraction $\lambda$) and $J$. The classical assumption of an invariant austenite/martens ite interface (habit plane) at zero stress has been used, under which the thermodynamic driving force related to (2) has been transformed to

$$f = -\partial\bar{\phi}_v / \partial\eta = \bar{\boldsymbol{\sigma}} \cdot \bar{\varepsilon}^t - \frac{1}{2}\bar{\boldsymbol{\sigma}} \cdot \mathbf{B}_a^T(\mathbf{M}_a - \mathbf{M}_m)\,\mathbf{B}_m\bar{\boldsymbol{\sigma}} - \Delta^{am}\phi_0 \tag{3}$$

with the second term proportional to the difference between the elastic compliance tensor for austenite $\mathbf{M}_a$ and the effective one for twinned martensite $\mathbf{M}_m$. Analytic expressions for $\mathbf{Q}$, $\mathbf{M}_m$, $\mathbf{B}_a$, $\mathbf{B}_m$ and further details are given in (Stupkiewicz and Petryk, 2002). The interfacial energy effect, not studied in the reference, is examined below.

## 3. Interfacial energy effect in a rank-two laminate

Consider again a representative volume element of a rank-two laminate but now in some more detail as illustrated in Fig.2. The twin spacing $h$ within a martensite plate is assumed unchanged up to a zigzag interface with austenite -possible "fingering" of twins is not examined here. Besides the bulk contribution (2) to the overall free energy, we take now into account also the contribution $\bar{\phi}_i$ of the free energy of the interfaces, per unit volume of a representative volume element of the material. Denote by $\gamma_{IJ} \equiv \gamma$ the

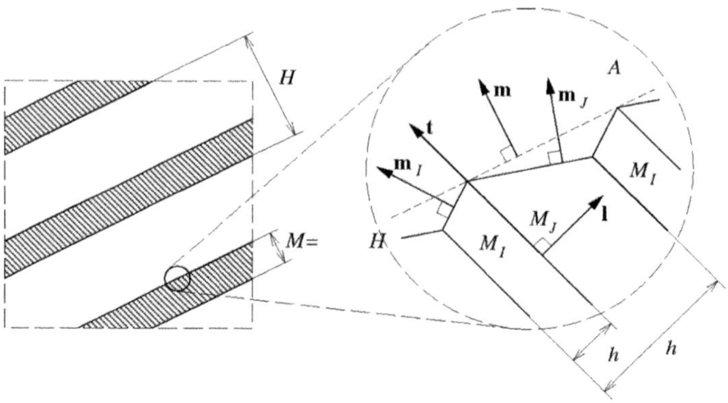


*Figure 2.* A representative volume element of the rank-two laminate with zigzag interfaces between austenite ($A$) and twinned martensite ($M_I/M_J$)

interfacial free energy, at the atomic level, of a coherent twin boundary that separates the martensite variants $I$ and $J$ within a martensite plate, per unit area of a twinning plane. An analogous energy density per unit area of a *smooth* interface portion between austenite and a *single* crystallographic variant $I$ of martensite is denoted by $\gamma_{aI}$. In a different scale, there is another interfacial energy term $\gamma_e$ provided by averaging the elastic strain energy in vicinity of the interface between the *twinned* martensite and austenite, per unit *nominal* area of the interface (i.e. per unit area of the habit plane). Disregarding interaction energy between such interfaces, straightforward dimensional analysis shows that the part of the elastic strain energy at the interface that corresponds to one pair of twinned layers of martensite is proportional to $h^2$ while the area of that part is proportional to $h$. Thus

$$\gamma_e = \Gamma h \tag{4}$$

where $\Gamma$ is a size-independent quantity, to be determined, that characterizes the microstructure of the austenite/martensite interface.

The total nominal area of the austenite/martensite interfaces and the total area of twinning planes within a representative volume $V$ are readily $S_{am} = 2V/H$ and $S_{IJ} = 2\eta V/h$, respectively, where $H$ is the spacing of martensite plates. The total true area of the interface between austenite and crystallographic variant $I$ of martensite is $S_{aI} = \lambda S_{am}$ $(\mathbf{m} \cdot \mathbf{t})(\mathbf{m}_I \cdot \mathbf{t})$, where $\mathbf{m}_I$ is the unit normal to this interface part, and $\mathbf{t} = \mathbf{m} - (\mathbf{m} \cdot \mathbf{l})\mathbf{l}$ is the projection of the habit plane normal $\mathbf{m}$ on the twinning plane. In general, $\mathbf{m}_I \neq \mathbf{m}_J$, but these unit normals are assumed here as fixed for each variant separately. Thus, the effective contribution of $\gamma_{aI}$ and $\gamma_{aJ}$ to the interfacial energy taken per unit *nominal* area of the austenite/mar tensite interface is

$$\gamma_{am} = (\mathbf{m} \cdot \mathbf{t})(\lambda\gamma_{aI}/(\mathbf{m}_I \cdot \mathbf{t}) + (1-\lambda)\gamma_{aJ}/(\mathbf{m}_J \cdot \mathbf{t})). \tag{5}$$

The total contribution $\bar{\phi}_i$ of the interfaces to the Helmoholtz free energy per unit representative volume is

$$\bar{\phi}_i = \frac{1}{V}(S_{am}(\gamma_e + \gamma_{am}) + S_{IJ\gamma}). \quad (6)$$

On substituting (4) and the expressions for the interfaces area into (6), it reads

$$\bar{\phi}_i = 2(\Gamma h / H + \eta\gamma / h + \gamma_{am} / H). \quad (7)$$

Introduce the working assumption that (i) the twin spacing $h$ is self-adjusting to minimize the interfacial energy $\bar{\phi}_i$, without any associated dissipation. The minimum of $\bar{\phi}_i$ with respect to $h$, if other parameters are fixed, $\bar{\phi}_i^* \equiv \bar{\phi}_i(h^*) \le \bar{\phi}_i(h)$ for any $h$, is obtained straightforwardly from $\partial\bar{\phi}_i / \partial h = 0$ at

$$h^* = \sqrt{\eta H \gamma / \Gamma} \equiv \sqrt{lM}, \quad M = \eta H, \quad l = \gamma / \Gamma. \quad (8)$$

Note that the ratio $\gamma/\Gamma$ provides a characteristic length $l$ such that the twin spacing $h^*$ represents geometric mean of $l$ and the nominal thickness $M$ of twinned martensite plates. This result is typical for laminated microstructures, cf. Khachaturyan (1983).

It is noteworthy that the first two terms on the righ-hand side of (7) dependent on $h$ become equal to each other for $h = h^*$, that is, the elastic energy at the austenite/twinned martensite interface becomes equal to the interfacial energy of the twinning planes. In particular, it is evident that

$$\bar{\phi}_i^* = 2(2\Gamma h^* + \gamma_{am}) / H = 2\eta(2\Gamma\sqrt{l/M} + \gamma_{am} / M). \quad (9)$$

In reference to the model outlined briefly in Section 2, the interfacial energy (7) or (9) affects the evolution rule (1) for the austenite/martensite laminate. This effect depends on how the transformation proceeds. Consider first the austenite-to-martensite transformation that proceeds by increasing proportionally $\eta$ and $M$ at fixed $H$ and $\lambda$, and consequently, at fixed $\Gamma$. According to the assumption (i), the corresponding increase in $h$ is governed by (8). Then, the thermodynamic driving force (in fact, resistance) due to the interfacial energy is

$$f_i^* = -\partial\bar{\phi}_i^* / \partial\eta = -2\Gamma\sqrt{l/M} = -2\Gamma h^* / M = -2\gamma / h^* \qquad \text{at fixed } H. \quad (10)$$

If the assumption (i) were replaced by the condition $h$ = const as $M$ is increasing then the corresponding force associated with (7) at fixed $H$ would be $f_i = -\partial\bar{\phi}_i/\partial\eta = -2\,\gamma/h$, which yields a higher resistance for $h < h^*$.

In turn, if the transformation proceeds by increasing the number of identical martensite plates, each of a fixed nominal thickness $M$ and a fixed twin spacing $h^*$ found from (8), then $\bar{\phi}_i^*$ is simply proportional to $\eta$ and the related resistance due to the increasing interfacial energy, *when smoothed out in time*, becomes

$$f_i^* = -4\Gamma\sqrt{l/M} - 2\,\gamma_{am}/M = -2(2\,\Gamma h^* + \gamma_{am})/M \qquad \text{at fixed } M. \tag{11}$$

This does not mean, however, that the growth of martensite plates at a fixed spacing $H$ must be energetically preferable since the increased resistance (11) at fixed $M$ can be balanced by lowering the elastic strain energy accumulated at the crystal boundary, cf. Section 4. Other transformation mechanisms can also be considered, especially at higher values of $\eta$.

In the evolution rule (1) for austenite-to-martensite transformation, the resistant force $f_i^* < 0$ can be taken as an interfacial energy contribution either to $f$ or with an opposite sign to $f_c$; these two options are, of course, mathematically equivalent. The accumulated interfacial energy density $\bar{\phi}_i^*$ is most likely dissipated as $\eta$ approaches 1 or, alternatively, during reverse transformation as $\eta$ decreases to 0. Accordingly, an interfacial energy release, contrary to accumulation, is not accounted for in the evolution rule (1), which means that there is a related difference between predicted stress levels for the forward and reverse transformations. This represents one of possible sources of a hysteresis loop in a cycle of transformation with intrinsic dissipation.

In the above formulae, there are two essential material parameters to be specified, $\gamma$ and $\Gamma$, if the influence of $\gamma_{am}$ is negligible when $\gamma_{am} \ll \gamma_e$. The value of interfacial energy density $\gamma$ on coherent twin boundaries is related to stacking fault energy, and its estimates can be found in the literature, e.g. (Porter and Easterling, 1992, Christian and Mahajan, 1995; Howe, 1997).

The proposed method of determining the parameter $\Gamma$ is outlined in Section 5. Then, the characteristic length $l = \gamma/\Gamma$ is estimated for twinned martensite plates in a CuAlNi shape memory alloy.

## 4. Boundary energy effect

Another free energy term, for brevity called the "boundary energy", is due to elastic interaction between the austenite/martensite laminate and the crystal boundary, for instance, at a stress free surface or at a high-angle grain boundary within a polycrystal. We restrict ourselves here to examine the case (ii) of small volume fraction of martensite ($\eta \ll 1$), so that the interactions between martensite plates can be neglected when estimating the boundary energy as follows

$$\phi_s = M^2 \int_L \Gamma_s \, dL, \tag{12}$$

where $\Gamma_s$ is another size-independent quantity that characterizes the boundary conditions on and the microstructure near the crystal surface, $L$ denotes the total length of the edges of martensite plates, $M$ is the nominal thickness of martensite plates far from the surface, and the factor $M^2$ follows from dimensional analysis.

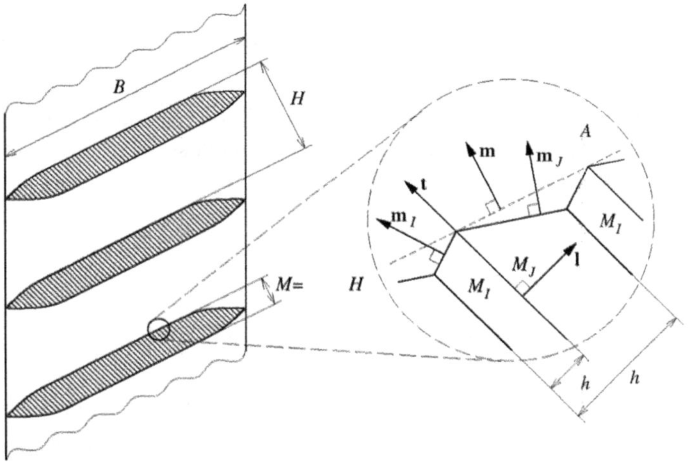


*Figure 3.* Schematic view of a rank-two laminate with the boundary effect

For simplicity we assume that (iii) the crystal is of the form of an infinite layer, with a distance $B$ between the boundaries measured within the habit plane (Fig. 3). Consequently, $L = 2V/BH$ in a representative volume $V$, and $\Gamma_s$ does not vary along $L$. A surface energy term analogous to (5) is neglected, which is admissible since the crystal boundary area does not vary. Then, the boundary energy contribution during transformation to the total Helmholtz free energy per unit representative volume is estimated as follows

$$\bar{\phi}_s = 2\Gamma_s M^2 / BH \tag{13}$$

This provides the third part of the total macroscopic Helmholtz free energy per unit volume, viz.

$$\bar{\phi} = \bar{\phi}_{\mathrm{v}} + \bar{\phi}_{\mathrm{i}} + \bar{\phi}_{\mathrm{s}}, \tag{14}$$

where from (7) and (13) the sum of size-dependent terms takes the form

$$\bar{\phi}_{\mathrm{is}} \equiv \bar{\phi}_{\mathrm{i}} + \bar{\phi}_{\mathrm{s}} = 2\eta(\Gamma_{\mathrm{s}} M / B + \gamma_{\mathrm{am}} / M + \Gamma h / M + \gamma / h). \tag{15}$$

The minimum of $\bar{\phi}_{\mathrm{is}}$ with respect to $M$, other parameters displayed in (15) being kept fixed, is reached for $M = \hat{M}$ given by

$$\hat{M} = \sqrt{l_{\mathrm{s}} B}, \quad l_{\mathrm{s}} = (\Gamma h + \gamma_{\mathrm{am}}) / \Gamma_{\mathrm{s}}, \tag{16}$$

where $l_{\mathrm{s}}$ is another characteristic length. It is worth mentioning that at the minimum point the first right-hand term in (15) becomes equal to the sum of the second and third terms. On substituting $h = h^*$ and $M = \hat{M} \equiv \eta \hat{H}$ into (15) and rearranging, we arrive at the following expression for the minimum interfacial/surface free energy per unit representative volume

$$\bar{\phi}_{\mathrm{is}}^{\min} = (6\Gamma h^* + 4\gamma_{\mathrm{am}}) / \hat{H}; \tag{17}$$

note the (perhaps unexpected) factors: 6 at $\gamma_{\mathrm{e}}^* = \Gamma h^*$ and 4 at $\gamma_{\mathrm{am}}$.

In particular, if the term with $\gamma_{\mathrm{am}}$ is disregarded then the remaining three interfacial energy contributions to $\bar{\phi}_{\mathrm{is}}^{\min}$ (from the twinning planes, twinned martensite interfaces and crystal boundary) become equal to each other.

In general, we cannot expect that the average spacing $H$ of martensite plates is self-arranged such that the equality $\bar{\phi}_{\mathrm{is}} = \bar{\phi}_{\mathrm{is}}^{\min}$ holds exactly true during transformation. Nevertheless, from the above analysis it follows that this may be a good approximation when the volume fraction $\eta$ of martensite, small enough, grows by increasing the number of martensite plates, each of fixed nominal thickness $M$ found from (16) and of fixed twin spacing $h^*$ determined from (8). It must be noted that creation of a new plate of a prescribed thickness $\hat{M}$ is associated with a finite jump in $\eta$, so that formal differentiation of (17) with respect to $\eta$ is, strictly speaking, not meaningful. However, when the number of plates is sufficiently large then the material response may be smoothed out in time, and the related resistance against martensitic transformation due to the interfacial and boundary energies follows from (17) as

$$f_{\text{is}}^{*}=-(6\Gamma h^{*}+4\gamma_{\text{am}})/\hat{M}=-6\,\Gamma\sqrt{l/\hat{M}}-4\gamma_{\text{am}}/\hat{M}\quad \text{at fixed}\,\hat{M}. \tag{18}$$

It is of interest to compare (18) with the driving force needed to increase $M$ at fixed $H$, analogous to (10) but with added contribution of $\bar{\phi}_{\text{s}}$. To calculate this, we cannot use the expression (17) since (16) no longer holds, but (15) with $h = h^*$ can be used instead. It yields

$$f_{\text{is}}^{*}=-\partial\bar{\phi}_{\text{is}}^{*}/\partial\eta=-4\Gamma_{\text{s}}M/B-2\Gamma\sqrt{l/M}\qquad \text{at fixed}\quad H \tag{19}$$

If $M = \hat{M}$ then $f_{\text{is}}^*$ in (18) and (19) coincide, cf. the remark that follows (16), but for $M > \hat{M}$ we have a higher resistance at fixed $H$.

From the above analysis one might conclude that under the assumptions introduced, especially those distinguished as (i)÷(iii), the minimization of the total interfacial energy $\bar{\phi}_{\text{is}}$ makes the increase of volume fraction of martensite by recursive formation of martensite plates of constant thickness $\hat{M}$ energetically preferable. However, $f_{\text{c}}$ might be lower for propagation of an existing transformation front than for formation of a new one, which in turn would prefer further growth of a martensite plate just created. This question is not pursued further here.

## 5. Elastic strain energy at twinned martensite interface

The problem of calculating the elastic strain energy in vicinity of the interface between the twinned martensite and austenite has been addressed in the literature in special cases, and certain analytic estimates are available (Khachaturyan, 1983; Sridhar et al., 1996; Roytburd, 1998). The present finite-element analysis goes beyond previous studies as it takes into account different elastic anisotropy of each constituent, the finite deformation effect at the micro-scale, and also the effect of a zigzag shape of the interface. Only a brief description is given below of the approach used; more details are provided in (Maciejewski *et al.*, to appear).

The value of $\Gamma$ is calculated here by the finite element method applied to a unit cell with periodic boundary conditions, Fig. 4(a), that contains a representative period of the interface between austenite and twinned martensite. The microstructure parameters (including **m**, **n** and $\lambda$) are determined beforehand in an independent way (from the crystallographic theory). The elastic strain energy stored in the unit cell at $\bar{\boldsymbol{\sigma}}=\mathbf{0}$ is calculated as the result of imposing transformation (Bain) strains $\varepsilon_I^{\text{t}}$ and $\varepsilon_J^{\text{t}}$ in domains

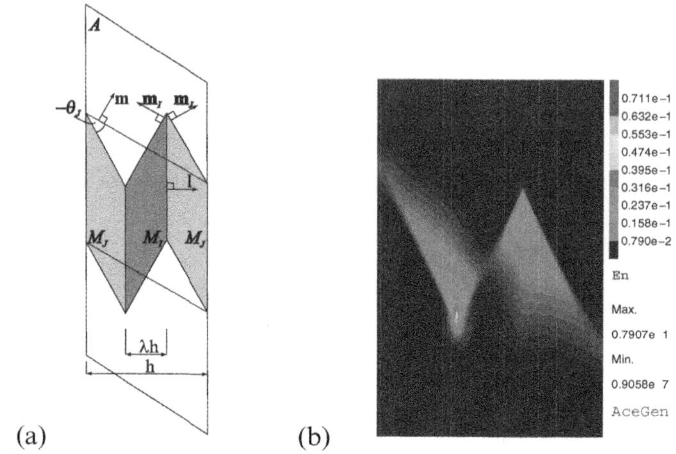


*Figure 4.* (a) A unit cell of the analyzed laminated microstructure (with vertical dimensions shortened), (b) elastic strain energy concentration near the zigzag interface

$M_I$ and $M_J$, respectively. Then, the last term in (2), if non-zero, is subtracted to obtain the elastic strain energy associated with twinned martensite interfaces. The ratio $h/M$ is taken sufficiently small to eliminate practically the interaction between interfaces (in accord with the Saint-Venant rule). Finally, the elastic strain energy calculated for the representative cell is divided by $S_{\mathrm{am}}h$ to obtain $\Gamma$. The value of $\Gamma$ depends on the angle $\theta_J$ between the vectors $\mathbf{m}$ and $\mathbf{m}_J$, negative in the reference configuration sketched in Fig. 4a, that represents a measure of the deviation of the twinned martensite interface from planar shape *before* imposing the transformation strains. An illustrative plot of the sensitivity of calculated $\Gamma$ on $\theta_J$ is presented in Fig. 5. It should be noted that this plot is *not* representative in the sense that for other microstructures of the examined martensite plates of the same alloy, the results are significantly different (Maciejewski *et al.*, to appear). Nevertheless, it can be concluded that the effect of a zigzag shape of the interface on the interfacial energy can be substantial, and that an optimal shape of the twinned martensite interface that minimizes the interfacial energy can be far from being planar.

## 6. Example: twinned martensite plates in CuAlNi crystal

The method outlined above has been applied to calculate $\Gamma$ for an interface between the twinned $\gamma_1'$ martensite and $\beta_1$ austenite after the cubic-to-orthorhombic transformation in a CuAlNi shape memory alloy. The microstructure parameters (including $\mathbf{m}$, $\mathbf{n}$ and $\lambda$) have been specified by solving the compatibility equations of the classical crystallographic theory of martensitic transformation (Wechsler et al., 1953; Ball and James, 1987). Consequently, the last term in (2) vanishes in this case. All the microstructures can be divided into four groups, each consisting of 24 crystallographically equivalent twinned martensite plates. The FE analysis has been carried out for four microstructures (one from each of the four groups) associated with the same $\varepsilon^t$ in the geometrically linear theory. The smallest value of $\Gamma$ has been estimated as $\Gamma = 3.2 \cdot 10^6$ Jm$^{-3}$ which corresponds to the minimum on the graph shown in Fig. 5 and to the elastic strain energy distribution visualized in Fig. 4(b).

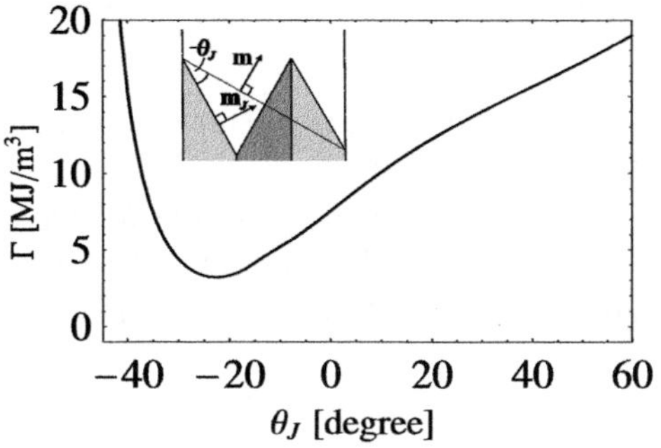


*Figure 5.* Interfacial elastic energy factor $\Gamma$ for a twinned $\gamma_1'$ martensite plate in CuAlNi

The value of interfacial energy density $\gamma$ on coherent twin boundaries is related to the stacking fault energy density and for Cu-based alloys can roughly be estimated as $\gamma \approx 0.02$Jm$^{-2}$ (Porter and Easterling, 1992; Christian and Mahajan, 1995; Howe, 1997).

This yields the estimate of the characteristic length $l = \gamma/\Gamma \approx 6$ nm.

Without calculating $\Gamma_s$, further estimates can be given in terms of the nominal thickness $M$ of martensite plates as a parameter. For instance, for $M = (0.6 \div 60)$ $\mu$m we obtain $M/h^* \approx (10 \div 100)$, $h^* \approx (60 \div 600)$ nm and $\gamma_e^* = h^*\Gamma \approx (0.2 \div 2)$ Jm$^{-2}$, respectively. Note that the influence of $\gamma_{am}$ on the total interfacial energy may be not negligible for thinner martensite plates.

If the influence of $\gamma_{am}$ is neglected then the resistant force (18) is estimated as $f_{is}^* \approx -(0.2 \div 2) \cdot 10^6$ Jm$^{-3}$ for $M = (60 \div 0.6)$ $\mu$m, respectively. The associated resistant force $f_i^*$ due to the interfacial energy without the boundary effect is estimated from (10) or (11) for $M = (60 \div 0.6)$ $\mu$m as $f_i^* \approx -(0.06 \div 0.6) \cdot 10^6$ Jm$^{-3}$ or $f_i^* \approx -(0.13 \div 1.3) \cdot 10^6$ Jm$^{-3}$ according to whether the austenite-to-martensite transformation proceeds at $H =$ const or $M =$ const, respectively. At least this part of $f_{is}^*$ can be one of possible sources of a hysteresis loop in a cycle of transformation with intrinsic dissipation, as mentioned at the end of Section 3.

It should be noted that the interfacial energy effect on the evolution rule (1) has been examined here under simplifying assumptions, e.g., (i)-(iii). The results may change if these assumptions are relaxed.

**Acknowledgement**

This work has been supported by the State Committee for Scientific Research (KBN), Poland, grant No. 4 T07A 032 26.

**References**

[1] J.M. Ball and R.D. James. Fine phase mixtures as minimizers of energy. *Arch. Ration. Mech. An.*, 100:13–50, 1987.

[2] K. Bhattacharya. *Microstructure of martensite: why it forms and how it gives rise to the shape-memory effect.* Oxford University Press, Oxford, 2003.

[3] J.W. Christian and S. Mahajan. Deformation twinning. *Progress Mater. Sci.*, 39:1–157, 1995.

[4] J.M. Howe. *Interfaces in Materials.* John Wiley & Sons, Inc., New York, 1997.

[5] A.G. Khachaturyan. *Theory of Structural Transformations in Solids.* John Wiley and Sons, New York, 1983.

[6] G. Maciejewski, S. Stupkiewicz, and H. Petryk. Elastic micro-strain energy at the austenite-twinned martensite interface (*to appear*).

[7] D.A. Porter and K.E. Easterling. *Phase Transformations in Metals and Alloys.* Chapman & Hall, London, 1992.

[8] A.L. Roytburd. Thermodynamics of polydomain heterostructures. II. Effect of microstresses. *J. Appl. Phys.*, 83(1):239–245, 1998.

[9] N. Sridhar, J.M. Richman, and D.J. Srolovitz. Twinning in thin films—I. Elastic analysis. *Acta Mater.*, 44(10):4085–4096, 1996.

[10] S. Stupkiewicz and H. Petryk. Modelling of laminated micro-structures in stress-induced martensitic transformation. *J. Mech. Phys. Solids*, 50:2303–2331, 2002.

[11] M.S. Wechsler, D.S. Lieberman, and T. A. Read. On the theory of the formation of martensite. *Trans. AIME J. Metals,* 197:1503–1515, 1953.

# MECHANISM-BASED STRAIN GRADIENT CRYSTAL PLASTICITY

C.-S. HAN[*◊], H. GAO[*], Y. HUANG[†], AND W.D. NIX[◊]
[*]*Max Planck Institute for Metals Research*
*Heisenbergstr. 3, D-70569 Stuttgart, Germany*
*Email: cshan@mf.mpg.de*
[◊]*Department of Materials Sciences and Engineering*
*Stanford University, Stanford, CA 94305, USA*
AND
[†]*Department of Mechanical and Industrial Engineering*
*University of Illinois, Urbana, IL 61801, USA*

## 1. Introduction

Size effects in plasticity at micron and submicron scales has been modeled by strain gradients by various authors, e.g. [2]. Analyzing micro-indentation experiments Nix & Gao [9] found that the measured indentation hardness strongly suggests a linear dependence between the square of plastic flow stress and strain gradient. This has led to the development of mechanism-based strain gradient (MSG) plasticity [3]. The core idea of MSG has been to incorporate the concept of geometrically necessary dislocations into the plastic formulation by the Taylor relation $\tau = \alpha\mu b\sqrt{\rho}$, where $\mu$ denotes the shear modulus, $b$ the Burgers vector length, $\alpha$ is an empirical coefficient usually taken to be 0.2–0.5, and $\rho$ the dislocation density.

Most of the theories on strain gradient plasticity have been developed for isotropic and polycrystalline materials. Crystal plasticity models, however, offer a higher resolution of plastic deformation at sizes comparable to single grains, which are often on the order of microns where significant size effects have been observed. The anisotropy of size effects has for example been observed in micro-indentation experiments of silver crystals in the (100) and (110) orientations [7]. In order to obtain a better resolution a MSG-type crystal plasticity model is presented.



*Q. P. Sun and P. Tong (eds.), IUTAM Symposium on Size Effects on Material and Structural Behavior at Micron- and Nano-Scales,* 163–171.

**Notation.** To set the stage, the notation has to be provided. The decomposition of the deformation gradient is adopted as, $\mathbf{F} = \mathbf{F}_e\mathbf{F}_p$, where the lattice rotation is contained in $\mathbf{F}_e$ and the plastic slip is included in $\mathbf{F}_p$. Further expressions in the intermediate configuration $\tilde{\mathcal{B}}$ are given in Table 1 (see [5] for details).

Table 1: Crystal plasticity relations in the intermediate configuration $\tilde{\mathcal{B}}$.

| | | |
|---|---|---|
| velocity gradient: | $\mathbf{l} = \mathbf{l}_e + \mathbf{F}_e \tilde{\mathbf{L}}_p \mathbf{F}_e^{-1}$ | (1) |
| | with $\mathbf{l}_e = \dot{\mathbf{F}}_e \mathbf{F}_e^{-1}$ | (2) |
| plastic velocity gradient: | $\tilde{\mathbf{L}}_p = \dot{\mathbf{F}}_p \mathbf{F}_p^{-1} = \sum \dot{\gamma}^\alpha \tilde{\mathbf{s}}^\alpha \otimes \tilde{\mathbf{m}}^\alpha$ | (3) |
| Lagrangian strain tensor: | $\tilde{\mathbf{E}} = \tilde{\mathbf{E}}_e + \tilde{\mathbf{E}}_p$ | (4) |
| | $\tilde{\mathbf{E}}_e = \frac{1}{2}(\mathbf{F}_e^T \mathbf{F}_e - \mathbf{1})$ and | |
| | $\tilde{\mathbf{E}}_p = \frac{1}{2}(\mathbf{1} - \mathbf{F}_p^{-T} \mathbf{F}_p^{-1})$ | |
| 2nd Piola-Kirchhoff stress: | $\tilde{\mathbf{S}} = \mathbf{F}_e^{-1} \boldsymbol{\tau} \mathbf{F}_e^{-T}$ | (5) |
| Mandel stress: | $\tilde{\mathbf{P}} = (\mathbf{1} + 2\tilde{\mathbf{E}}_e)\tilde{\mathbf{S}} = \tilde{\mathbf{C}}_e \tilde{\mathbf{S}}$ | (6) |
| elastic relation: | $\tilde{\mathbf{S}} = \tilde{\mathbb{C}}_e \tilde{\mathbf{E}}_e$ | (7) |
| shearing rate: | $\dot{\gamma}^\alpha = \dot{\gamma}_o \left\| \frac{\tau^\alpha}{\mathcal{G}^\alpha} \right\|^n sign(\tau^\alpha)$ | (8) |
| linear hardening in silp resistance: | $\mathcal{G}^\alpha = \mathcal{G}_o + c_h \sum \|\gamma^a\|$ | (9) |
| resolved shear stress: | $\tau^\alpha = \tilde{\mathbf{P}} \cdot (\tilde{\mathbf{s}}^\alpha \otimes \tilde{\mathbf{m}}^a)$ | (10) |

## 2. Dislocation density tensor

Conventional crystal plasticity assumes that the lattice remains essentially undistorted in the deformation process. Lattice distortions can be described with the dislocation density tensor which can be associated with the concept of geometrically necessary dislocations and size dependent deformation [1, 4]. In the following *Grad, Div,* and *Curl* denote gradient, divergence, and curl differential operators with respect to a material point in the reference configuration $\mathcal{B}_o$ and *grad, div,* and *curl* denote operators with respect to a material point in the current configuration $\mathcal{B}$. The compatibility conditions Curl $\mathbf{F} = \mathbf{0}$ and curl $\mathbf{F}^{-1} = \mathbf{0}$, yield [4]

$$\tilde{\mathbf{A}}_e = det(\mathbf{F}^e)\mathbf{F}^{e-1}\,(\mathrm{curl}\,\mathbf{F}^{e-1})^T \text{ and } \tilde{\mathbf{A}}_p = \frac{1}{det(\mathbf{F}^p)}\mathbf{F}^p(\mathrm{Curl}\,\mathbf{F}^p)^T \tag{11}$$

where $\tilde{\mathbf{A}} = \tilde{\mathbf{A}}_e = \tilde{\mathbf{A}}_p$ is defined as the dislocation density tensor in $\tilde{\mathcal{B}}$.

**Relation of $\tilde{\mathbf{A}}$ to dislocation theory**. The dislocation density tensor $\tilde{\mathbf{A}}$ has been defined above in the continuum theory of crystal plasticity. It is helpful to investigate the correlating expressions in the discrete dislocation theory. For a dislocation segment, $\tilde{\mathbf{A}}$ is defined by

$$\tilde{\mathbf{A}}\;dv = \tilde{\mathbf{l}} \otimes \tilde{\mathbf{b}}\;dl, \tag{12}$$

where $\tilde{\mathbf{l}}$ denotes the unit tangent vector and $dl$ the infinitesimal element of the dislocation line, $\tilde{\mathbf{b}}$ the Burgers vector, and $dv$ the associated volume element. In the case of multiple dislocations linear combination yields $\tilde{\mathbf{A}} = \sum_k \eta_k^A\ \tilde{\mathbf{l}}_k \otimes \tilde{\mathbf{b}}_k$, where $\tilde{\mathbf{b}}_k$ and $\tilde{\mathbf{l}}_k$ are unit vectors of dislocation $k$, i.e., $|\mathbf{b}_k| = |\mathbf{l}_k| = 1$. A coordinate expansion [1] yields $\tilde{\mathbf{A}} = \sum_{ij} \eta_{ij}^A\ \tilde{\mathbf{e}}_i \otimes \tilde{\mathbf{e}}_j$, with an orthonormal basis set $\tilde{\mathbf{e}}_i$. The dimensions of $\eta^A = |\tilde{\mathbf{A}}|$, $\eta_{ij}^A$, $\eta_k^A$ in continuum crystal plasticity are 1/m because the length of the Burgers vector is incorporated. The usual dislocation density can be recalculated with $\rho = \eta^A/b$.

## 3. Peach-Koehler and related forces in continuum plasticity

An essential concept in dislocation theory is the Peach-Koehler force. The Peach-Koehler force on a dislocation line element in the intermediate configuration is defined here by [5]

$$\tilde{\boldsymbol{\xi}}_L = -\tilde{\mathbf{l}} \times \tilde{\mathbf{P}}^T\,\tilde{\mathbf{b}}, \tag{13}$$

where $\tilde{\mathbf{P}}$ represents the applied stress field. The Peach-Koehler force describes the the total energetic force acting on a dislocation line segment. For multiple dislocations the Peach-Koehler force on a dislocation $k$ can be described by $\tilde{\boldsymbol{\xi}}_L^k = -\tilde{\mathbf{l}}^k \times (\tilde{\mathbf{P}}^T + \sum_{j\neq k} \tilde{\mathbf{P}}_j^T)\ \tilde{\mathbf{b}}^k$ which has been derived in [10] via superposition of external stress fields $\tilde{\mathbf{P}}$ and stress fields associated with the $j$ dislocation $\tilde{\mathbf{P}}_j$. Note that this is in a strict sense only valid for small strains. The component $\xi_L^k$ of $\tilde{\boldsymbol{\xi}}_L^k$ in the direction $\tilde{\mathbf{l}}^k \times \tilde{\mathbf{m}}^k$, i.e. normal to the dislocation line direction and parallel to the $\alpha$ slip plane is $\xi_L^k = \tilde{\mathbf{m}}^k \cdot (\tilde{\mathbf{P}}^T + \sum_{j\neq k} \tilde{\mathbf{P}}^T_j)\ \tilde{\mathbf{b}}^k$, which determines the force and motion of the glide dislocation in its slip plane. The net Peach-Koehler force of a volume element can be written as $\tilde{\boldsymbol{\xi}}_L^{net} = \sum_k \tilde{\boldsymbol{\xi}}_L^K$.

**Corresponding forces in continuum plasticity**. In view of (12) and (13), the corresponding force induced by geometrically necessary dislocations in a crystalline continuum – referred to in the following as the density force – can be defined by

$$\tilde{\boldsymbol{\zeta}} = -\tilde{\mathbf{A}} \times \tilde{\mathbf{P}}^T, \tag{14}$$

where $[\mathbf{A} \times \mathbf{B}]_k = A_{mi} B_{ni} e_{mnk}$ defines a tensor cross product with the permutation tensor $e_{ijk}$. $\tilde{\zeta}$ can be understood as the force acting on geometrically necessary dislocations - described by $\tilde{\mathbf{A}}$ -because of the applied stress $\tilde{\mathbf{P}}$. Note that $\tilde{\zeta}$ in (14) is identical to $\tilde{\boldsymbol{\xi}}_L$ in (13) for a single dislocation and can be directly derived on the basis of (12) [8]. In the general case of multiple dislocations the dislocation-dislocation interactions are however not reflected in (13).

**Resolved density force on a slip system**. $\tilde{\zeta}$ is in general not parallel to the slip plane and the relevant components of $\tilde{\zeta}$ have to be projected. For a given slip system, the shear rate $\dot{\gamma}^\alpha$ is related to the plastic velocity gradient by (3). The associated density force of its thermodynamic conjugate, $\tau^\alpha$, resolved onto the slip system $\alpha$ is defined as [5]

$$\tilde{\boldsymbol{\zeta}}^\alpha = -\tau^\alpha\, \tilde{\mathbf{A}} \times (\tilde{\mathbf{s}}^\alpha \otimes \tilde{\mathbf{m}}^\alpha)^T. \tag{15}$$

The norm yields

$$\left|\tilde{\boldsymbol{\zeta}}^\alpha\right| = \left|T^\alpha\right| \eta_G^\alpha, \quad \text{with} \quad \eta_G^\alpha = \left|\tilde{\mathbf{A}} \times (\tilde{\mathbf{s}}^\alpha \otimes \tilde{\mathbf{m}}^\alpha)^T\right|, \tag{16}$$

where $\eta_G{}^\alpha$ can be seen as a projection of the dislocation density tensor $\tilde{\mathbf{A}}$ into the slip system $\alpha$.

In dislocation theory the dislocation velocity is a direct function of $\xi_L{}^k$ and is therefore directly linked to the plastic strain rate via Orowan relation $\dot{\varepsilon}_p = b\rho_m u_m$, where $\rho_m$ denotes the mobile dislocation density and $v_m$ the average velocity of the dislocations contained in $\rho_m$.

The density force $\tilde{\zeta}^\alpha$ in continuum plasticity can be interpreted as the force acting on the geometrically necessary dislocations because of the shear stress $\tau^\alpha$. The dislocations of the density $\eta_G{}^\alpha$ are in essence however immobile because of their geometrically necessary characteristics. The dislocations of $\eta_G{}^\alpha$ will therefore hinder the dislocation motion in the $\alpha$ slip plane.

## 4. Hardening description with projected dislocation density

The relation $\tau = \alpha\mu b\sqrt{\rho}$ links the dislocation density $\rho$ to the shear flow stress. It should be noted that this relation has been experimentally verified for different deformation stages, materials, e.g. single crystals in single and double slip, and polycrystals, and has been found to agree well for dislocation densities in the range of $10^9$ to $10^{15}\frac{1}{m^2}$. Dividing $\rho$ into the density of statistically stored $\rho_S$ and geometrically necessary dislocations $\rho_G$ the Taylor relation has been suggested in the MSG approach [3] as

$$\tau = \alpha\mu b\sqrt{\rho S + \rho G}. \tag{17}$$

Following MSG, $\rho_S$ is related to the strain hardening which is described in crystal plasticity by the slip resistance $g^\alpha$. With the notion that the lattice is unaffected by the deformation in conventional crystal plasticity we can write $g^\alpha = \alpha\mu b\sqrt{\rho_S^\alpha}$ which yields the density $\rho_S{}^\alpha$ effective on slip system $\alpha$ as $\rho_S^\alpha = \left(\frac{g^\alpha}{\alpha\mu b}\right)^2$.

The basic assumptions of the suggested adaptation of the MSG approach to crystal plasticity are: (A) the hardening mechanisms of dislocation–dislocation interactions can be approximated by (16) and (B) the anisotropy of size effects can be described by the projected density of geometrically necessary dislocations $\eta_G{}^\alpha$. With these assumptions the total slip resistance effective on slip system $\alpha$ can be described by

$$g_T^\alpha = g_o\sqrt{(g^\alpha / g_o)^2 + \ell\eta_G^\alpha} \tag{18}$$

where $g_o$ denotes a reference slip resistance and $l = \alpha^2\mu^2 b / (g_o)^2$ the intrinsic length scale [5]. Assuming $b$ to be around one tenth of a nanometer and $\mu/g_o \approx 100$, the intrinsic length is estimated on the order of microns. The shear rate of a rate-dependent material is then modified to

$$\dot{\gamma}^\alpha = \dot{\gamma}_o^\alpha \left(\frac{\tau^\alpha}{g_T^\alpha}\right)^n \operatorname{sign}\left(\tau^\alpha\right). \tag{19}$$

It should be noted that also other projections of $\tilde{\mathbf{A}}$ could be applied. The projection $\eta_G{}^\alpha$ incorporates the property that gradients normal to the slip plane are not contained (this is actually also true for $\tilde{\mathbf{A}}$, however only for the single slip case). Forest dislocations intersecting a slip plane $\alpha$ can for example expressed by $\eta_F^\alpha = \left|\tilde{\mathbf{A}}^T \tilde{\mathbf{m}}^\alpha\right|$ whose hardening mechanisms are not accounted for in the suggested material model.

## 5. Examples

For the case of small deformation and negligible elastic strains, two examples are briefly discussed in the following. For a detailed discussion of these problems it is referred to [5].

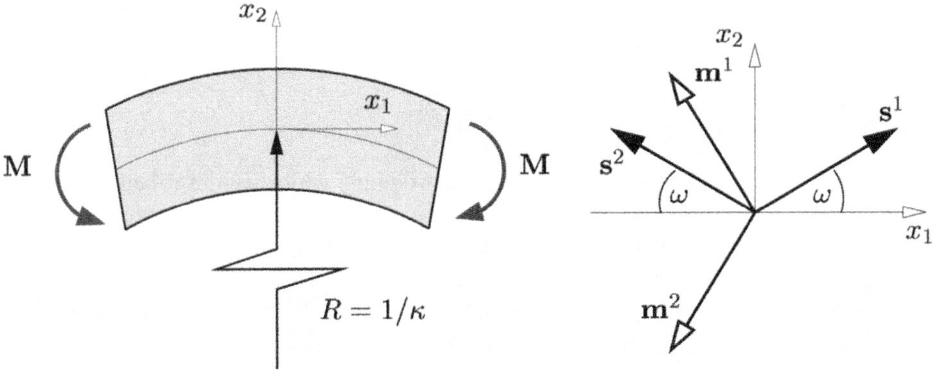


*Figure 1.* Micro-bending of a thin crystalline beam with two active slip systems.

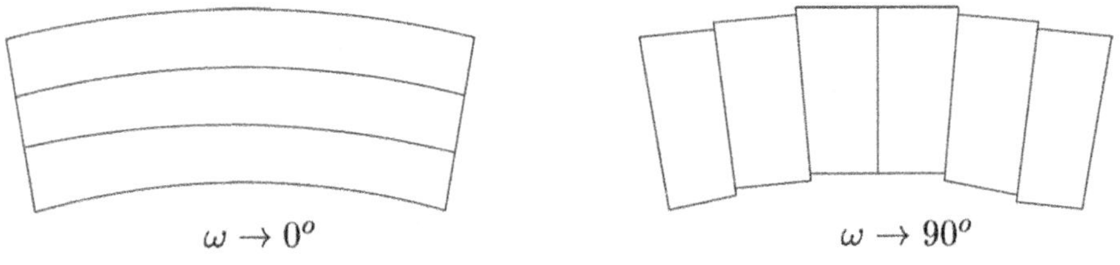


*Figure 2.* Lattice distortion and orientation of slip systems.

**Micro-bending of a beam in plane strain.** The first problem is concerned with micro-bending of a thin beam under plain strain, Fig. 1 (left). The strain components are obtained by assuming the plain strain condition $\varepsilon_{i3} = 0$, $i = 1,2,3$ and deformation by pure bending $\varepsilon_{11} = kx_2$. The remaining strain components are obtained by the incompressibility of plastic deformation yielding $\varepsilon_{22} = -kx_2$ and the Kirchhoff beam theory assuming $\varepsilon_{12} = 0$. Two active slip systems are assumed, Fig. 1 (right). After some algebra the non-zero components of the dislocation density tensor are obtained as $A_{31} = -k$ and are therefore independent of the orientation of slip systems. The resolved strain gradient measure yields $q^\alpha = k \cos \omega$.

To illustrate how the anisotropy of gradient hardening affects the deformation of the beam, the bending moment is considered. The exponent $n$ in (19) with a given $\dot{\gamma}_o$ is assumed to be large enough for the relation $\tau^\alpha = g_T{}^\alpha$ to be valid. Thus rate effects are assumed to be negligible. The normalized moments $M/M_o$, where $M_o = \tau_o h^2/4$, are plotted in Fig. 3 for $\omega = 15°, 75°$ and several values of $\beta = l/h$. Here $\beta = 0$ corresponds to macroscopic values of the beam thickness $h$ and is equivalent to solutions without strain gradients. With decreasing size $h$, the normalized moment $M/M_o$ required to deform the beam increases. The effect of the orientation of the slip systems can be assessed by comparing normalized moments for $\omega = 15°$ and $75°$. While the moment to initiate deformation is the same the hardening induced by the resolved dislocation density shows significant differences with increasing deformation. This is consistent with the notion of lattice distortions induced by geometrically necessary dislocations, as indicated in Fig. 2.

**Shear in constrained half-space.** A constant shear stress over a crystalline bi-material is now considered, Fig. 4. The slip systems are oriented with an angle $\omega$ in the same way as in the previous example. The lower half space of the structure is assumed to be rigid. The domain is subjected to a constant shear stress $\tau_{12} = \tau_o$. Continuous double slip is again assumed $\gamma^1 = \gamma^2$. Considering the limiting case, $n \to \infty$, of (19), the condition $\tau^\alpha = g_T{}^\alpha$ is applied. Since the substrate is rigid, slip vanishes $\gamma^\alpha = 0$ at $x_2 = 0$.

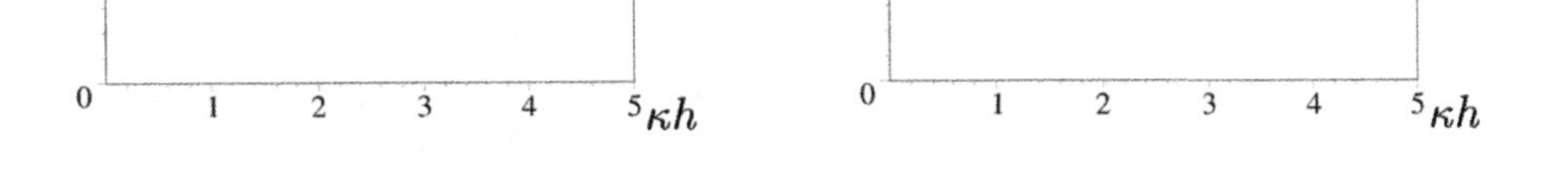


*Figure 3.* Bending moments versus $kh$ for $\omega = 15°$ and $75°$, and several values of $\beta = l/h$.

The influence of the slip system orientation on the plastic deformation can be seen in Fig. 5 (left). The shear strain is plotted with respect to the $x_2$ coordinate normalized by the length scale $l$ for $\tau^\alpha/g_o = 1.1$, $c_h/g_o = 1/5$ and various $\omega$. For these plots $\tau^\alpha$ has

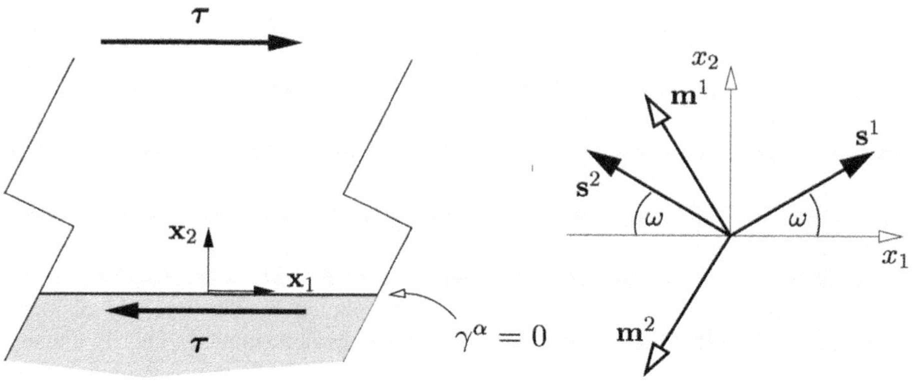


*Figure 4.* Shear in constrained half-space.

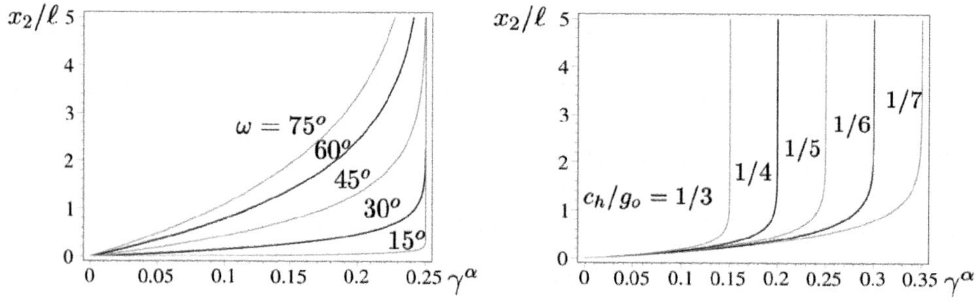


*Figure 5.* $\gamma^\alpha$ for $\tau^\alpha/g_o = 1.1$, $c_h/g_o = 15$ (left) and for $\tau^\alpha/g_o = 1.1$, $\omega = 30°$ (right).

been chosen to be the same for all $\omega$. The actual applied stress is obtained as $\tau_o = \tau^\alpha/\cos 2\omega$. Geometrically necessary dislocations will not be present for $\omega \to 0°$ since the domain can be deformed homogeneously. The hardening due to geometrically necessary dislocations should however increase as $\omega$ approaches 90°, which is in accordance to the predictions shown in Fig. 5 (right). The crystalline anisotropy of size dependent hardening is therefore reflected in the qualitatively correct way. For this problem, back stresses are expected in the interface area and are not explicitly incorporated in the suggested material description. In Fig.5 (right), the variation of the shear strain $\gamma^\alpha$ with respect to the $x_2$ coordinate is presented for $\tau^\alpha/g_o = 1.1$, $\omega = 30°$, and various ratios of $c_h/g_o$. With decreasing strain hardening expressed by $c_h$, the interface spreads its influence into the interior of the material. Thus the stronger the strain hardening, the less pronounced the boundary layer effect will be.

**Acknowledgements**

The authors gratefully acknowledge support from an NSF-NIRT Project: "Mechanism Based Modeling and Simulation in Nanomechanics," through grant No. NSF CMS-0103257, under the direction of Dr. Ken Chong. Helpful discussions with Alexander Hartmaier are also gratefully acknowledged.

## References

1. Arsenlis, A., Parks, D.M.: Crystallographic aspects of geometrically necessary and statistically stored dislocation density, *Acta Mater.*, **47** (1999), 1597–1611.
2. Fleck, N.A., Muller, G.M., Ashby, M.F., Hutchinson, J.W.: Strain gradient plasticity: theory and experiment, *Acta Mater.*, **42** (1994), 475.
3. Gao, H., Huang, Y., Nix, W.D., Hutchinson, J.W.: Mechanism-based strain gradient plasticity – I. Theory, *J. Mech. Phys. Sol.*, **47** (1999), 1239.
4. Gurtin, M.E.: A gradient theory of single-crystal viscoplasticity that accounts for geometrically necessary dislocations, *J. Mech. Phys. Sol.*, **50** (2002), 5.
5. Han, C.-S., Gao, H., Huang, Y., Nix, W.D.: Mechanism-based strain gradient crystal plasticity -I. Theory, submitted (2004).
6. Huang, Y., Gao, H., Nix, W.D., Hutchinson, J.W.: Mechanism-based strain gradient plasticity – II. Analysis, *J. Mech. Phys. Sol.*, **48** (2000), 99.
7. Ma, Q., Clarke, D.R.: Size dependent hardness of silver single crystals, *J. Mater. Res.*, **10** (1995), 853.
8. Mura, T.: *Micromechanics of Defects in Solids,* Martinus Nijhoff, Dordrecht (1987).
9. Nix, W.D., Gao, H.: Indentation size effects in crystalline materials: A law for strain gradient plasticity, *J. Mech. Phys. Sol.*, **46** (1998), 411.
10. Van der Giessen, E., Needleman, A.: Discrete dislocation plasticity: A simple planar model, *Model. Sim. Mater. Sci. Eng.*, **3** (1995), 689.

# SIZE EFFECT DOWN TO NANOSCALES: FUNCTION OVER STRENGTH

**WANLIN GUO[1]*， YUFENG GUO[1]**

[1]*Institute of Nano Science, Nanjing University of Aeronautics and Astronautics, Nanjing, 210016, P. R. China*

**Corresponding Author: wlguo@nuaa.edu.cn*

**Abstract**

Exceptional physical and biochemical functions of matter and devices become predominant when size down to nanoscale, although strength of solids may also have strong size effects. The emerging of the molecular physical mechanics as the theoretical fundamental of the new interdisciplinary fields is discussed to meet the requirement of development in science and technology. New examples in nanoscale tribology, thermal dynamic phenomena, nano intelligent systems and structural growth are presented to shown the mainpoints.

## 1. Introduction

Strength is an old problem more than 2500 years back to the ancient age of the Greek civilization. Especially in mechanics, strength is a vital issue for materials and structures, and study on the size-dependence of strength of solids can be traced back to the renaissance. Leonardo Da Vinci found that the strength of iron wires increases with decreasing length (Leonardo *et al.*, 1972) and Galileo analysed mathematically the strength of materials and invented the concept of stress (Galile *et al.*, 1638). A concise review of the history of strength of solid from the ancient philosophical concepts to modern atomic and quantum theories has been made by Malescio (Gianpietro *et al.*, 2003).



*Q P. Sun and P. Tong (eds.), IUTAM Symposium on Size Effects on Material and Structural Behavior at Micron- and Nano-Scales,* 173–186.

The size effect raised by Da Vinci is a problem of scaling, which is central to every physical theory. In applied mechanics research, the problem of scaling continuously played a prominent role for over a hundred years. Nevertheless most of size effect problems in macroscopic scale are related to the strength of materials such as materials' failure, fracture and fatigue (Bazant *et al.*, 1999). Many size effect theories based on traditional mechanics are developed to solve various problems in engineering. However, when size decreases to atomic level, these theories and methods unfortunately may be invalid and incorrect.

The strength of nano materials as well as the related mechanical properties is studied extensively and many extraordinary properties are discovered due to the strong size effect. For example, a bulk nanocrystalline pure copper with high purity and high density is found possessing an extreme superplastic extensibility without a strain hardening effect at room temperature (Lu *et al.*, 2000).

However, the more exciting and exceptional discoveries and properties in nanoscale caused by the size effect are the advanced functional nano-elements and intelligent structures with superior exceptional physical, chemical and biological characteristics and functions (Baughman *et al.*, 1999; Thomas *et al.*, 2000; Michael *et al.*, 2000; Craighead 2000; Shankar *et al.*, 2003; Gudiksen *et al.*, 2002; Krusin *et al.*, 2001). Recently most of researches are focused on the investigation of functions, intelligences and coupled behaviors of nano- and biomaterials and structures (Feng *et al.*, 2003; Mark *et al.*, 2001; Kenichiro *et al.*, 2001; Loscertales *et al.*, 2002). The results of these researches exhibit a surprising world for nano- and biotechnology and may lead to great advances in potential applications and fabrications. Therefore it is indicated that the functions and intelligences are more important and attractive than strength when size down to the nanoscale. In this paper, we will focus at the function aspect of the size problem and suggest molecular physical mechanics as the theoretical basis and tool.

## 2. Molecular Physical Mechanics

The philosophy of atomics together with the Newton's mechanical principles and differential-integral theory provide the fundamental for the discipline of continuum mechanics, which are the theoretical basis of applied mechanics. In the history of science and technology and the development of human civilization, mechanics has made great contributions and many milestones have been created by scientists in this field. Traditional mechanics in our thought is mainly the applied mechanics branch in the great family of mechanics that has played important role in the machining revolution age, especial after the two world wars. Applied mechanics in the construction of modern industry, transportation and civil engineering system, which provide us present environment for living in, are mainly related to mechanical movement induced by force, thermal transition and fluid behaviors in large scale, or in the frame of continuum mechanics. With continuum assumption the mechanics has made the best use of calculus theory and is capable to perform beautiful mathematical calculations in modeling complex engineering problems. On the other hand, philosophy of atomic concepts and modern physical researches lead to the establishment of quantum mechanics in the first quarter of last century. Quantum mechanics soon become the theoretical fundamental of modern physics, chemistry and molecular biology and forms another main branch of mechanics, as shown by Fig.1. Due to the disciplinary boundary and subject limitations, quantum mechanics and applied mechanics are mainly applied in different research fields.

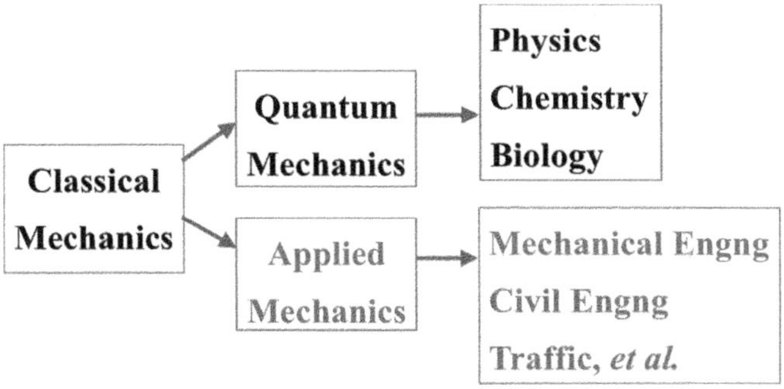


*Figure 1.* Disciplinary divarication of classical mechanics.

However, when the technology go down to the nano scale with the fast development of information technology and peering and digging into the molecular level of life systems, the traditional mechanics meets problems and the need to reunite with quantum mechanics and molecular mechanics is increasing. With solid background in large scale continuum mechanics and wide knowledge in macro world of things, mechanical scientists face the challenges to know the micro world, where the discrete characteristic becomes significant and the mechanical motion is closely related to physical and chemical effects. Consequentially, the so-called bottom-up methodology is proposed which need the combination and fusion of the two main branches in mechanics. This leads to a new interdisciplinary field, the molecular physical mechanics to meet the requirement of science and technology development.

The relationship of molecular physical-mechanics with other theories and fields is plotted in Figure 2. Down to nanoscale, all disciplines are highly interlinked and united, so the molecular physical- mechanics as a powerful tool can be applied not only in the traditional fields such as information and aerospace but also in the newest directions and fields of nano- and biotechnology. For the nanoscopic size effect, it is apparently a good way to use molecular physical-mechanics to explore the functions and intelligence of nano systems and structures.

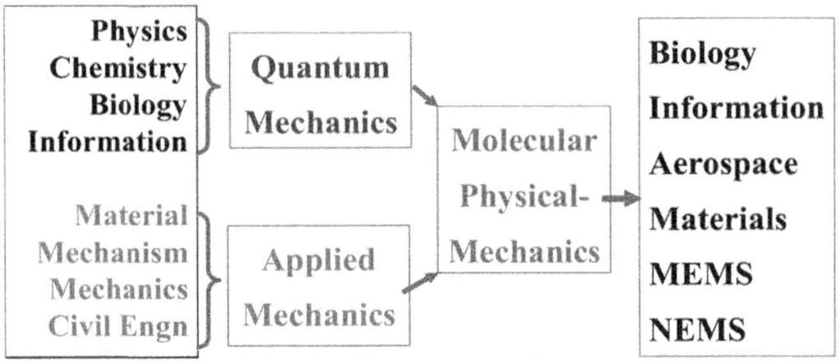


*Figure 2.* Inter-Disciplines and fusion.

### 3. Methodology

Atomistic simulations as one of the methods in molecular physical-mechanics have been played important roles in scaling down to nanoscale and can help in the elucidation of their properties and in the development of new methods for their fabrications and applications. In order to display how useful in computations and simulations of nanoscale science, some atomistic methods including molecular mechanics and quantum mechanics and the comparison with continuum theory are introduced and discussed in this paper.

Carbon nanotubes (CNTs), a novel form of carbon structures, have been rapidly recognized as particularly important nanoscopic systems. Their exceptional mechanical and physical functions and properties are related to their remarkable and significant size effect due to the one-dimensional nano structure (Robertson *et al.*, 1992; Falvo *et al.*, 1997; Wong *et al.*, 1997; Yu *et al.*, 2000; Lourie *et al.*, 1998), so carbon nanotubes systems are good proofs and representations for the topic of function over strength. The following parts in this paper introduce the investigations of energy dissipation and temperature controlling in nanoscale, same chirality growth of carbon nanotubes. Some new mechanisms and important findings are presented and summarized for carbon nanotube systems to show the size effect of the functions.

To solve various physical and mechanical problems from macroscopic to mesoscopic scale, different theory and methods such as continuum mechanics, molecular dynamics and quantum mechanics are developed in the corresponding research fields. When size goes down to nanoscale, quantum mechanics becomes the fundamental of some basic disciplines while continuum theory maybe not suitable and correct in these scales. Therefore many atomistic methods based on molecular mechanics, molecular dynamics simulation and *ab initio* calculations are chosen and developed to investigate the functions and properties of nanoscale materials. Combination of atomistic methods and continuum mechanics will produce a new concept: molecular physical-mechanics, which can be widely used for nanoscale researches. There are some examples outlined below for the explanation of how to combine the atomistic methods and classic mechanics.

Carbon nanotubes due to their exceptional electronic and mechanical properties have been widely studied by atomistic methods (Yakobson *et al.*, 1996; Rochefort *et al.*, 1999; Lier *et al.*, 2000; Zhou *et al.*, 2001). However, many atomistic methods applied to CNTs are limited by the computational cost, spatial and temporal scales. Therefore many scientists have tried to find approximate simulations based on continuum mechanics to study larger systems as requested. For example, large deformation behavior and post-buckling modes of single-walled carbon nanotubes are studied by using traditional continuum shell theory and the results show that the continuum shell theory is convenient and efficient in predicting the post-buckling behavior of the nanotubes subjected to axial compression, torsion and bend loads (Li *et al.*, 2003).

The more exciting potential uses of CNTs are related to their exceptional electronic and chemical properties, but in this field continuum mechanical theory become unsuitable and invalid. Thus atomistic methods based on molecular mechanics and quantum mechanics are widely and extensively used to solve different kinds of problems and phenomena in nanoscale. Completely quantum mechanics such as *ab initio* Hatree-Fock and density function calculations can be extremely expensive in terms of the computer resources required while the approximate and semi-empirical quantum mechanical methods require less computational resources. Moreover, by incorporation of parameters derived from experimental data some semi-empirical methods can calculate some properties more accurately than even the highest level of *ab initio* methods. Therefore according to the actually requirements of calculations and simulations, we use different atomistic methods to investigate the physical properties and behaviors of CNTs system. For example, firstly a semi-empirical quantum mechanical method PM3 (Stewart 1989) is adopted to study the coupled mechanical and electronic behaviors of single-walled carbon nanotubes under applied electric field and tensile loading (Guo *et al.*, 2003; Guo *et al.*, 2002). Then we use more accurate Hatree-Fock quantum mechanical calculation to obtain an exceptional characteristic of large axial electrostrictive deformation of SWNTs. For the essential physical properties, accurate density function technique is employed to achieve the energy gap and electronic structure of SWNTs (Guo and Guo, 2003).

## 4. Size-Dependent Energy dissipation at nanoscale

Using molecular statics and dynamics methods, coupled defects-size effects on interlayer friction in multiwalled carbon nanotubes are systematically studied. The 5-7-7-5 type defects shown in Figure 3 are created on the core and outer tubes respectively. The defects in bi-walled carbon nanotubes are shown to have strong effect on interlayer friction. The variations of the normalized amplitude of potential energy with the time for the different lengths of bi-tube system are plotted in Figure 4. The oscillating energy dissipation rate increases monotonically in a bi-tube system with increasing tube length and defect density. The results indicate that the defects as well as the size effect can greatly change the functions of the whole carbon nanotube systems. This coupled defect-size effect also can explain how an ultrasmooth nano bi-tube system leads to rough longer tube system and provide a new mechanism for multiscale tribology (Guo *et al.*, 2003; Guo *et al.*, 2005).

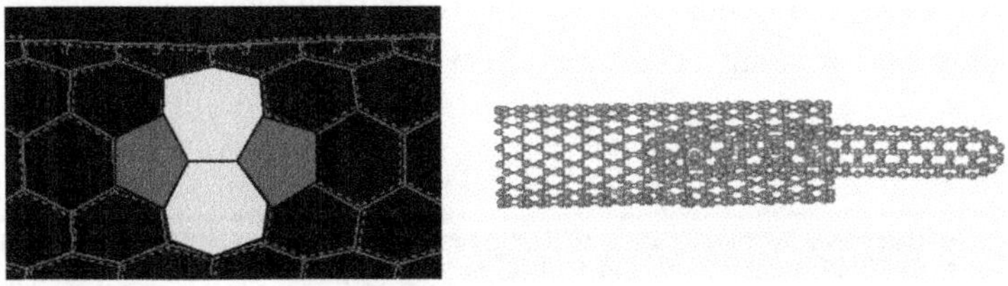

*Figure 3*. The structure of 5-7-7-5 defects on the carbon nanotubes.

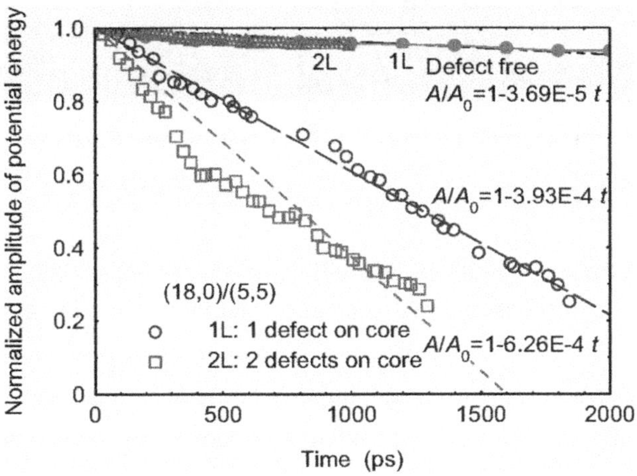


*Figure 4*. The variations of the normalized amplitude of potential energy with the time for the different lengths of bi-tube system.

## 5. Temperature at nanoscale

Temperature is always a vital issue in physical processes. When size goes down to nanoscale, dissipation and temperature in non-equilibrium systems become more challenging, especially in nano electromechanical systems where mechanical behaviors are the first important. In continuum mechanics, temperature is taken as a parameter independent of mechanical movements, but at atomistic scale, the mechanical movement of a NEMS always merges with the thermal vibration of atoms. As shown in Figure 5, the thermodynamic process of the multi-walled carbon nanotubes systems is not a random motion at a temperature of 500 K, and the temperature drives the inner tubes oscillating and rotating. The size effect on the temperature controlling is so obvious and strong in nanoscale.

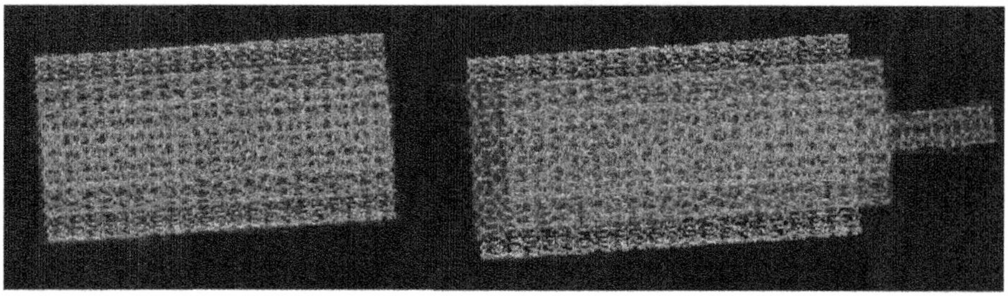

*Figure 5.* Thermodynamics motion of MWNT system at the temperature of 500K.

In common-purpose atomistic simulation codes, kinetic temperature controls are widely used, in which the isothermal simulation can be realized by re-scaling the atom velocity using some standard schemes, but the motion of thermodynamics and regular mechanical components is not divided in this kind of temperature control method. Using atomistic methods and the conceptions of thermodynamic temperature and kinetic temperature, isothermal simulations can be realized by constant thermodynamics temperature control of a system, while constant kinetic temperature control scheme yields misleading results for mechanical motions of the system. Figure 6 shows the variations of temperatures with time for the (10,10)(5,5) carbon nanotube commenurate system. The isothermal simulations yield similar trends as an adiabatic approximation (Guo and Zhu, *et al.*,).

## 6. Growth of carbon nanotubes

To investigate the chiral effect on the growth of multi-walled carbon nanotubes, several zigzag carbon nanotubes are selected as the core and template tubes. The single-walled carbon nanotube is indexed by a pair of integers (n,m) which define the circumferential direction in graphitic lattice coordinates (White *et al.*, 1993). Assuming n $\geq$ m, the SWNT

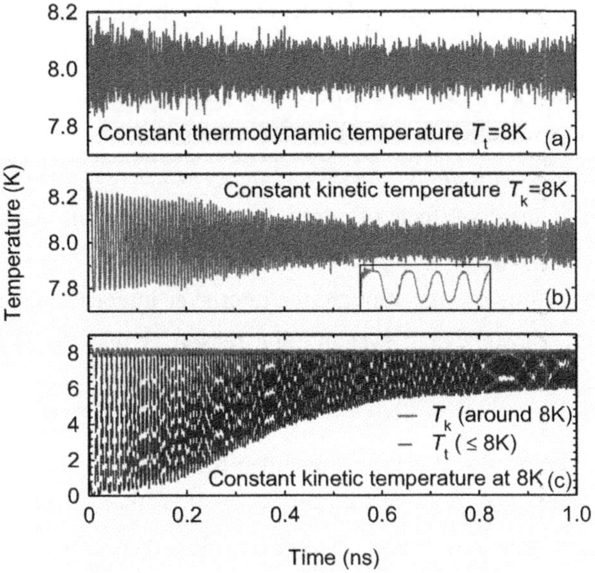


*Figure 6.* Variation of temperatures with time for the (10,10)(5,5) carbon nanotubes systems. (a) The thermodynamic temperature under constant thermodynamic temperature control at 8K. (b) The kinetic temperature under constant kinetic temperature control at 8K with initial temperature of 8K; inset is the kinetic temperature varied with time in for time from 0 to 50ps. (c) Comparison of the kinetic temperature $T_k$ with the thermodynamic temperature under constant kinetic temperature control, in the same case as (b).

radius is given by $R=\frac{\sqrt{3}a_{c-c}}{2\pi}\sqrt{n^2+m^2+nm}$, where $a_{c-c}$ is the length of the C-C bond which is taken as 1.42Å. According to the experiment result, the interlayer distance for the MWNTs is 3.38Å, then in our calculations the outer tubes' radius and corresponding index (n,m) are determined from the condition $R_c+3.23$ Å$\leqslant R_o \leqslant R_c+3.53$ Å, where subscript "o" denotes the outer shell and "c" the inner core tube. The interaction between outer tube and inner tube is described by using K-C potential (Kolmogorov and Crespi, 2000). For these zigzag core tubes, all possible orientations of the outer tubes whose diameters are in the given range are selected to calculate the interlayer potential energy. Fig.7 shows the variations of the minimum interlayer potential energy for the different outer tubes when m/n ranges from 0 to 1. As shown in these curves due to slight changing in diameters of outer tubes, the interlayer energy is varying with the m/n of outer tubes. However, when both outer and inner tubes are zigzag tube with m/n = 0, the minimum interlayer energy is much lower than others in the all-possible outer tubes arrangements. This indicates that zigzag/zigzag system possess more stable structure than other tube assembles.

The variation of the minimum interlayer energy with core tube radius is plotted in Fig.8 to study the diameters effects of core tubes. The minimum interaction energy of outer tube decreases remarkably with the increasing of inner tube radius. However, when the core tube radius is large enough, the interlayer interaction energy is gradually to be the same for zigzag/zigzag system. The reason is that the two carbon sheets become more and more flat and the effect of curvature radius of carbon nanotubes is not so significant.

According to the results, it can be concluded that zigzag (n,0) core tubes prefer to be with zigzag outer tubes to form multi-walled carbon nanotubes due to the lowest minimum interlayer potential energy. Moreover, this theory prediction can give a reasonable interpretation for the growth mechanism of experimental synthesized zigzag carbon cones (Zhang *et al.*, 2003; Zhang *et al.*, 2005) as well.

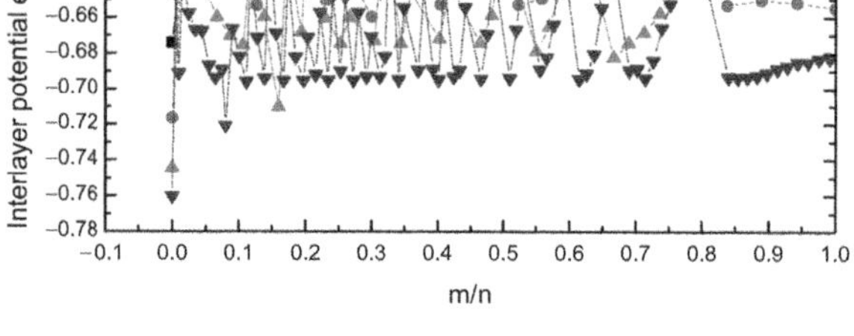


*Figure 7.* Variations of the minimum interlayer interaction energy with outer tubes' chiralities m/n for the (26,0), (50,0), (100,0) and (200,0) core tubes, where integers n and m denote the orientation of the outer tube.

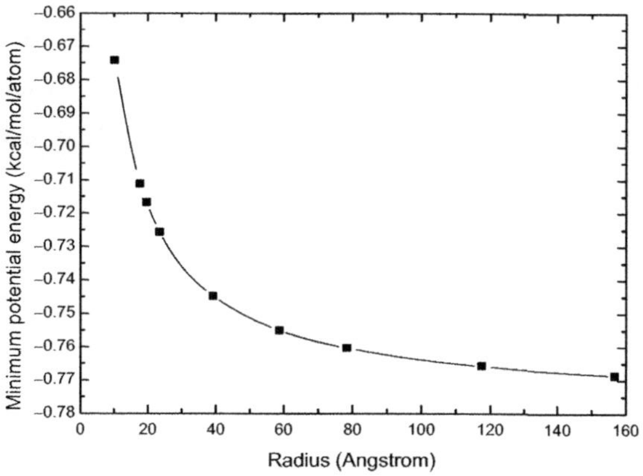


*Figure 8*. Variation of the minimum interlayer interaction energy of zigzag outer tube with the radius of zigzag core tubes.

## 7. Summary

When size goes down to nanoscale, the functions and intelligences of smart materials and nano elements are more attractive and significant in the future applications. Compare to the macroscopic systems, the size effect in nanoscale not only influences and changes the strength of the nano systems and structures but aslo essentially exhibits the surprising functions and intelligences and leads to many exciting and exceptional properties and characteristics. The coupled mechanical, physical and chemical properties and behaviors are most important and vital for the nano systems. Various kinds of methods including continuum mechanics and atomistic simulations are developed to study the size effect and new phenomena in nano scale. Then a new concept molecular physical-mechanics based on these theories and methods is established to solve the problems of size effect in nanoscale. Combination of molecular mechanics, molecular dynamics and quantum mechanics is a powerful tool in nano researches, which provides the wide background for not only the nanotechnology, but also the nano systems with surprised functions.

CNTs are good examples and proofs for 'function over strength when size effect down to nanoscales' due to their interesting structures and exceptional properties. The energy dissipation, temperature control of CNTs systems and growth of carbon nanotubes are investigated respectively by using quantum mechanics and molecular mechanics. All the given results present a clear picture that functions and intelligences originated from the size effect are more important and necessary in nano materials and structures.

## References


Bazant, Z. P. (1999) Size effect on structural strength: a review, *Archive of Applied Mechanics* **69**, 703-725.

Baughman, R. H. *et al.*, (1999) Carbon Nanotube Actuators, *Science* **284**, 1340-1344.

Craighead, H. G. (2000) Nanoelectromechanical Systems, *Science* **290**, 1532-1535.

Feng, P. L. *et al.*, (2003) A Two-State DNA Lattice Switched by DNA Nanoactuator, *Angew. Chem. Int. Ed.* **42**, 4342-4346.

Falvo, M. R. *et al.*, (1997) Bending and buckling of carbon nanotubes under large strain, *Nature* **389**, 582-584.

Galilei, G., Discorsi, E. (1638) Dimostrazioni matematiche, Elsevier Leiden.

Gianpietro, M. (2003) Intermolecular potentials: past, present, future, *Nature Materials* **2**, 501-503.

Gudiksen, M. S. *et al.*, (2002) Growth of nanowire superlattice structures for nanoscale photonics and electronics, *Nature* **415**, 617-620.

Guo, Y., and Guo, W. (2003) Mechanical and electrostatic properties of carbon nanotubes under tensile loading and electric field, *J.Phys. D: Appl. Phys.* **36**, 805-811.

Guo, W., and Guo, Y. (2002) Molecular Dynamics and Quantum Mechanics Investigation on Mechanic-electric Behaviors of Nanotubes, *Int. J. Nonlinear Science and Numerical Simulation* **3**, 469-472.

Guo, W., and Guo, Y. (2003) Giant Axial Electrostrictive Deformation in Carbon Nanotubes, *Phys. Rev. Lett.* **91**, 115501 1-4.

Guo, W., Guo, Y., Gao, H., Zheng, Q., and Zhong, W. (2003) Energy Dissipation in Gigahertz Oscillators from Multiwalled Carbon Nanotubes, *Phys. Rev. Lett.* **91**, 125501 1-4.

Guo, W., Zhong, W., Dai, Y., and Li, S. (2005) Coupled Defect-Size Effects on Interlayer Friction in Multiwalled Carbon Nanotubes, *Phys. Rev. B* **72**, 075409 1-10.

Guo, W., Zhu, C., Yu, T., and Woo, C. H. Isothermal Atomistic Simulations of Nano electromechanical Systems, submitted to *Phys. Rev. Lett.*

Krusin, E. L. *et al.*, (2001) Stable ultrahigh-density magneto-optical recordings using introduced linear defects, *Nature* **410**, 444-446.

Kenichiro, K., Gao, G. T., Hideki, T., and Zeng, X. C. (2001) Formation of ordered ice nanotubes inside carbon nanotubes, *Nature* **412**, 802-805.

Kolmogorov, A. N., and Crespi, V. H. (2000) Smoothest Bearings: Interlayer Sliding in Multiwalled Carbon Nanotubes, *Phys. Rev. Lett.* **85**, 4727-4730.

Leonardo, D. V., and Libri, D. M. (1972) Reconstructed from the priginal notes by Arturo Uccelli, Kraus Reprint, Nendeln Liechtenstein.

Lu, L., Sui, M. L., and Lu, K. (2000) Superplastic Extensibility of Nanocrystalline Copper at Room Temperature, *Science* **287**, 1463-1446.

Loscertales, I. G. *et al.*, (2002) MicroNano Encapsulation via Electrified Coaxial Liquid Jets, *Science* **295**, 1695-1698.

Lourie, O., Cox, D. M., and Wagner, H. D. (1998) Buckling and Collapse of Embedded Carbon Nanotubes, *Phys. Rev. Lett.* **81**, 1638-1641.

Li, C., and Guo, W. (2003) Continuum Mechanics Simulation of Post-Buckling of Single-Walled Nanotubes, *Int. J. Nonlinear Science and Numerical Simulation* **4**, 387-393.

Lier, G. V., Alsenoy, C. V., Doren, V. V., and Geerlings, P. (2000) Ab initio study of the elastic properties of single-walled carbon nanotubes and graphene, *Chem. Phys. Lett.* **326**, 181-185.

Michael, M. and Uzi, L. (2000) Formation, Stability, and Breakup of Nanojets, *Science* **289**, 1165-1169.

Mark, S. P., and Philip, C. B. (2001) Biophysics: Water at the nanoscale, *Nature* **414**, 156-159.

Robertson, D. H., Brenner, D. W., and Mintmire, J. W. (1992) Energetics of nanoscale graphitic tubules, *Phys. Rev. B* **45**, 12592-12595.

Rochefort, A., Avouris, P., Lesage, F., and Salahub, D. (1999) Electrical and mechanical properties of distorted carbon nanotubes, *Phys. Rev. B* **60**, 13824-13830.

Shankar, G., Sood, A. K., and Kumar, N. (2003) Carbon Nanotube Flow Sensors, *Science* **299**, 1042-1044.

Stewart, J. J. P. (1989) Optimization of parameters for semiempirical methods. I. Method, *J. Comput. Chem.* **10**, 209-220.

Thomas, R. *et al.*, (2000) Carbon Nanotube-Based Nonvolatile Random Access Memory for Molecular Computing, *Science* **289**, 94-97.

Wong, E., Sheehan, P., and Lieber, C. (1997) Nanobeam Mechanics: Elasticity, Strength, and Toughness of Nanorods and Nanotubes, *Science* **277**, 1971-1975.

White, C. T., Robertson, D. H., and Mintmire, J. W. (1993) Helical and rotational symmetries of nanoscale graphitic tubules, *Phys. Rev. B* **47**, 5485-5488.

Yu, M. F., and Lourie, O. (2000) Strength and Breaking Mechanism of Multiwalled Carbon Nanotubes Under Tensile Load, *Science* **287**, 637-640.

Yakobson, B., Brabec, C., and Bernholc, J. (1996) Nanomechanics of Carbon Tubes: Instabilities beyond Linear Response, *Phys. Rev. Lett.* **76**, 2511-2514.

Zhou, G., Duan, W., and Gu, B. (2001) First-principles study on morphology and mechanical properties of single-walled carbon nanotube, *Chem. Phys. Lett.* **333**, 344-349.

Zhang, G. Y., Jiang, X., and Wang, E. N. (2003) Tubular Graphite Cones, *Science* **300**, 472-474.

Zhang, G. Y., Bai, X. D., Wang, E. G., Guo, Y., and Guo, W. (2005) Identical Chirality of Tubular Graphite Cones Controlled by Radial Layer-by-Layer Growth, *Phys. Rev. B* **71**, 113411 1-4.

# MICROMECHANICS OF MICROPOLAR COMPOSITES

Gengkai Hu

*Department of Applied Mechanics, Beijing Institute of Technology 100081, Beijing, China*

## 1. Introduction

The objective of micromechanics is to establish a relation between the macroscopic properties and the microscopic information for heterogeneous materials. There are different length scales involved in micromechanical modeling, for example, the structural characteristic size $L$, the size of the representative volume element (RVE) $l$, the characteristic length scale of the reinforced phase $A$ and the characteristic length scale of the matrix $l_m$. The classical micromechanical method applies for the following length scale condition $L >> l >> A >> l_m$ (Forest and Sab, 1998; Zaoui, 2003; Hu *et al.*, 2004); in this case both constituents and the homogenized material are idealized as Cauchy material model without any inner microstructure. However when the structure size $L$ is small (MEMS, thin films), the previous length scale condition is not fulfilled; in another situation, for metal matrix composites, nanocomposites, the size of the filler can reach the same order as the characteristic length scale of the matrix material ($l_m$). Under these length scale conditions, the nonlocal nature of either the homogenized macro-element (RVE) or the constituent materials must be considered in a proper high order continuum theory. Homogenization strategies and the corresponding length scale conditions are presented in table 1 (Forest and Sab,1998; Hu *et al.*, 2004).



*Q. P. Sun and P. Tong (eds.), IUTAM Symposium on Size Effects on Material and Structural Behavior at Micron- and Nano-Scales,* 187–194.

TABLE 1. Length scale condition and homogenization method

| Constituent model | length scale condition | | homogenized model |
|---|---|---|---|
| Cauchy model | $A >> l_m$ | $L >> l >> A$ | Cauchy model |
| High order continuum | $A \approx l_m$ | | |
| Cauchy model | $A >> l_m$ | $L \approx l \approx A$ | High order continuum |
| High order continuum | $A \approx l_m$ | | |

Discussions on high order continuum theory can be found in the monographs given by Eringen (1999) and Nowacki (1986). For composite materials, Wei (2001), Chen and Wang (2002) have used high order continuum models to describe the size-dependence of the overall nonlinear stress and strain relations, however their methods are basically numerical. In this paper, we are concerned with the following length scale condition $L >> l >> A \approx l_m$, locally the constituents are idealized as micropolar material model, however the homogenized macro-element can be still considered as a Cauchy material due to the small size of the RVE compared to the structural size. The paper will be arranged as follows: A brief recall of the micropolar elasticity and plasticity theory will be presented in section 2; the micro-macro transition method and the method to evaluate the effective moduli of the composite will be given in section 3; in section 4, the secant moduli method will be extended to micropolar composites to predict the effective nonlinear stress and strain relation.

## 2. Micropolar elasticity and plasticity

In the micropolar theory, three rotation angles $\phi$ are introduced in addition to the macro-displacements $u_i$ of a material point. The kinetic, equilibrium and constitutive equations for an isotropic centro-symmetric micropolar material are (without body force and couple)

$$\begin{cases} \varepsilon_{ij} = u_{j,i} - e_{kij}\phi_k \\ k_{ij} = \phi_{j,i} \end{cases}, \quad \begin{cases} \sigma_{ij,j} = 0 \\ m_{ij,j} + e_{jik}\sigma_{ik} = 0 \end{cases}, \quad \begin{cases} \sigma_{ij} = C_{ijkl}\varepsilon_{kl} \\ m_{ij} = D_{ijkl}k_{kl} \end{cases} \tag{1}$$

The stress $\sigma_{ij}$ and the strain $\varepsilon_{ij}$ are not symmetric in the micropolar theory, $m_{ij}, k_{ij}$ are the couple stress and the torsion respectively. For an isotropic micropolar material, we have

$$C_{ijkl} = \lambda\delta_{ij}\delta_{kl} + (\mu+\kappa)\delta_{jk}\delta_{il} + (\mu-\kappa)\delta_{ik}\delta_{jl} \tag{2a}$$

$$D_{ijkl} = \alpha\delta_{ij}\delta_{kl} + (\beta+\gamma)\delta_{jk}\delta_{il} + (\beta-\gamma)\delta_{ik}\delta_{jl} \tag{2b}$$

where $\mu, \lambda$ are the classical Laméconstants, $\kappa, \gamma, \beta, \alpha$ are new material constants introduced in the micropolar theory, $\delta_{ij}$ is the Kronecker delta. With $\sigma'_{(ij)}$, $\sigma'_{<ij>}$, $\sigma$ and $\varepsilon'_{(ij)}$, $\varepsilon'_{<ij>}$, $\varepsilon$ denoting separately the deviatoric symmetric, anti-symmetric and hydrostatic parts of the stress and the strain tensors, and similar notations for the couple-stress and torsion tensors, the well-established elastic constitutive relations for a linear isotropic micropolar material can be rewritten as (Nowacki, 1986):

$$\sigma'_{(ij)} = 2\mu\varepsilon'_{(ij)}, \sigma'_{<ij>} = 2\kappa\varepsilon'_{<ij>}, \sigma = 3K\varepsilon\, ; m'_{(ij)} = 2\beta k'_{(ij)}, m'_{<ij>} = 2\gamma\varepsilon'_{<ij>}, m = 3Pk \tag{3}$$

where $K = \lambda + 2/3\,\mu$ is the bulk modulus, $P = \alpha + 2/3\,\beta$ can be regarded as the corresponding stiffness measure for the torsion, and symbols ( ) and < > in the subscript denote the symmetric and anti-symmetric parts of a second order tensor, respectively.

The two distinct sets of moduli: ($\mu, \lambda, \kappa$) and ($\gamma, \beta, \alpha$) define three intrinsic characteristic lengths for an isotropic elastic micropolar material (Liu and Hu,2005):

$$l_1 = (\gamma/\mu)^{1/2}, l_2 = (\beta/\mu)^{1/2}, l_3 = (\alpha/\mu)^{1/2} \tag{4}$$

For simplicity, in the following discussion, we set $l_1 = l_2 = l_3 = l_m$.

For a nonlinear micropolar material, the generalized equivalent stress may be defined as (Lippmann, 1995; Fleck and Hutchinson,1997; Liu and Hu,2005):

$$\bar{\sigma}_e = \sqrt{\frac{3}{2}\sigma'_{(ij)}\sigma'_{(ij)} + \frac{3}{2}l_m^{-2}(m'_{(ij)}m'_{(ij)} + m_{<ij>}m_{<ij>})} \tag{5}$$

Following the same idea as in a Cauchy material model, the stress potential for a nonlinear micropolar material may then be written as:

$$w = w_0(\bar{\sigma}_e) + \frac{1}{6\kappa}\sigma^2_{<e>} + \frac{1}{18K}\sigma^2 + \frac{1}{18P}m^2 \tag{6}$$

where $w_0(\overline{\sigma}_e)$ has the same form as for a classical material:

$$w_0 = \frac{\overline{\sigma}_e^2}{6\mu} + \frac{n}{n+1}\frac{1}{H^{1/n}}\left(\overline{\sigma}_e - \sigma_y\right)^{\frac{n+1}{n}} \tag{7}$$

where $\sigma_y$ and $n, H$ are the initial yield stress, and hardening parameters of the micropolar material respectively, and $\sigma_{<e>}^2 = 3\,\sigma_{<ij>}\sigma_{<ij>}/2$.

From the deformation theory in plasticity, the secant moduli of a nonlinear micropolar material at the stress state $\overline{\sigma}_e$ can be defined from equation(6) as

$$\mu^s = \frac{1}{1/\mu + 3[(\overline{\sigma}_e - \sigma_y)/H]^{1/n}/\overline{\sigma}_e}, \quad \kappa^s = \kappa, \quad K^s = K \tag{8a}$$

$$\beta^s = l_m^2\mu^s, \quad \gamma^s = l_m^2\mu^s, \quad P^s = P \tag{8b}$$

These secant moduli will be used to determine the effective nonlinear stress and strain relation for composite materials.

## 3. Micro-macro transition method and effective moduli

Under the length scale condition $L >> l >> A \approx l_m$, the RVE is sufficiently small and and it can still be considered as a material point in the structure, so the classical boundary condition is applied: $u_i = E_{(ij)}x_j$, $\phi_i = 0$ or $F_i = \Sigma_{(ij)}n_j$, $m_{ij}n_j = 0$, where $n_i$ is the unit normal to the boundary of the RVE. The average of the local energy density over RVE can be written as:

$$<\boldsymbol{\sigma}:\boldsymbol{\varepsilon} + \boldsymbol{m}:\boldsymbol{k}> = \boldsymbol{E}^{sym}:\overline{\boldsymbol{C}}^{sym}:\boldsymbol{E}^{sym} = \boldsymbol{\Sigma}^{sym}:\overline{\boldsymbol{M}}^{sym}:\boldsymbol{\Sigma}^{sym} \tag{9}$$

Equation (9) provides a method to evaluate the effective stiffness $\overline{\boldsymbol{C}}^{sym}$ or compliance tensor $\overline{\boldsymbol{M}}^{sym}$ for micropolar composites. To compute the average strain and torsion in the different phases, the average equivalent inclusion method will be utilized. It has been shown by Chen and He (1995) that for an infinite micropolar matrix, the strain and torsion produced by a spherical region with a uniform eigenstrain $\boldsymbol{\varepsilon}^*$ and a uniform eigentorsion $\boldsymbol{k}^*$ are

$$\boldsymbol{\varepsilon} = \boldsymbol{A}(\boldsymbol{x}):\boldsymbol{\varepsilon}^* + \boldsymbol{B}(\boldsymbol{x}):\boldsymbol{k}^*;\quad \boldsymbol{k} = \hat{\boldsymbol{A}}(\boldsymbol{x}):\boldsymbol{\varepsilon}^* + \hat{\boldsymbol{B}}(\boldsymbol{x}):\boldsymbol{k}^* \tag{10}$$

where $\boldsymbol{A}(\boldsymbol{x})$, $\boldsymbol{B}(\boldsymbol{x})$, $\hat{\boldsymbol{A}}(\boldsymbol{x})$ and $\hat{\boldsymbol{B}}(\boldsymbol{x})$ are Eshelby-like tensors, their expressions are given in Chen and He(1995). For a spherical inclusion, it can be shown that $<\boldsymbol{B}(\boldsymbol{x})>_I=0$ and $<\hat{\boldsymbol{A}}(\boldsymbol{x})>_I=0$, where $<\bullet>_I$ means the volume average of the considered quantity over the spherical region. The analytical expressions for $<\boldsymbol{A}(\boldsymbol{x})>_I$ and $<\hat{\boldsymbol{B}}(\boldsymbol{x})>_I$ are given in Liu and Hu (2004). For a spherical inclusion with a uniform eigenstrain and a uniform eigentorsion, the induced strain and torsion inside of the inclusion are not uniform, so the equivalent inclusion method widely used for classical micromechanics has to be applied in an average sense in the case of micropolar composites (Xun *et al.*, 2004). With help of the average equivalent inclusion method, we can follow exactly the same method as in classical micromechanics to determine the effective moduli. For a two-phase particulate composite, $a$ and $c_1$ denoting the radius and the volume fraction of the particle respectively, the effective shear and bulk moduli estimated by Mori-Tanaka's method are:

$$\frac{\overline{\mu}}{\mu_0} = 1 + \frac{c_1}{(1-c_1)\xi_1 + \mu_0/(\mu_1-\mu_0)},\quad \frac{\overline{K}}{K_0} = 1 + \frac{c_1}{(1-c_1)\xi_2 + K_0/(K_1-K_0)} \tag{11}$$

where $\xi_1 = \dfrac{6(K_0+2\mu_0)}{5(3K_0+4\mu_0)} - \dfrac{6\kappa_0}{5(\kappa_0+\mu_0)}G(\eta)$, $\xi_2 = \dfrac{3K_0}{3K_0+4\mu_0}$, and

$G(\eta) = e^{-\eta}(\eta^{-2}+\eta^{-3})[\eta\cosh\eta - \sinh\eta], \eta = a/h, h^2 = (\mu_0+\kappa_0)(\gamma_0+\beta_0)/4\mu_0\kappa_0$.

When $a/h \to \infty$, the classical results can be recovered, as expected. However when the particle size is reduced, the effective shear modulus increases, as shown in figure 1 ($\kappa_0/\mu_0 = 0.5$, $K_0/\mu_0 = 2.53$, $\mu_1/\mu_0 = K_1/K_0 = 10$). However the effective bulk modulus will not depend on the particle size.

## 4. Effective nonlinear stress-strain relation

We can follow the main idea of secant moduli method as in the classical micromechanics to determine the overall nonlinear property for a micropolar composite. The key point is to evaluate the secant moduli of the nonlinear micropolar matrix as function of the applied load $\boldsymbol{\Sigma}^{sym}$. To this end, we consider a linear comparison micropolar composite $\overline{\boldsymbol{M}}_s^{sym}$ characterized locally by the compliance tensors $\boldsymbol{M}^s, \boldsymbol{R}^s$. Consider local material compliances variations $\delta\boldsymbol{M}^s$, $\delta\boldsymbol{R}^s$, this will lead to the variations of the local stress, couple stress and the macroscopic compliance of the composite $\delta\boldsymbol{\sigma}$, $\delta\boldsymbol{m}$ and $\delta\overline{\boldsymbol{M}}_s^{sym}$ respectively. From equation (9), we can show that

$$< \boldsymbol{\sigma} : \delta\boldsymbol{M}^s : \boldsymbol{\sigma} + \boldsymbol{m} : \delta\boldsymbol{R}^s : \boldsymbol{m} >= \boldsymbol{\Sigma}^{sym} : \delta\overline{\boldsymbol{M}}_s^{sym} : \boldsymbol{\Sigma}^{sym} \tag{12}$$

Consider the following variations $\delta\mu_0^s, \delta\beta_0^s, \delta\gamma_0^s$ separately (the superscript $s$ here denotes a quantity related to the reference material, and according to the secant moduli method the reference material is assumed to have the secant moduli of the actual nonlinear phase), the average equivalent stress of the matrix in the linear comparison micropolar composite defined by equation (5) can then be calculated by

$$\overline{\sigma}_e^2 = \frac{1}{1-c_1}\boldsymbol{\Sigma}^{sym} : \left[ -3\mu_0^{s\,2} \left( \frac{\partial \overline{\boldsymbol{M}}_s^{sym}}{\partial \mu_0^s} + \frac{\partial \overline{\boldsymbol{M}}_s^{sym}}{\partial \beta_0^s} + \frac{\partial \overline{\boldsymbol{M}}_s^{sym}}{\partial \gamma_0^s} \right) \right] : \boldsymbol{\Sigma}^{sym} \tag{13}$$

The effective moduli of the linear comparison composite can be estimated by equation (11), so the average equivalent stress of the matrix can be evaluated without any difficulty. When the matrix material has undergone a plastic deformation, the secant moduli method can be utilized to determine the overall nonlinear stress-strain relation of the composite.

The comparison between the prediction and the experimental results conducted by Kouzeli and Mortensen (2002) is shown in figure 2 for different particle sizes and volume concentrations. The following material constants are used in the modeling: $K_1/K_0 = 3.71$, $\mu_1/\mu_0 = 6.52$, $K_0/\mu_0 = 2.61$, $\kappa_0/\mu_0 = 0.5$, $l_m = 4.2\mu m$, $H = 110Mpa$, $\sigma_y = 25Mpa$, $n = 0.255$. Four sets of composite samples are examined, with the particle diameter $2a$ and the volume fraction $c_1$ separately given by: (4.5 $\mu m$, 0.39), (9.3 $\mu m$, 0.54), (29.2 $\mu m$, 0.461) and (58 $\mu m$, 0.475). A good correlation is found between the predictions and the experimental values.

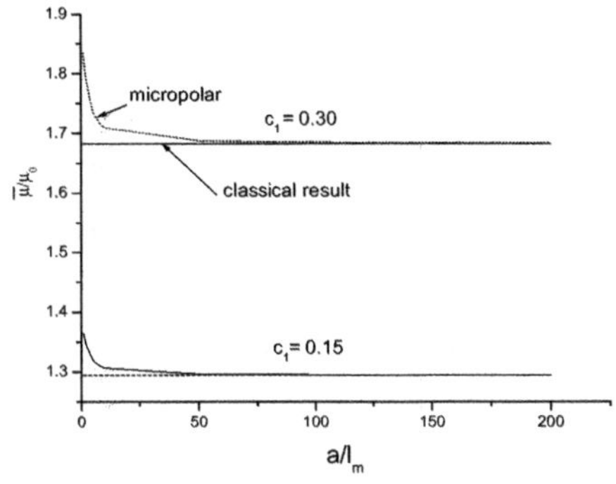


*Figure 1.* Effective shear modulus as a function of particle size

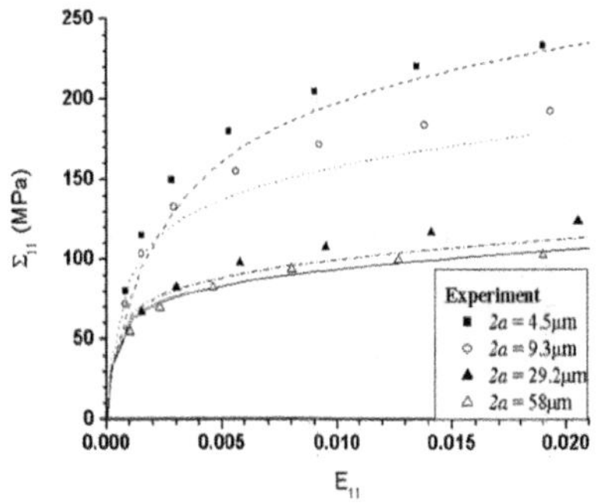


*Figure 2.* Comparison between theory and experiment

Similar to the case of classical micromechanics (Hu, 1996), the proposed secant moduli method based on second order stress and couple stress moments can also have a variational interpretation of the Ponte Castañda's type (Ponte Castañda, 1991), which can be written in the following form

$$\overline{W}_{eff} \geq \overline{W}_{eff}^{-} \equiv \sup_{\forall \boldsymbol{M}^s, \boldsymbol{R}^s} \left[ \overline{W}^s(\boldsymbol{M}^s, \boldsymbol{R}^s, \boldsymbol{\Sigma}^{sym}) - \sum_{r=1}^{N} c_r \sup_{\forall \boldsymbol{\sigma}, \boldsymbol{m}} \left[ w_r^s(\boldsymbol{\sigma}, \boldsymbol{m}) - w_r(\boldsymbol{\sigma}, \boldsymbol{m}) \right] \right] \tag{15}$$

where $w_r$ is the stress potential of the nonlinear phase $r$, $w_r^s$ is the stress potential for a linear micropolar material associated with the phase $r$, $c_r$ is its volume fraction. $\overline{W}^s$ and $\overline{W}_{eff}$ are the effective stress potentials of the linear comparison composite and the nonlinear composite, respectively.

## 5. Summary

We have presented a homogenization method, in which the nonlocal nature of the matrix due to its coarse-grain structure is characterized by a micropolar material model. Mori-Tanaka's method and the secant moduli method are extended to micropolar composites, the particle size-dependence of the overall nonlinear stress-strain relation can be well predicted analytically. It is also shown that the proposed method has also a variational interpretation of Ponte Castañeda's type.

**Acknowledgement**: this work is supported by NSFC through grant 10325210, 10332020

## 6. References

Chen, S., Wang T. C. (2002) Size effects in the particle-reinforced metal-matrix composites, Acta Mechanica 157(1-4), 113-127.

Cheng, Z. Q, He, L. H. (1995) Micropolar elastic fields due to a spherical inclusion. Int. J. Eng. Sci., 33, 389–397

Eringen, A. C. (1999) Microcontinuum field theory. Springer.

Fleck, N. A., Hutchinson, J. W. (1993) A phenomenological theory for strain gradient effect in plasticity. J. Mech. Phys. Solids 41(12), 1825-1857.

Forest, S., Sab, K. (1998) Cosserat overall modeling of heterogeneous materials, Mech. Res. Comm. 25, 449-454.

Hu, G. K. (1996) A method of plasticity for general aligned spheroidal void or fiber-reinforced composites. Int. J. Plasticity 12, 439-449.

Hu, G. K. Liu, X., Xun, F. (2004) Micromechanics of heterogeneous micropolar mediums. Advance in Mechanics (in press, in Chinese)

Kouzeli, M., Mortensen, A. (2002) Size dependent strengthening in particle reinforced aluminium. Acta Materialia 50, 39-51.

Lippmann, H., (1995) Cosserat plasticity and plastic spin. ASME Appl. Mech. Rev. 48, 753-762.

Liu, X. N., Hu, G. K. (2005) A Continue Micromechanical Theory of Overall Plasticity For Particulate Composites Including Particle Size Effect, Int. J. of Plasticity 21, 777-799.

Nowacki, W. (1986) Theory of asymmetric elasticity. Pergamon press.

Ponte Castañda, P. (1991) The effective mechanical properties of nonlinear isotropic composite. J. Mech. Phys. Solids 39, 45-71.

Wei, Y. G. (2001) Particulate size effects in the particle-reinforced metal-matrix composites. Acta Mech. Sinica. 17, 45-58.

Xun, F., Hu, G. K., Huang, Z. P. (2004) Effective in-plane moduli of composites with a micropolar matrix and coated fiber. Int. J. of Solids and Structures 41, 247-265.

Zaoui, A. (2002) Continuum micromechanics: Survey. Journal of Engineering Mechanics-ASCE 128, 808-816.

# ASYMPTOTIC MODEL EQUATION FOR PHASE TRANSITIONS IN A SLENDER CIRCULAR ELASTIC CYLINDER

HUI-HUI DAI
*Department of Mathematics*
*City University of Hong Kong*
*Kowloon, Hong Kong*
*Email: mahhdai@cityu.edu.hk*

**Abstract**

In the literature, people have used *pure* one-dimensional theories to study boundary-value problems of phase-transforming materials. In this paper, we show that to model phase transitions physically and mathematically it is essential to consider the influence from the other dimension(s). For a slender circular cylinder, by taking into account the effects due to the radial deformation and traction-free boundary conditions, we establish the proper model equation through an asymptotic approach, which shows that the problem is a singular perturbation one. The *pure* one-dimensional model equation used in the literature is only the leading order equation valid in the outer regions. Various possible boundary conditions for our model equation are discussed. It is known that the *pure* one-dimensional model equation cannot produce the unique solution. With a proper set of boundary conditions, the model equation provided here may yield the unique solution.

# 1 Introduction

Phase-transforming materials have many applications and many authors have studied various aspects of these types of materials (e.g., Boullay et al. 2002, Ahluwalia and Ananthakrishna 2001, Barsch and Krumhansl 1984, Bales and Gooding 1991, Kartha et al. 1995). In the continuum scale, these materials may be modeled by strain-energy functions with



*Q. P. Sun and P. Tong (eds.), IUTAM Symposium on Size Effects on Material and Structural Behavior at Micron- and Nano-Scales,* 195–204.

multiple local minima (Ball and James 1992) or in general by not-strictly convex strain-energy functions. It is well-known that one important and difficult issue in the mathematical modeling of phase-transforming materials is the nonuniqueness of solutions. In a recent article by Abeyaratne et al. (2001), an elegant review on this and other aspects was given. In the classical work by Ericksen (1975), the *pure* one-dimensional model was used to study a two-end boundary value problem. It was found that in the case of the given end displacement within a certain interval there are two solutions. (**Remark:** According to the model used by Ericksen, mathematically there are also solutions with multiple phase boundaries (a family of them), although only the solution with one phase boundary was reported in that paper.) As pointed out by Abeyaratne et al. (2001) the lack of uniqueness of solutions is usually due to the incompleteness in the model. So, what physically important effects are neglected in the *pure* one-dimensional model?

Our view is that the influence from the other dimension must be considered (see Dai 2004). Consider the axial equilibrium equation in an axially symmetrical static problem

$$\frac{\partial S_{zZ}}{\partial Z} + \frac{\partial S_{zR}}{\partial R} + \frac{S_{zR}}{R} = 0, \tag{1.1}$$

where $S_{zZ}$ and $S_{zR}$ are the components of the first Piolar-Kirchhoff stress tensor $\mathbf{S}$, and $(r, \theta, z)$ and $(R, \Theta, Z)$ are the current and reference cylindrical coordinates, respectively. For a thin wire (in experiments, the ratio of the radius and length is about $\frac{1}{60}$), one might think that the last two terms in (2.1) (representing the influence of the radial deformation in the axial direction) are very small and can be neglected. For standard elastic materials, indeed such an approximation is valid (actually, $\frac{\partial S_{zR}}{\partial R}$ and $\frac{S_{zR}}{R}$ are exponentially small for a neo-Hookean material; cf. Dai and Bi (2001)). However, for a phase-transforming material, the stress-strain curve typically has a peak-valley combination. In the loading process, as the external stress approaches the peak value $\sigma_p$, say, it is equal to $\sigma_p^-$. In this case, $\frac{\partial S_{zZ}}{\partial Z} = \frac{\partial S_{zZ}}{\partial \gamma}\gamma_Z$, where $\gamma$ is the axial strain, is exactly equal to zero. Then, the terms, $\frac{\partial S_{zR}}{\partial R}$ and $\frac{S_{zR}}{R}$, no matter how small they are, are dominant terms and cannot be neglected! This implies there must be a radial deformation in the process of phase transformation. In other words, there do not exist one-dimensional stress problems for phase transitions. Thus, to model phase transitions, the influence of the radial deformation should be taken into account.

There are also experimental evidences that the radial deformation is important.

In experiments (Shaw and Kyriakides 1995,1997, Li and Sun 2002), it was observed that after the nucleation stress was reached the nucleation process began and accompanying with it there was a radial contraction (necking). Also, after the two-phase state was formed, the deformation was inhomogeneous, one part being thin and one part being thick. Thus, to model the nucleation process and the inhomogeneous deformation of different thicknesses, one should consider the radial deformation. Indeed, the solutions provided by Ericksen (1975) based on one-dimensional stress model cannot describe the nucleation process.

In the experiments on seven specimens with different diameters by Sun et al. (2000), some strong size effects were observed: The smaller the diameter is, the larger the nucleation stress is; also, the smaller the diameter is and the sharper the stress drop (taking place after the nucleation stress) is. Thus, to capture these size effects one has to consider the diameter, which implies the consideration of the radial deformation.

In this paper, we shall also give the asymptotic model equation for phase transitions in a slender circular cylinder composed of a phase-transforming material, which takes into account the influence of the radial deformation and traction-free boundary conditions. The model equation shows that the problem is a *singular* perturbation one and the *pure* one dimensional model used in the literature is only valid in the outer region at the leading order. Many people (e.g., Miyazaki et al. 1981, Shaw and Kyriakides 1995 & 1997) conducted experiments on thin wires composed of phase-transforming materials and for a given end displacement only a unique deformation state was observed. We believe that the model equation provided in this paper will lead to the unique solution with proper boundary conditions. A complete mathematical study on this nonlinear equation under various boundary conditions and the constitutive stress-strain curves as well as the structures of the phase boundaries will be carried out in the near future.

# 2 Three-dimensional Field Equations

Based on the results given in section 1, it can be seen that to model phase transitions in a slender cylinder, it is essential to take into account the influence of the radial deformation. Here, we consider the axisymmetrical deformation of a circular cylinder of radius $a$ and length $l$ ($a/l$ is small) composed of a phase-transforming material whose strain energy function is nonconvex. The material is also assumed to be isotropic and isothermal and thus the strain energy function $\Phi = \Phi(I_1, I_2, I_3)$, where $I_i$ $(i = 1, 2, 3)$ are the principal invariants of the Cauchy-Green strain tensor.

In the case of a uniform stretch of the cylinder it is possible to write the average strain energy function $\frac{1}{\pi a^2}\int_0^{2\pi}\int_0^a \Phi(I_1, I_2, I_3) R dR d\Theta = \bar{\Phi}(\bar{\gamma})$ and the average axial stress $\bar{\sigma} = \frac{\partial \bar{\Phi}}{\partial \bar{\gamma}}$, where $\bar{\gamma}$ is the axial strain. So, for a given $\Phi$ the form of the function $\bar{\Phi}(\cdot)$ can be determined. As a result of the nonconvexity of the strain energy function, the $\bar{\sigma} - \bar{\gamma}$ curve has a peak-valley combination. The governing equations according to the three-dimensional theory are

$$\mathrm{Div}\mathbf{S} = 0. \tag{2.1}$$

In the above, one of the equations is just the axial equilibrium equation (1.1) and another is the radial equilibrium equation. We shall also consider the case that the lateral surface is free of traction. Thus, we have the boundary conditions:

$$S_{rR}|_{R=a} = 0, \quad S_{rZ}|_{R=a} = 0. \tag{2.2}$$

For a given strain energy function $\Phi$ the first Piolar-Kirchhoff stress tensor can be calculated through $\mathbf{S} = \frac{\partial \Phi}{\partial \mathbf{F}}$, where $\mathbf{F}$ is the deformation gradient. Then, for small strains it is possible to expand this relation up to any order (see Fu and Ogden 1999). Then, two equations together with two boundary conditions for the two unknowns $U$ (radial displacement) and $W$ (axial displacement) can be explicitly written out.

# 3 Asymptotic Model Equation

Mathematically, it will be too hard to study the two-dimensional field equations directly. In this section, we shall use the smallness of the radius-length ratio to obtain the asymptotic equation which takes into account the influence of the radial deformation and the traction-free boundary conditions.

An approach, which is very similar to that used in Dai and Huo (2002) and Dai and Fan (2004) for nonlinear dispersive waves in rods composed of standard nonlinear elastic materials, can be adopted to deduce the asymptotic equation. Here, we shall only describe the main steps (the lengthy expressions, many are as long as one page, are omitted). The first step is to introduce an essential change of variables for the radial displacement $U$ and the radial variable $R$ by

$$U = RV, \quad Y = R^2. \tag{3.1}$$

Consequently, we find that in the resulting equations and boundary conditions $R$ does not appear explicitly or implicitly in the form of $\sqrt{Y}$. As a result, the two unknowns $V$ and the axial displacement $W$ can be regarded as functions of $Z$ and $Y$. To conduct an asymptotic analysis, it is needed to identify the small quantities, which characterize this problem, through a nondimensionalization process. By a careful analysis, we introduce that

$$W = hw, \quad Z = lx, \quad Y = l^2 s, \quad V = \frac{h}{l} v, \tag{3.2}$$

where $h$ is a characteristic axial displacement. Consequently, two small parameters $\nu = a^2/l^2$ and $\epsilon = h/l$ (usually the maximum displacement among the material points in the specimen is substantially smaller than the total length) and a small variable $s = Y/l^2$ arise. We note that $0 \leq s \leq \nu$. These three small quantities play essential roles in the asymptotic analysis. So, we can write

$$w = w(x, s; \nu; \epsilon), \qquad v = v(x, s; \nu; \epsilon). \tag{3.3}$$

As $s$ is a small variable, we shall assume that the unknowns have Taylor series in $s$, i.e.,

$$w = w_0(x) + s w_1(x) + s^2 w_2(x) + \cdots, \tag{3.4}$$

$$v = v_0(x) + s v_1(x) + s^2 v_2(x) + \cdots. \tag{3.5}$$

(**Remark:** In the *pure* one-dimensional stress theory, $w = w_0(x)$ and $w_1 = w_2 = \cdots = 0$ and $v_1 = v_2 = \cdots = 0$.) Then, we substitute the above two equations into equations (2.1). For the axial equilibrium equations, by setting the coefficients of $s^0$ and $s^1$ to be zero, two equations containing $w_0, w_1, w_2, v_0, v_1$ are obtained. For the radial equilibrium equation, setting the coefficient of $s^0$ to be zero, one equation containing $w_0, w_1, w_2, v_0, v_1$ is obtained. The boundary conditions (2.2) on the lateral surface provide additional two equations. Thus, we have five equations for five unknowns $w_0, w_1, w_2, v_0, v_1$. If we further use the smallness of $\epsilon$, it is possible to represent $w_1$, $w_2$ and $v_1$ in terms of $w_0$ and $v_0$. Substituting these into the last two equations arisen due to the lateral boundary conditions, two equations for $w_0$ and $v_0$ can be obtained. Upon using the smallness of $\nu$ and neglecting terms of $O(\nu\epsilon)$ (for the purpose to get the leading term due to the radial deformation, $O(\nu\epsilon)$ terms, representing the coupling effects of geometry and material nonlinearity, can be neglected), these two equations can be combined to give (after retaining the dimensional variables)

$$\frac{\partial}{\partial Z}[\frac{\partial \bar{\Phi}(\gamma)}{\partial \gamma}] - \frac{a^2}{4} E\gamma_{ZZZ} = 0 \tag{3.6}$$

or

$$[\bar{\sigma}(\gamma)]_Z - \frac{a^2}{4} E\gamma_{ZZZ} = 0, \tag{3.7}$$

where $\gamma = W_{0Z}$, $E$ is the Young's modulus. Equation (3.6) or (3.7) is the proper model equation for phase transitions in a slender circular cylinder. The above equation can also be obtained through the following argument (this is very similar to an argument used to study nonlinear waves in fluids; see Whitham 1974). The idea to obtain the model equation is to consider which terms should be present. For a slender cylinder, the dominant equation is the axial equation. For that equation, obviously the term $[\bar{\sigma}(\gamma)]_Z$ should be present (we require the model equation to be able to produce the correct solution for a uniform stretch of the cylinder before the phase transformation takes place). Other term(s) should represent the influence due to the radial deformation. The leading term due to the effect of the radial deformation in the axial deformation should be a linear term. Thus, to deduce that term is sufficient to work with linear three-dimensional field equations. In Dai and Huo (2002) and Dai and Fan (2004), we have worked with nonlinear (up to the second-order material nonlinearity) three-dimensional field equations. Thus, as long as we neglect nonlinear terms in that two papers, we can get the required term. Indeed, from (35*a*) to (36*b*) of Dai and Fan (2004), we can easily deduce that the required term is $-\frac{a^2}{4}E\gamma_{ZZZ}$. However, the complete derivation as described above can have the advantage to give the precise relations how $w_1, w_2, v_0$ and $v_1$ depend on $w_0$. These are very important for boundary value problems, since the proper boundary conditions may have to be proposed according to $w_1, w_2, v_0$ and $v_1$.)

From (3.7), we can immediately observe the defect of the *pure* one-dimensional model $[\sigma(\gamma)]_Z = 0$ used in the literature. For small $a$, (3.7) represents a singular perturbation problem (in the front of the highest derivative term there is a small parameter). The equation $[\sigma(\gamma)]_Z = 0$ is the correct equation in the outer region valid to the leading order. No matter how small $a$ is, in the inner regions (boundary layer regions, which should contain at least the region containing the phase boundary), the highest derivative term plays an essential role and cannot be neglected.

Equation (3.6) (or (3.7)) is a fourth-order nonlinear ODE for $W_0$ and four boundary conditions are required. A variety of boundary conditions are possible in reality. Here, we list some possibilities: 1. Given both $W_0$ and $V_0$ at two ends. 2. Given $W_0$ and $\gamma_Z = 0$ (i.e., the free end condition) at two ends. 3. Given $W_0$ at $Z = 0$, the axial resultant and $V_0$ at two ends. 4. Given $W_0$ at $Z = 0$, the axial resultant and $W_{0Z} = C$ at two ends (it is expected that depending on the value of $C$ (whether it is in the austensite, martensite or the softening part corresponding to the constitutive curve) completely different results will be obtained). 5. Given $W_0$ and $V_0$ at $Z = 0$ and $W_0$ and $W_{0Z}$ at $Z = l$.

It should be noted that (3.7) is a fully (not simply second order) nonlinear equation and a slight change of boundary conditions can result in completely different situations. We also point out that not all the boundary conditions can yield the unique solution. It is expected that in the cases of the boundary conditions based on the axial deformation alone there are no unique solutions. To get unique solutions, at least one boundary condition should be based on the radial displacement (or radial strain) at one end. In the cases of unique solutions, the results may be used to explain some phenomena observed in experiments analytically. (**Remark:** In particular, the first set of boundary conditions corresponds to the real situations in experiments and we also expect that the unique solution will be obtained by using it.) Complete studies on this model equation together various stress functions $\bar{\sigma}(\gamma)$ and boundary conditions will be carried and the results will be reported in the near future.

# 4 Conclusions

Due to the non-existence of one-dimensional stress problems for phase-transforming materials, it is essential to consider the influence of the radial deformation for phase transition problems in a circular rod (bar) and as well the traction-free boundary conditions at the lateral surface. We have adopted an asymptotic approach to obtain the model equation for static problems, which takes into account the effect of the radial deformation. It turns out that the model equation represents a singular perturbation problem and the pure one-dimensional model is only the leading order equation in the outer regions. We expect that with proper boundary conditions this model equation will yield the unique solution. For more difficult dynamical problems in phase transitions, the uniqueness of solutions is still a problem which has not been solved completely. By considering the effects from other dimension(s) in the axial equation, it may be possible to derive an additional condition besides the two jump conditions across the phase boundary. The results for dynamical problems of a slender cylinder composed of a compressible phase-transforming material will be reported elsewhere.

Finally, we point out that our conclusion on the non-existence of one-dimensional stress problems for phase-transforming materials is purely based on the fact that there is a maximum/minimum in the constitutive stress-strain curve. It seems that for other physical problems if the constitutive relations have the same critical feature, the non-existence of one-dimensional problems might also hold.

## References

Abeyaratne, R., Bhattacharya K., and Knowles, J.K. (2001) **in:** *Nonlin ear Elasticity: Theory and Applications* (Fu,Y., and Ogden, R. (eds.)), pp. 433-490, Cambridge University Press: Cambridge.

Abeyaratne, R., and Knowles, J.K. (1991) *Arch. Rational Mech. Anal.* **114**, 119-154.

Abeyaratne, R., and Knowles, J.K. (1993) **in:** *Shock Induced Transitions and Phase Structures in General Media* (Dunn, J.E., Fosdick, R.L., and Slemrod, M. (eds.)), Springer-Verlag IMA Volume 52, 3-33.

Abeyaratne, R., and Knowles, J.K. (2000) *J. Appl. Phys.* **87**, 1123-1134.

Ahluwalia, R., and Ananthakrishna, G. (2001) *Phys. Rev. Lett.* **86**, 4076-4079.

Bales, G.S., and Gooding, R.J. (1991) *Phys. Rev. Lett.* **67**, 3412-3415.

Ball, J.M., and James, R.D. (1992) Phil. Trans. *R. Soc. Lond. A* **338**, 389-450.

Barsch, G.R., and Krumhansl, J. (1984) *Phys. Rev. Lett.* **53**, 1069-1072.

Boullay, Ph, Schryvers, D., and Kohn, R.V. (2002) *Phys. Rev. B* **64**, paper 144105.

Dai, H.-H. (2004) *C. R. Adca. Paris Ser. I* **338**, 981-984.

Dai, H.-H., and Q. Bi, Q. (2001) *J. Mech. Appl. Math.* **54**, 39-56.

Dai, H.-H., and Fan, X. (2004) *Math. Mech. Solids.* bf 9, 61-79.

Dai, H.-H., and Huo, Y. (2002) *Acta Mech.* **157**, 97-112.

Ericksen, J.L. *J. Elasticity* **5**, 191-201.

Favier, D., Liu, Y., Orgeas, L., and Rio, G. (2001) **in:** *Proceedings of the IUTAM Symposium on Mechanics of Martensitic Phase Trans formation in Solids* (Sun, Q.P. (ed.)), pp. 205-212, Kluwer Academic Publishers:Boston.

Fu, Y., and Ogden, R.W. (1999) *Continuum Mech. Thermodn.* **11**, 141-172.

Kartha, S., Krumhans, J., Sethna, J.P., and Wickham, J. (1995) *Phys. Rev. B* **52**, 803-822.

Knowles, J.K. (2002) *SIAM J. Appl. Math.* **62**, 1153-1175.

Miyazaki, S., Imai, T., Otsuka, K., and Suzuki, Y. (1981) *Sripta Met.* **15**, 853-856.

Li, Z.Q., Sun, Q. P. (2002) *Inter J. Plasticity* **18**, 1481-1498.

Shaw, J.A., and Kyriakides, S. (1995) *J. Mech. Phys. Solids* **43**, 1243-1281.

Shaw, J.A., and Kyriakides, S. (1997) *Acta Materialia* **45**, 683-700.
Shaw, J.A., and Kyriakides, S. (1998) *Int. J. Plasticity* **13**, 837-871.
Sun, Q.P., Li, Z.Q., and Tse, K.K., (2000) **in**: *Proc. of IUTAM Symposium on Smart Structures and structure systems* (U. Gabbert, U., and Tzou, H. S. (eds.)), pp. 1-8, Kluwer Academic Publishers: Boston.
Whitham, G.B. (1974) *Linear and Nonlinear Waves*, Wiley: New York.

# EFFECT OF PHASE TRANSITION ON THE UNUSUAL MICROWEAR BEHAVIOR OF SUPERELASTIC NITI SMA

LINMAO QIAN[1#], QINGPING SUN[1*] and XUDONG XIAO[2]
[1]Department of Mechanical Engineering, The Hong Kong University of Science and Technology, Hong Kong, China
[2]Department of Physics and Institute of Nano Science and Technology, The Hong Kong University of Science and Technology, Hong Kong
* To whom correspondence should be addressed. E-mail: meqpsun@ust.hk

## 1. Introduction

Nickel titanium (NiTi) shape memory alloy (SMA) can exhibit the well-known shape-memory effect and superelastic behavior through its unique reversible phase transition between the low-temperature martensite phase and the high-temperature austenite phase [1]. Owing to its large recoverable deformation, long fatigue life and high work output per unit volume, NiTi SMA has attracted strong interest in the field of micro-electromechanical systems (MEMS) [2]. The understanding and control of tribological behaviors of the material become an important issue of concern [2-4]. In most applications of NiTi, the stress-induced phase transition process is unavoidable and therefore may play an important role in deformation and wear of materials. Research on wear behavior of NiTi SMA started to gain momentum only in recent years and some important results have been obtained [5-7], with more journal papers on the wear behavior of NiTi appearing in the literature [8-9]. However, most of the previous studies were mainly conducted by traditional pin-on-disk type machines under a relatively large normal load; the quantitative role of phase transition as well as its interplay with plasticity in deformation and wear at micrometer or submicrometer scales are still not entirely clear.

This paper focuses on the microwear behavior of NiTi and it is organized as follows. In section 2, the materials and testing methods are described. In section 3, the temperature dependent constitutive relations of the material during loading which involves both stress-induced phase transition and plasticity are first characterized. Detailed experimental results for the wear performance and hardness of the material measured at different temperatures are reported. The observations are analyzed in terms of the intrinsic temperature dependent constitutive law of the material, where the role of phase transition and its interaction with plasticity in the observed unusual wear

[#] Also with the Department of Physics, The Hong Kong University of Science and Technology, Hong Kong, China. Permanent address: Tribology Research Institute, Southwest Jiaotong University, Chengdu 610031, China.



*Q. P. Sun and P. Tong (eds.), IUTAM Symposium on Size Effects on Material and Structural Behavior at Micron- and Nano-Scales,* 205–216.

performance is emphasized. The results are further quantified in a simple contact model for indentation and wear. The final conclusions are given in section 4.

## 2. Materials and Testing Methods

Commercial 0.3 mm thick superelastic NiTi polycrystalline cold-rolling sheets were purchased from Shape Memory Applications, Inc. (San Jose, CA, USA). The nominal alloy composition was Ni-56.4% Ti-43.6% (weight %). The size of the grains is about 50~100 nm as observed by transmission electron microscopy [10]. With a differential scanning calorimeter (DSC 92, SETARAM, France), the characteristic transformation temperatures indicate that the material is in the austenite phase at room temperature [5]. The tensile tests of the sheet sample were performed by a Universal Testing Machine with a temperature chamber (MTS SINTECH 10/D, USA). For wear tests at microscale, a NiTi plate was cut into 8 × 8mm pieces by a diamond saw. A series of silicon carbon and aluminum oxide sand papers were used to polish the samples until a satisfactory surface quality was obtained.

All the indentation and wear tests were conducted on a triboindenter (Hysitron Inc., Minneapolis, USA) at temperatures ranging from 22 to 120°C with Berkovich diamond tips. For a given load and temperature, the microwear was performed by 200 cycles of scanning-scratch over a fixed 5μm × 5μm area on the surface of sample. One cycle of scanning-scratch consisted of 256 lines of scratch with a 10μm/s scratching speed. After every five cycles' of scanning-scratch, one in-situ AFM image was taken over a 10μm × 10μm area with the wear mark in the center. The wear depth was measured as the average height difference between the worn area and its surrounding flat surface.

## 3. Results and Discussion

### 3.1 PHASE TRANSITION AND PLASTICITY IN THE CONSTITUTIVE RESPONSE OF NITI

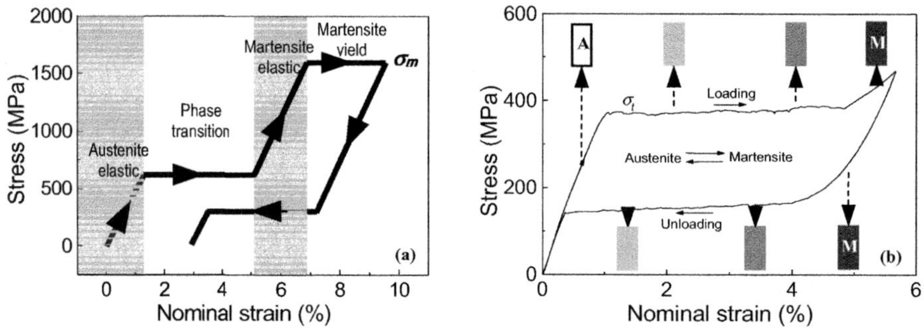


*Figure 1.* (a) Schematic illustration of the four deformation process during loading. (b) Typical measured recoverable stress-strain curve of NiTi at room temperature without plastic deformation.

Fig. 1(a) schematically shows a typical stress versus strain curve of NiTi under tensile loading and unloading test. If the specimen is unloaded soon after the completion of the phase transition, both the elastic and phase transition deformations will recover during unloading after a hysteresis as shown in Fig. 1(b).

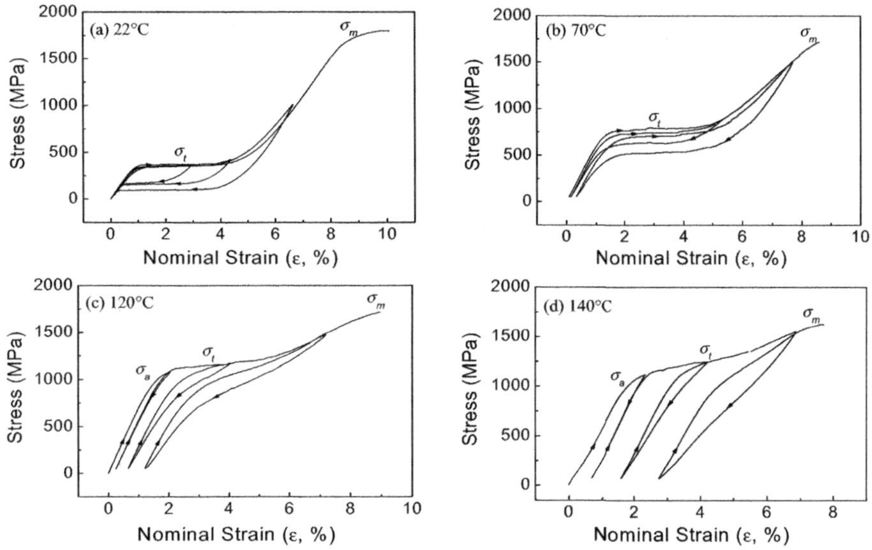


*Figure 2.* Repeated stress versus strain curves at various peak tensile strains measured at different ambient temperatures: (a) 22°C, (b) 70°C, (c) 120°C and (d) 140°C.

Since the stress-induced phase transition in NiTi is thermoelastic in nature [1], there should be a clear temperature dependence of the transformation stress of the material. At each temperature, the tensile loading and unloading were repeated with different maximum strains to study the contribution of plasticity and phase transition to the total strain. At 22°C and 70°C, no residual strain was observed at the first tensile loading/unloading cycle as shown in Fig. 2 (a) and (b), which indicates that the phase transition stress was below the austenite plastic yield stress. At higher temperatures such as 120°C and 140°C (see Fig. 2 (c) and (d)), since the phase transition stress became higher than the austenite plastic yield stress, plastic deformation occurred prior to the start of the phase transition. The residual strain was immediately observed when the stress is over the linear elastic limit at the first and second cycles (Fig. 2 (c) and (d)).

The tensile stress-strain curves for the first loading-unloading cycle at various temperatures are plotted in Fig. 3 (a). It was found that both the phase transition stress plateau and austenite elastic modulus increase with temperature. Furthermore the plastic deformation of austenite started to evolve with phase transition at about 90 °C (also from Fig. 2 (c) and (d)). Fig. 3 (b) shows that while the austenite and martensite plastic yield stresses are almost kept constant in the entire temperature range, the phase transition stress increases linearly with temperature.

To summarize, the stress-induced phase transition process in NiTi is not only strongly temperature dependent but also intertwined with plasticity. As will be unfolded in detail in the following sections, this feature has important implications and may play a key role in the unusual temperature dependence of the hardness and wear performance of the material during indentation and repeated scanning-scratch.

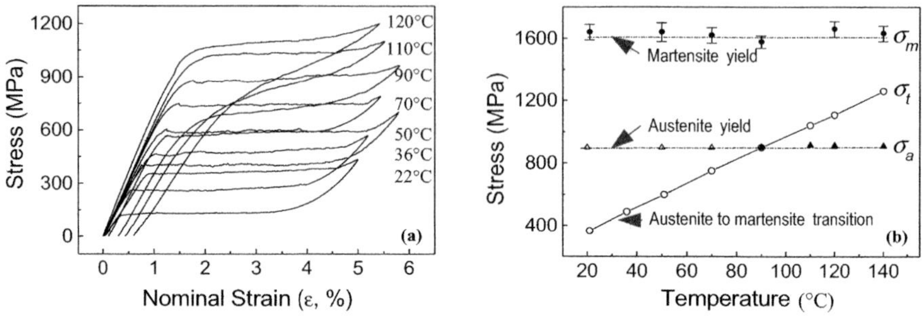


*Figure 3.* (a) The tensile stress-strain curves for the first loading-unloading cycle at various temperatures. (b) Temperature effect on the measured martensite transition stress, martensite and austenite yield stress.

## 3.2 MICROWEAR PROPERTIES OF NITI AT ROOM TEMPERATURE — THRESHOLD LOAD AND WEAR MODE

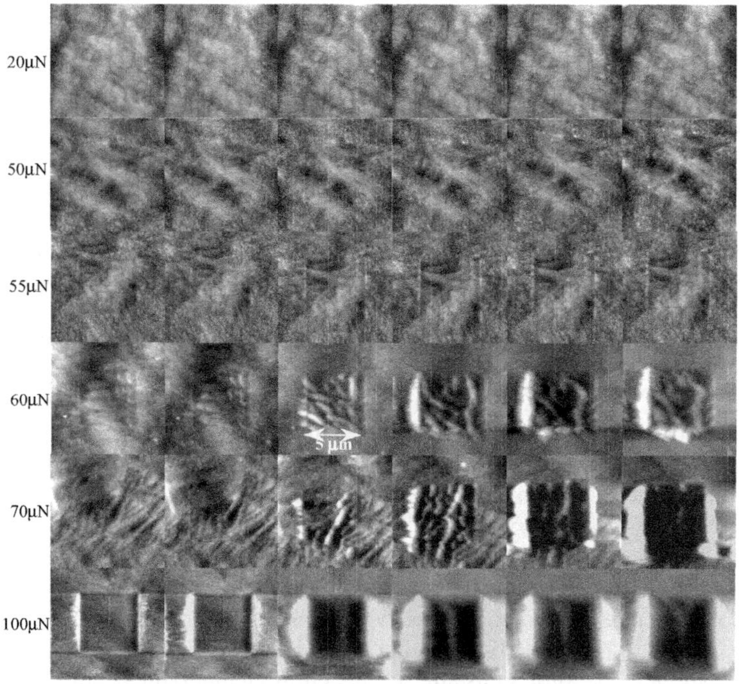


*Figure 4.* In-situ AFM images of the wear marks on NiTi corresponding to various scanning-scratch cycles (5~200 cycles) and normal loads (20~100μN). The microwear is conducted by repeated scanning-scratch cycles over a 5μm × 5μm area. The image size is 10μm × 10μm and the tip curvature radius is 5.7μm.

Figure 4 records the typical wear process of NiTi by in-situ AFM images corresponding to various scanning-scratch cycles and loads. It is seen that there existed an apparent *threshold load* for NiTi. When the normal loads fell below 55μN, even after 200 cycles of scanning-scratch, no obvious material loss happened. However, when the load was increased to above 60μN, serious wear occurred. According to the plough mechanism of abrasive wear, the so-called *threshold load* for wear should be related to the condition for the occurrence of plasticity in the indenter tip contacted region of the sample. Since the stress in the contact region is strongly influenced by the tip shape and the amount of stress-induced phase transformation during indentation (which in turn depends on the temperature and applied load), it must be emphasized that the *threshold load* should not be treated as an intrinsic material property, instead it is a comprehensive nominal wear property of the material. For a given tip shape, the temperature dependence of the phase transition stress implies that the *threshold load* is also temperature dependent. This will be elucidated in section 3.4.

The above discussion on the influence of phase transition to the threshold load can be further supported by comparing the wear test results of NiTi at room temperature with that of stainless steel 304. Indentation result shows that NiTi has the same hardness as that of stainless steel. However, NiTi exhibits much better microwear resistance than steel as shown in Fig. 5. This phenomenon is anomalous according to the existing wear theory which states that wear property should be similar for metal materials with the same hardness [11]. It is speculated that the unusual superior wear performance of NiTi over stainless steel may be attributed to the superelastic behavior of the material due to its unique phase transition property.

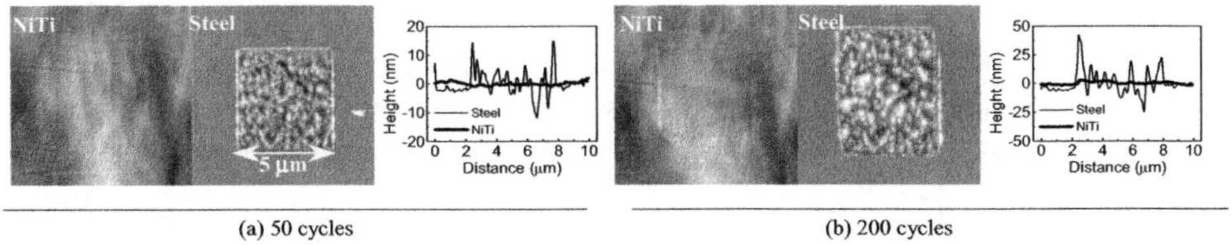


*Figure 5.* Comparison of the microwear property of NiTi and stainless steel-304. The normal load is 40μN and the tip curvature radius is 5.2μm. (a) The comparison of the wear marks on NiTi and stainless steel after 50 scanning-scratch cycles. The right figure shows the cross section profile of the wear marks at the center line. (b) Comparison of the wear marks after 200 scanning-scratch cycles.

## 3.3 ROLE OF PHASE TRANSITION AND PLASTICITY IN ROOM TEMPERATURE WEAR — CONTACT MECHANICS ANALYSIS

Since wear originates from the accumulation of plastic deformation and damage in the contact region, to obtain a quantitative estimation of the threshold wear load $P_c$, we apply the Hertzian contact theory to steel and NiTi, respectively.

For the stainless steel wear test the Berkovich tip's curvature $R$ is 5.2μm and the diamond tip is assumed to be rigid, setting the maximum stress in the contact region to be the plastic yield stress $\sigma_y$, we can obtain $P_c$ as [12]

$$P_c = 4.43R^2 \frac{\sigma_y^3}{E^2} \cdot \qquad (1)$$

The elastic modulus $E$ measured is 200GPa and $\sigma_y$ is roughly 1/3 of its indentation hardness [13] and therefore is about 1330MPa. From Eq. (1) $P_c$ for the steel is 6.5μN. This $P_c$ is far below the experimentally applied load (=40μN) for the wear test in Fig. 5, thus it is not surprising that serious wear occurs immediately after 50 cycles of scanning-scratch on the steel.

For NiTi, the phase transition stress is lower than the austenite plastic yield stress and NiTi will be transformed to martensite before austenite plastic yield (Fig. 2(a)) at room temperature [5]. From Eq. (1), the load corresponding to the start of phase transition is only 10μN. If we assume no phase transition before plastic yield, the plastic yielding (i.e., wear) is expected to start at a load of 40μN. However, no obvious wear is observed at 40μN on NiTi even after 200 cycles of scanning-scratch in the experiments (Fig. 5). Therefore the great increase in $P_c$ (here the measured value is 50 μN at 22°C) is mainly due to the additional shielding effect by the reversible austenite-martensite phase transition in the near tip region.

Due to the phase transition process during the indentation, the projected area $A_c$ of NiTi (with the area due to linear elastic unloading deducted as defined [14]) can be divided into two regions (inset of Fig. 7): the phase transition region $A_t$ and the martensite plastic yield region $A_m$ [5] (notice that $A_t$ overlaps and is larger than $A_m$). To estimate the contribution of the phase transition to the measured hardness of NiTi, we introduce the recovery ratio $\eta$, defined as the ratio of $A_t/A_c$. The areas $A_t$ and $A_m$ can then be related to $\eta$ by:

$$A_t = \eta A_c, \qquad A_m = (1-\eta)A_c \cdot \qquad (2)$$

During indentation, the normal pressure on $A_t$ is about three times the uniaxial compressive transition stress $\sigma_t'$ and on $A_m$ is about three times the uniaxial compressive plastic yield stress $\sigma_m'$ of martensite [13]. In the triboindenter, hardness $H$ is defined as:

$$H = \frac{P_{\max}}{A_c}, \qquad (3)$$

where $P_{max}$ is the peak load which can be expressed by the rule of mixture as [15]

$$P_{\max} = 3\sigma_t' A_t + 3\sigma_m' A_m \cdot \qquad (4)$$

Here, because of the asymmetry of martensite deformation during compression and tension, $\sigma_t'$ and $\sigma_m'$ are about 1.5 times of tensile transition stress $\sigma_t$ and tensile plastic yield stress $\sigma_m$, respectively [16]. Combining Eqs. (2)-(4), we obtain

$$H = 4.5\eta\sigma_t + 4.5(1-\eta)\sigma_m \cdot \qquad (5)$$

There are two ways to estimate the value of the geometric quantity $\eta$. One is from the indentation force versus indentation depth curve [5] and the other is by the dimensional analysis and the concept of geometric similarity for indentation [12]. Here we use the second one. For the Berkovich indenter used (or any sharp indenter such as a pyramidal or conical indenter) the total strain distribution in NiTi under indentation is determined by the face angle and independent of indentation load or depth [12]. Therefore we would have the same strain distribution after scaling even if the material experiences two stage (phase transition followed by plastic yielding) deformations during indentation under various loads. For a given point in the material whether it is at the phase transition stage (plateau) or enters into plastic yielding is completely determined by its strain state (or equivalent strain), so the geometrical quantity $\eta$ will be the same under different indenter loads (depth). With the measured values of $H$, $\sigma_t$ and $\sigma_m$ at room temperature, $\eta$ equals to 0.59 by Eq. (5), which indicates that the phase transition area is about 60% of the project contact area. Again, that $\eta$ is independent of depth was affirmed by the measured values of $H$ which is almost a constant (for indenter depths > 50 nm, plateau part of Fig. 6). In summary, the above analysis shows that the measured nominal indentation hardness of the superelastic NiTi alloy is no longer a measure of a single material property such as plastic yield stress; rather it has two contributions (two plateaus in the stress-strain curve of NiTi) and thus is a comprehensive indicator of the material's ability against plastic deformation and phase transition.

## 3.4 HARDNESS AND MICROWEAR OF NITI AT HIGH TEMPERATURES — THE INTERPLAY OF PHASE TRANSITION AND PLASTICITY

Since phase transition property is temperature dependent, to further explore and reveal the effect of phase transition on wear of NiTi, it would be ideal to perform the wear and hardness tests of NiTi at various temperatures. The results of such tests are summarized as follows.

### 3.4.1 *Hardness Variation with Temperature*

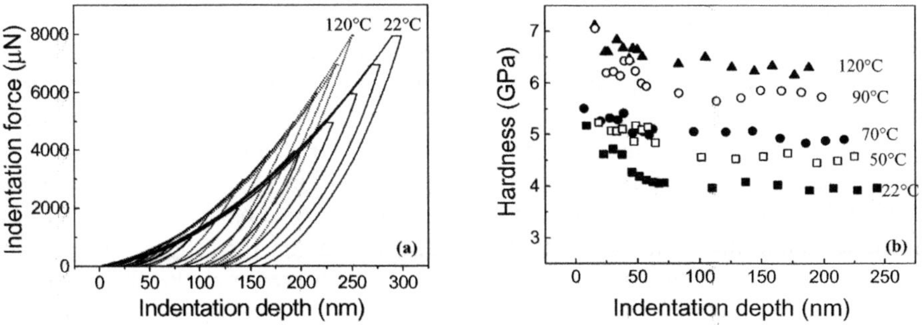


*Figure 6.* (a) Indentation curves on NiTi at various peak loads measured at 22°C and 120°C. (b) The hardness of NiTi as a function of indentation depth at various temperatures (22°C ~120°C).

The hardness of NiTi was measured by indentations at loads ranging from 100μN to 8000μN and at temperatures from 22°C to 120°C. Fig. 6(a) shows the indentation curves with various peak loads at 22°C and 120°C, respectively. From these indentation curves, hardness can be deduced and the results are collected in Fig. 10(b). It is clear that for all indentation depths the hardness increased almost linearly with temperature.

It is well known that the hardness of normal metals usually decreases with an increase in temperature because of the decreasing critical stress required for dislocation motion [17]. Here the trend for NiTi is exactly the opposite. Again we may directly link this phenomenon with the unique temperature dependence of the stress-induced phase transition process involved in the indentation.

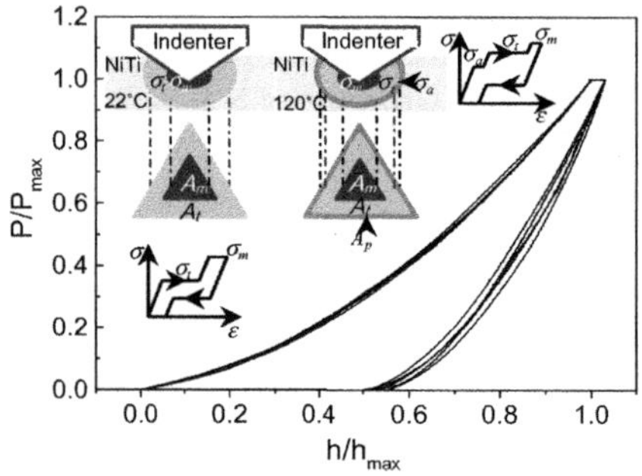


*Figure 7.* The non-dimensional indentation curves on NiTi at various temperatures (22°C ~120°C). The inset on the upper left is a schematic showing indentation in NiTi at 22°C, where the contact region consists of the martensite plastic yield zone $A_m$ and phase transition zone $A_t$. The inset on the upper right is a schematic showing indentation in NiTi at 120°C. Besides $A_m$ and $A_t$, the contact region also includes austenite yield zone $A_p$.

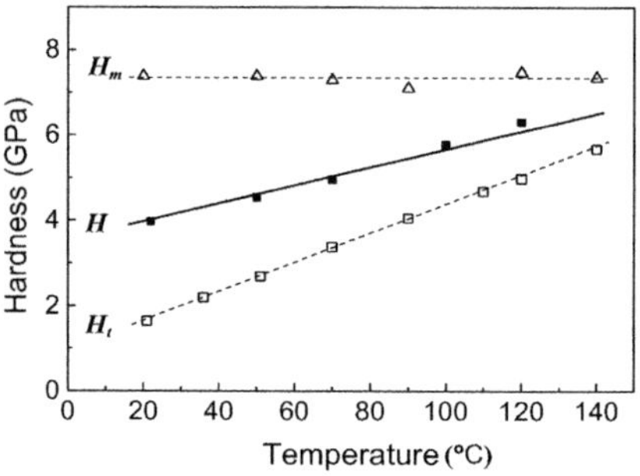


*Figure 8.* The hardness $H_m$ corresponding to plastic yield of martensite (open triangle point), the hardness $H_t$ corresponding to phase transition stress $\sigma_t{}'$ (open square point) and NiTi indentation hardness $H$ (solid square) as a function of temperature. The solid line shows the calculation of the hardness of NiTi by eq. (8). The dot lines are the linear fit of $H_m$ and $H_t$.

We must be aware that, with an increase in temperature, there is an interchange in the order of the two physical processes (i.e., austenite → martensite phase transition versus plastic deformation) in the indenter region. As discussed in section 3.3, at 22°C, the wear occurs only in the core plastic deformed region inside the transformed martensite phase region (as shown in Fig. 7). This situation will change when the test temperature is moved to a high temperature regime, for example, at or above 70°C, since the austenite yield stress is lower than the transition stress, austenite will enter plastic yield before phase transition (Fig. 7 and Table 1). As a result, the wear will start from the austenite phase at 70°C and 120°C and the region beneath the indenter is a fully plastic deformed region. Now the projected contact area includes three parts: the only austenite plastic yield area $A_p$, the phase transition area $A_t$ and the martensite plastic yield area $A_m$. Setting $A_p = \lambda A_c$ and $A_t = \eta A_c$, then $A_m = (1-\lambda-\eta)A_c$. Therefore, the hardness $H$ of NiTi can be determined by:

$$H = \lambda H_a + \eta H_t + (1-\lambda-\eta)H_m \,. \tag{6}$$

Here, $H_a$ is the hardness corresponding to the austenite plastic yield stress under compression ($H_a = 3\sigma_a'$), $H_t$ is the hardness corresponding to the phase transition stress under compression ($H_t = 3\sigma_t'$) and $H_m$ is the hardness corresponding to the martensite plastic yield stress under compression ($H_m = 3\sigma_m'$), respectively. This contact model and eq. (6) is valid for the entire temperature range. When the temperatures are below 70°C there is no austenite plastic yield before phase transition, so $\lambda$ equals to zero and eq. (6) becomes eq. (5). When the temperature is above 70°C, $A_p$ can be ignored in the calculation of NiTi's hardness because the austenite plastic yield is a work-hardening process at higher temperatures. Therefore, for temperatures between 22°C~120°C, in a first approximation, eq. (6) can be simplified as:

$$H = \eta H_t + (1-\eta) H_m. \tag{7}$$

Since $\sigma_t'$ increases almost linearly with temperature, $H_t$ in eq. (7) is a linear function of temperature. The dimensionless indenter load-depth curves obtained by scaling the load and indentation depth by their respective maximum values for different temperatures almost collapsed onto a single curve as shown in Fig. 7, which implies that $\eta$ is almost independent of temperature (~0.6) for the present test temperature range. With the independently measured values of $\sigma_t$ and $\sigma_m$ (Fig. 3(b)), the calculated values of the hardness of NiTi by eq. (7) are plotted as the solid line and compared with the values by directly measurement (solid square) in Fig. 8. It is seen that the agreement is very good therefore we can conclude that the increase in hardness with temperature is mainly due to the increase in the transition stress since the martensite plastic yield stress is almost temperature independent (Fig. 3(b)) [5].

TABLE 1. Comparison of the calculated NiTi hardness by eq. (8) with the measured values.

| Temperature (°C) | 22 | 50 | 70 | 100 | 120 |
|---|---|---|---|---|---|
| Transition $\sigma_t$ (GPa) | 0.37 | 0.59 | 0.75 | 0.97 | 1.11 |
| Austenite $\sigma_a$ (GPa) | 0.90 | 0.90 | 0.90 | 0.90 | 0.90 |
| Martensite $\sigma_m$ (GPa) | 1.60 | 1.60 | 1.62 | 1.66 | 1.63 |
| Measured $H$ (GPa) | 3.93 | 4.52 | 5.03 | 5.71 | 6.22 |
| Calculated $H$ (GPa) | 3.93 | 4.51 | 4.98 | 5.64 | 5.95 |

### *3.4.2 Wear Test at Various Temperatures*

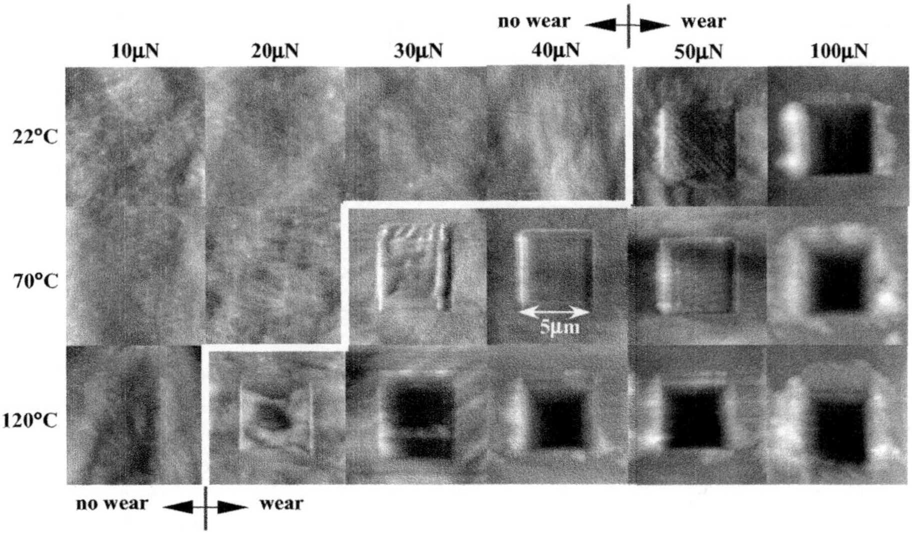


*Figure 9.* Images of the wear marks on NiTi after 200 cycles of scanning-scratch corresponding to various normal loads (10~100μN) and temperatures (22°C~120°C).

At 22°C, 70°C and 120°C, by using a diamond Berkovich tip with 5.2μm tip radius, wear tests were made at loads ranging from 10μN to 100μN. The wear marks after 200 cycles of wear at various loads and temperatures are summarized in Fig. 9, which clearly show that there exists a threshold load at a given temperature.

To quantitatively describe the wear properties of NiTi, the wear depth versus load curves at various temperatures are plotted in Fig. 10(a). The results show the following: (1) Beyond a threshold load value the wear depth increases with load; (2) The threshold load decreases with increasing temperature; (3) For a given load the wear depth/volume of NiTi increases (or the wear resistance decreases) with the increasing temperature. Combing (3) with the hardness results in 3.4.1, it turns out that a higher wear resistance actually corresponds to a lower hardness as shown in Figs. 11 and 12. This anomalous wear-hardness relationship cannot be explained by the existing theory that has been supported by most experimental work for metals [11]. The above temperature effect on the wear performance of NiTi may again be attributed to the unique temperature dependence of phase transition and elastic properties of NiTi. It is necessary to have a detailed discussion as follows.

TABLE 2. Calculated threshold load by Eq. (2) and the measured values.

| Temperature (°C) | 22 | 70 | 120 |
|---|---|---|---|
| Austenite $E$ (GPa) | 45 | 56 | 64 |
| Austenite $\sigma_a$ (GPa) | 0.9 | 0.9 | 0.9 |
| Calculated $P_c$ (μN) | 40 | 27.9 | 21.4 |
| Measured $P_c$ (μN) | 50 | 30 | 20 |

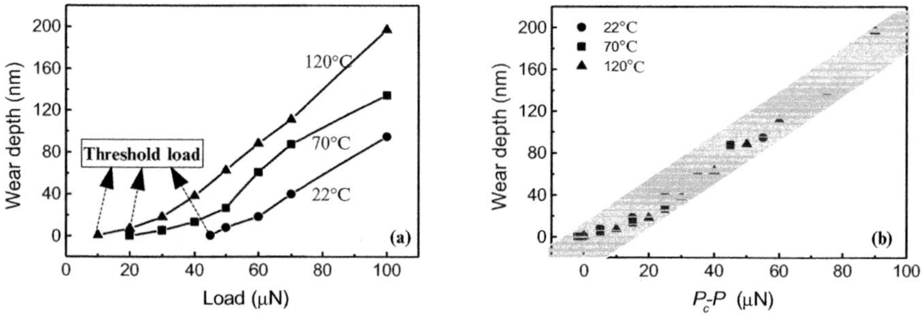


*Figure 10.* (a) Wear depths of the wear marks in Fig. 9 versus normal load at various temperatures. (b) Replot of the wear depth as a function of the "effective normal load" ($P$-$P_c$).

Firstly, the interchange in the order of the two physical processes with temperature in the indenter region (i.e., A→M phase transition versus plastic deformation) plays an important role in the observed change in the threshold loads. As discussed in section 3.3 and 3.4.1, at 22°C, the wear occurs only in the core plastic deformed martensite phase region (as shown in Fig. 7). Compared to the case of no phase transition, the lower austenite modulus and the lower phase transition stress plateau (with a 5% reversible strain) relaxed the stress concentration in the indented region, which provided a stress shielding effect in the near tip region and thus can effectively shift the threshold wear load for the start of plastic flow to a higher value. At high temperatures above 70°C, on the one hand the austenite modulus increased considerably and on the other hand the above transformation shielding effect is much weaker because austenite will enter plastic yield before phase transition (Fig. 7 and Table 1) because the austenite yield stress is lower than the transition stress. As a result, the wear will happen in the austenite phase and the region beneath the indenter will first experience the plastic deformation and then phase transition as shown in Fig. 7. The threshold load $P_c$ of plastic deformation at these temperatures can then be directly estimated by using the Hertzian contact theory (eq. (1)). As shown in Table 2, when the temperature increases from 70°C to 120°C, $P_c$ decreases mainly due to the increase in austenite modulus $E$ (with a constant $\sigma_a$). Thus the decrease in the threshold load with increase in temperature can be well explained by the contact mechanics analysis incorporating the roles of both plasticity and phase transition.

Secondly, when examining the observed apparent anomalous wear-hardness relationship in Fig. 10 (a) and Fig. 11 we noticed that beyond $P_c$ the volume of plastically deformed austenite region increased with the load and therefore the wear increased as shown. This at least means that the actual amount of wear should monotonically increase with the effective normal load ($P$-$P_c$) instead of $P$. If Fig. 10 (a) is re-plotted by wear depth versus "effective normal load" ($P$-$P_c$) in Fig. 10 (b) it becomes clear that all the data points under different temperatures almost fall on a single narrow band. Therefore, we can conclude that the observed apparent anomalous wear-hardness relationship is actually caused by the decrease in the threshold load with temperature (Fig. 12) due to two key factors. One is the increase in Young's modulus of austenite with temperature and the other is the temperature dependent interplay of stress-induced phase transition and plasticity in the contact region of NiTi.

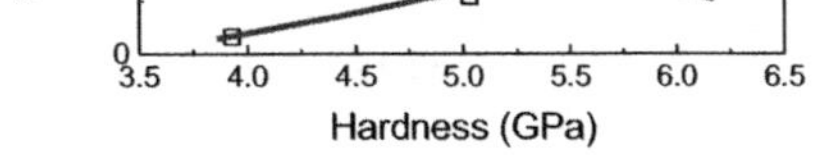


*Figure 11.* Wear depth of the wear marks in Fig. 9 as a function of hardness at various normal loads (50~100μN).

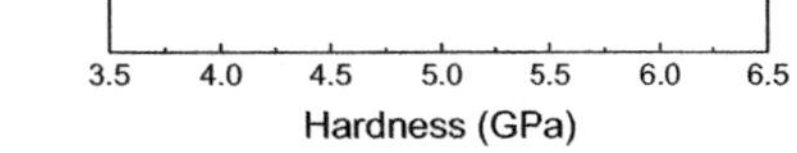


*Figure 12.* The threshold load of the wear of NiTi as a function of hardness.

## 4. Conclusions

Microwear tests on superelastic NiTi SMA were performed with a triboindenter. Excellent microwear performance of the material is reported in this paper where the role of stress-induced phase transition in the deformation and microwear behavior of NiTi has been emphasized and investigated by wear and hardness tests at various temperatures. The obtained results are analyzed in terms of the intrinsic constitutive relations of the NiTi material and by the contact mechanics theory. The microwear of NiTi is the result of the competition between the plasticity and phase transition. The unique reversible transformation that dominates the material's property at room temperature plays a key role in the unusual excellent microwear behavior of the superelastic NiTi alloy. Our experimental results suggest that the lower the transformation stress, the better the wear performance. Therefore "Transition before plastic yielding" becomes the key mechanism in controlling the wear behavior of NiTi.

## Acknowledgment

The authors are grateful for the financial support from the Research Grants Council of Hong Kong (Project No. HKUST 6199/03E and project No. N_HKUST601/01). QLM also would like to acknowledge the partial support from the Sichuan Youth Science & Technology Foundation of China (Project No. 04ZQ026-004) and the Chinese Natural Science Foundation (Project No. 50305029). The authors also wish to acknowledge the financial support of the Hong Kong University of Science and Technology through the William Mong Solid State Cluster Laboratory and the Institute of Nano Science and Technology.

## References

[1] Duerig, T., Melton, K.N., Stockel, D., Wayman, C.M. (1990) *Engineering Aspect of Shape Memory Alloys*, Butterworth-Heinemann, Boston.

[2] Fu, Y.Q., Du, H.J., Huang, W.M., Zhang, S., Hu, M. (2004) TiNi-based thin films in MEMS applications: a review, *Sensor. Actuat. A-Phys.* **112**, 395.

[3] Maboudian, R. (1998) Adhesion and friction issues associated with reliable operation of MEMS, *MRS Bulletin* **23**, 47.

[4] Romig, A.D., Dugger, M.T., McWhorter, P.J. (2003) Materials issues in microelectromechanical devices: science, engineering, manufacturability and reliability, *Acta. Mater.* **51**, 5837.

[5] Qian, L.M., Xiao, X.D., Sun, Q.P., Yu, T.X. (2004) Anomalous relationship between hardness and wear properties of a superelastic nickel–titanium alloy, *Appl. Phys. Lett.* **84**, 1076.

[6] Ni, W., Cheng, Y.T., Grummon, D.S. (2002) Recovery of microindents in a nickel–titanium shape-memory alloy: A "self-healing" effect, *Appl. Phys. Lett.* **80**, 3310.

[7] Liu R. and Li, D.Y. (2000) Experimental studies on tribological properties of pseudoelastic TiNi alloy with comparison to stainless steel 304, *Metall. Mater. Trans. A* **31A**, 2773.

[8] Shida, Y., Sugimoto, Y. (1991) Water jet erosion behavior of Ti-Ni binary alloys, *Wear* **146**, 219.

[9] Clayton, P. (1993) Tribological behavior of a titanium-nickel alloy, *Wear* **162-164**, 202.

[10] Ng, K.L. (2002) *Stress-induced Phase Transformation and Reorientation in NiTi Tubes*, Master Thesis, the Hong Kong University of Science and Technology, Hong Kong.

[11] Bhushan, B. (1999) *Principles and Applications of Tribology*, John Wiley & Sons, Inc., New York.

[12] Johnson, K.L. (1985) *Contact Mechanics*, Cambridge University Press, London.

[13] Tabor, D. (1951) *The Hardness of Metals*, Oxford University Press, London.

[14] Oliver, W.C., Pharr, G.M. (1992) An improved technique for determining hardness and elastic modulus using load and displacement sensing indentation experiements, *J. Mater. Res.* **7**, 1564.

[15] Ekvall, J.C. (1965) Structure Behavior of Monofilament Composites, *Proc. AIAA $6^{th}$ Structures and Materials Conf.*, AIAA, New York, April.

[16] Orgeas, L. and Favier, D. (1998) Stress-induced martensitic transformation of a NiTi alloy in isothermal shear, tension and compression, *Acta Mater.* **46**, 5579.

[17] Szymanski, A., Szymanski, J.M. (1989) *Hardness Estimation of Minerals Rocks and Ceramic Materials*, Elsevier, New York.

# HOMOGENIZATION METHOD FOR STRENGTH AND INELASTIC BEHAVIOR OF NANOCRYSTALLINE MATERIALS

L. CAPOLUNGO[a,b], M. CHERKAOUI[a] and J. QU[b]
[a] *LPMM-CNRS University of Metz, Ile du Saulcy, 57045, Metz, Cedex1 France*
[b] *George W.Woodruff School of Mechanical Engineering, Atlanta Georgia 30332-0405 USA*

## Abstract

A model based on a self consistent scheme is developed to account for the plastic behavior of f.c.c. nanocrystalline materials. The material is assimilated as a composite with two phases: the inclusion phase which represents the grain cores, and the matrix phase which represents both grain boundaries and triple junctions. Both phases exhibit an elastoplatic behavior. In the case of the inclusion phase, the glide of dislocations pilots the plastic deformation, while Coble creep and the mechanism of grain boundary sliding are responsible for the elastoplastic behavior of the matrix phase.

## 1. Introduction

The possibility of improving of the mechanical properties of materials following a refinement in their structure is such that considerable effort has been put into the study of nanostructured materials. In the case of pure polycrystalline materials, a decrease in the grain size is known to engender an increase in the yield stress. This is the well known Hall Petch law which is given by:

$$\sigma_y = \sigma_0 + \frac{K^{HP}}{\sqrt{d}} \tag{1}$$

Here $\sigma_0$ is the friction stress, d is the grain size and $K^{HP}$ is the Hall Petch slope.

Various models have been proposed to account for the evolution of the yield stress with grain size. The first type of model is based on the pile up of dislocations (Hall 1951; Petch 1953). Plasticity occurs when the pile up of dislocations at the grain boundaries is such that it creates a local stress larger than the strength of the grain boundaries. The stress concentration can activate Frank and Read sources in adjacent grains whose orientation is less favorable for plasticity to happen. Materials with b.c.c. structures exhibit the same size dependence of the yield stress. Dislocation pile up does not occur



*Q. P. Sun and P. Tong (eds.), IUTAM Symposium on Size Effects on Material and Structural Behavior at Micron- and Nano-Scales,* 217–226.

in b.c.c. material which suggests that the pile up model is not adequate to describe the phenomenon. The second type of model is based on the density of dislocations in the grain cores (Ashby 1970). The yield stress depends on the density of immobile dislocations also called stored dislocations (Taylor 1934). The latter increases the stress necessary to obtain a motion of the glide plans. In most models no difference is made between mobile dislocations and immobile ones. The total dislocation density evolves dynamically with strain and is related to the grain size via statistics (Kocks 1976). The third type of model is based on the emission of dislocations by grain boundary ledges (Li 1963). The emitted dislocations are supposed to create a Taylor type forest. The density of dislocations is proportional to the angle of misfit between the grain boundaries. The latter is then related to the ledge density. Even though the model was intended to account for the Hall Petch law when a pile up of dislocations is not possible, the influence of the ledge density on the yield stress in f.c.c. materials has been revealed to be non negligible by experimental studies (Murr 1981).

When decreasing the grain size to the nano range, the Hall Petch law exhibits a zero slope or even a negative slope (Conrad et al. 2000; Schuh et al. 2002). This is the so called breakdown of the Hall Petch law. The critical threshold at which the phenomenon occurs is approximately 30nm. The breakdown was very quickly attributed to a change in the structure of grain boundaries. The latter were thought to exhibit an amorphous structure (Birringer 1989). Later, experimental studies have shown that grain boundaries exhibit the same facetted structure without any dependence on grain size (Kumar et al. 2003). Though, triple junctions have been reported to have a less organized structure (Ranganathan et al. 2000).

The current mainstream accounts for the change in the Hall Petch slope in terms of a change in the mechanism dominating the deformation. Hence, most of the current research focuses on identifying the nature of those mechanisms.

In coarse grain materials with an f.c.c. structure, plasticity can be described with the mechanism of glide of dislocations. It is a thermally activated mechanism. When the local stress in the grain cores is larger than the Taylor flow stress, the glide of dislocation can happen, provided thermal fluctuation brings the energy required for the glide to take place. The critical stress to be overcome depends on the dislocation density which evolves via two phenomena: the statistical storage of dislocation which creates work hardening, and the annihilation of dislocations which engenders a recovery of the material. The latter was first referred to as work softening (Stokes et al. 1954).

When the grain size is in the ultrafine range ($d < 100$nm), molecular dynamics (MD) simulations have exhibited the emergence of a different mechanism; the emission of dislocations by grain boundaries (Van Swygenhoven et al. 1999; Yamakov et al. 2001). The simulations also revealed that when the grain size is on the order of 10nm, grain boundaries emit only single partial dislocations, leaving stacking faults in the material. The previous results are now believed to be an artifact of MD simulations, and grain boundaries are supposed to have a perfect dislocation activity (Kumar et al. 2003). Even though MD simulations are characterized by unrealistically high strain rates, the emission of dislocations by grain boundaries is thought to occur and to contribute to the deformation of NC materials even at more common strain rates. It has been shown that the emission of dislocations by grain boundaries can be energetically favorable (Gutkin et al. 2003). Also three types of grain boundary sources have already been identified (1) low angle grain boundaries (2) grain boundary ledges and (3) disclinations in motion.

MD simulations have also shown the possibility of a contribution of the motion of twin boundaries even in high stacking fault energy (SFE) materials (Yamakov et al. 2002). These results must be taken with caution. Twinning usually occurs at very high strain rates and is not supposed to operate in high SFE materials. In contrast, twins have been experimentally observed.

When the grain size is very small, typically $d < 10$nm, dislocation activity ceases. Two types of mechanisms have been proposed to dominate the deformation: (1) rigid motion of grains and (2) diffusion based mechanisms. Grain boundary sliding (GBS) has been identified as a mode of deformation in various MD simulations and experimental studies (Kumar et al. 2003). Grain boundary sliding usually occurs in superplasticity. This mechanism could either be accommodated by diffusion phenomena or not be accommodated. MD simulations predict an accommodation of GBS while experimental results do not. Finally, diffusion phenomena have been proposed to be operating in NC materials. Various paths for vacancies have been examined. Coble creep, which corresponds to the diffusion of vacancies at the interface between grain boundaries and the grain cores, has been experimentally observed at room temperature.

Most models assimilate the NC material as a composite with two phases and are based on simple rules of mixtures (Carsley et al. 1995; Konstantinidis et al. 1998; Kim et al. 2000). Even though it simplifies the analytical development, it also leads to a loss of information.

In the present study, a model based on a self consistent scheme is developed. The material is assimilated as a composite with two phases; (1) an inclusion phase which represents the grain cores and (2) a matrix phases including both grain boundaries and triple junctions. The dislocation glide mechanism is incorporated in the constitutive behavior of the inclusion phase while Coble creep and GBS are incorporated in the constitutive behavior of the matrix.

## 2. The Model

The model is based on a self consistent scheme developed by Cherkaoui et al. (Cherkaoui et al. 2000). Usually, self consistent schemes are based on Eshelby's Solution of the ellipsoidal inclusion. In the latter, the solution of the inclusion problem is solved by taking into account the difference in the plastic strains of the matrix and of the inclusion. In the extended model proposed by Cherkaoui et al., the differences in the elastic constants as well as the one in the inelastic deformation are taken into account.

Both phases are taken as isotropic and not compressible. We suppose perfect bonding between the inclusions and the matrix.

In the following, $\Sigma_{ij}$ and $\dot{\Sigma}_{ij}$ denote the global stress tensor and its rate, $E_{ij}$ and $\dot{E}_{ij}$ denote the global strain tensor and its rate, $\sigma_{ij}$ and $\dot{\sigma}_{ij}$ denote the microscopic stress tensor and its rate, $\varepsilon_{ij}$ and $\dot{\varepsilon}_{ij}$ denote the microscopic strain tensor and its rate. The superscript I will refer to the inclusion phase while the superscript M will refer to the matrix phase.

## 2.1. THE INCLUSION PHASE

Although not necessary, it is assumed here for convenience that the crystals in the NC materials are all spherical. The volume fraction of inclusions is given by

$$f(d) = \left(\frac{d-w}{d}\right)^3 \tag{2}$$

where d is the grain size and w is the grain boundary thickness.

The inclusion phase has an elastoplastic behavior given by:

$$\dot{\sigma}_{ij}^{I} = C_{ijkl}^{I} \cdot (\dot{\varepsilon}_{kl}^{I} - \dot{\varepsilon}_{kl}^{vpI}) \tag{3}$$

Here $C_{ijkl}^{I}$ is the elastic tensor of the inclusion phase. $\dot{\varepsilon}_{kl}^{vpI}$ is the viscoplastic part of the strain rate of the inclusion phase. It is given by the modified Prandtl Reuss flow rule:

$$\dot{\varepsilon}_{ij}^{I,VP} = \frac{3}{2}\left(\frac{\dot{\varepsilon}_{eq}^{I,VP}}{\sigma_{eq}^{I}}\right) s_{ij}^{I} \tag{4}$$

Here, $\sigma_{eq}^{I}$ is the equivalent stress in the inclusion phase, $s_{ij}^{I}$ is the deviatoric part of the stress tensor in the inclusion phase and $\dot{\varepsilon}_{eq}^{I,VP}$ is the equivalent viscoplastic strain rate. The latter takes into account the contribution of the dislocation glide mechanism. Is is a thermally activated mechanism that can be written as the ratio of the applied stress over the critical flow stress $\sigma_{0,I}$. The latter is proportional to the square root of the normalized dislocation density Z. Hence, the equation describing the dislocation glide is written as follows:

$$\dot{\varepsilon}_{eq}^{I,VP} = \dot{\varepsilon}_{I}^{*}\left(\frac{\sigma_{eq}^{I}}{\sigma_{0,I}}\right)^{m} Z^{-\frac{m}{2}} \tag{5}$$

Here m is inversely proportional to the Temperature. This is the term that accounts for the activation energy. The density of dislocations evolves with the strain; its evolution is written as follows (Kim et al. 2000):

$$\frac{dZ}{d\varepsilon_{dis}} = C + C_1\sqrt{Z} - C_2 Z \tag{6}$$

Equation 6 accounts for the fact that the dislocation density evolves via two phenomena (1) the statistical storage of dislocations and (2) the annihilation of dislocation. The statistical storage of dislocations engenders work hardening and is described by the first two terms in equation 8. $C_1$ is a numerical constant while C is given by (Kim et al. 2000):

$$C = M\frac{b}{d}\left(\frac{M\alpha G}{\sigma_{0,I}}\right)^2 \tag{7}$$

Here M is the Taylor factor, b is the Burger vector, G is the shear modulus, $\alpha$ is a numerical constant. One can notice the dependence on the grain size. The annihilation of dislocations is also a thermally activated mechanism whish is ruled by the following equation

$$C_2 = C_{20}\left(\frac{\dot{\varepsilon}_I}{\dot{\varepsilon}_0}\right)^{-\frac{1}{n}} \tag{8}$$

Here n, is a constant which is inversely proportional to the temperature, $C_{20}$ is a constant as well as $\dot{\varepsilon}_0$.

## 2.2. THE MATRIX PHASE

The mechanisms of Coble creep and of GBS are taken into account in the description of the behavior of the matrix phase. The latter has an elastoplastic constitutive law written as follows:

$$\dot{\sigma}_{ij}^M = C_{ijkl}^M\left(\dot{\varepsilon}_{kl}^M - \dot{\varepsilon}_{kl}^{M,VP}\right) \tag{9}$$

Here $C_{ijkl}^M$ is the elastic tensor of the matrix phase and $\dot{\varepsilon}_{kl}^{M,VP}$ is the viscoplastic part of the strain rate tensor of the matrix phase. It is related to its equivalent quantity via the modified Prandtl Reuss Flow rule and takes into account the contribution of both GBS and Coble creep.

$$\dot{\varepsilon}_{eq}^{M,VP} = \dot{\varepsilon}_{co} + \dot{\varepsilon}_{gbs} \tag{10}$$

Here the subscripts 'co' and 'gbs' respectively refer to Coble and grain boundary sliding. The effect of the diffusion of vacancies along the grain boundaries is taken into account. As a first attempt in this model, the steady state equation for the flow of vacancies is taken into account (Coble 1963). One obtains:

$$\dot{\varepsilon}_{eq}^{M,VP} = \frac{14 \cdot \pi \cdot \Omega_c \cdot w \cdot D_{bd} \cdot \sigma_{eq}^{M}}{k \cdot T \cdot d^3} \tag{11}$$

where $D_{bd}$ is the self-diffusion coefficient of the boundary phase, k is the Boltzmann constant, T is the temperature, d is the grain size, w, is the grain boundary thickness $\Omega_c$ is the atomic volume.

Grain boundary sliding occurs during superplasticity. It can be accommodated by either a diffusion of vacancies through the grain cores or through the grain boundaries. It is given by (Luthy et al. 1979):

$$\dot{\varepsilon}_{gbs} \approx 10^8 \frac{D_{gb}}{d^3} b \left( \frac{\sigma_{eq}^{I}}{E} \right)^2 \tag{12}$$

Here E is the elastic modulus of the inclusion phase, d is the grain size, $D_{gb}$ is the diffusion coefficient of the grain boundaries.

## 2.3. HOMOGENIZATION

The global properties of the material are now obtained with the use of a homogenization technique developed by Cherkaoui et al. (Cherkaoui et al. 2000). In self consistent schemes, the global response of the material is obtained by embedding the two phases in an effective matrix. The solution of the problem of the inclusion is usually provided by Eshelby's results (Eshelby 1957). In the latter the sole effect of the difference in the plastic strains of both phases is taken into account. In our case, the extension of Eshelby's problem takes also into the difference in the elastic deformation as well as the one in the inelastic deformation. The global behavior of the material is given by:

$$\dot{\Sigma}_{ij} = C_{ijkl} \left( \dot{E}_{kl} - \dot{E}_{kl}^{VP} \right) \tag{13}$$

Here, $\dot{E}_{kl}^{VP}$ is the viscoplastic strain rate tensor. The local strain rates and stress rates are related to the global ones via equation 14 and 15.

$$\dot{\varepsilon}_{ij}(r) = A_{ijkl}(r) \dot{E}_{kl} + a_{ij}(r) \tag{14}$$

$$\dot{\sigma}_{ij}(r) = B_{ijkl}(r)\dot{\Sigma}_{kl} + b_{ij}(r) \tag{15}$$

The concentration tensors denoted by $A_{ijkl}$ and $B_{ijkl}$ are obtained with the use of the self consistent scheme. These fourth order tensors take into account the heterogeneity of the tangent modulus. The second order tensors $a_{ij}$ and $b_{ij}$ take into account the heterogeneity in the plastic deformation of the different phases. The self consistent scheme used provides a complete set of equation giving the global response of the material (Cherkaoui et al. 2000).

$$A^I_{ijkl} = \langle A_{ijkl}(r)\rangle_{V^I} = \left[I_{ijkl} + S_{ijmn} C^{-1}_{mnpq}\left(C^I_{pqkl} - C_{pqkl}\right)\right]^{-1} \tag{16}$$

$$A^M_{ijkl} = \langle A_{ijkl}(r)\rangle_{V^M} = \left[I_{ijkl} + S_{ijmn} C^{-1}_{mnpq}\left(C^M_{pqkl} - C_{pqkl}\right)\right]^{-1} \tag{17}$$

$$a^I_{ij} = \langle a_{ij}(r)\rangle_{V^I} = A^I_{ijkl} S_{klmn} C^{-1}_{mnpq}\left(C^I_{pqrs}\dot{\varepsilon}^{I,VP}_{rs} - C_{pqrs}\dot{E}^{VP}_{rs}\right) \tag{18}$$

$$a^M_{ij} = \langle a_{ij}(r)\rangle_{V^M} = A^M_{ijkl} S_{klmn} C^{-1}_{mnpq}\left(C^M_{pqrs}\dot{\varepsilon}^{M,VP}_{rs} - C_{pqrs}\dot{E}^{VP}_{rs}\right) \tag{19}$$

$$\dot{E}^{VP}_{kl} = C^{-1}_{ijkl}\left(f \cdot C^I_{klmn} A^I_{mnpq}\dot{\varepsilon}^{I,VP}_{pq} + (1-f)C^M_{klmn}A^M_{mnpq}\dot{\varepsilon}^{M,VP}_{pq}\right) \tag{20}$$

In the above, $S_{klmn}$ denotes Eshelby's tensor.

## 3. Results

The model presented in the above was applied to pure copper. The numerical data used in the simulations are the following; $\mu_I = 21.45 GPa.$, m = 230, n = 8.25, $\sigma_{0,I} = 220 MPa.$, $\dot{\varepsilon}^*_I = 0.005/s$, M = 3.06, $\alpha = 0.33$, G = 42.1GPa., $C_1 = 52.86$, $C_{20} = 18.5$, $\dot{\varepsilon}_0 = 1/s$, $\Omega_c = 1.18e-29$, $D_{bd} = 2.6e-20$, T = 300K, E = 130GPa., w = 1.5nm.

In figure 1, one can see a comparison between the predictions of the model, represented by lines, and various experimental studies (Sanders et al. 1996; Youngdahl et al. 1997), represented by dots. Stress strain curves were plotted for various grain sizes. A decrease in the grain size engenders an increase in the strain hardening. However when the grain size reaches 20 nm, strain hardening is less than when d = 26 nm. These results are explained in terms of a change in the mechanism dominating the deformation. When d > 20nm., the glide of dislocation dominates the deformation. When d < 20nm diffusion related mechanisms take over the dislocation glide engenders a decrease in the strain hardening. The model is in agreement with experimental results in the first stage of deformation. When the strain is larger than a few percent, the model underestimates the global stresses.

In figure 2, one can see the evolution of the yield stress with the grain size. The predictions of the current model are represented by lines while experimental results are represented by dots. The yield stress increases linearly down to a critical grain size which is approximately equal to 40nm. Once this critical threshold is reached, the breakdown of the Hall Petch law occurs. The model predicts values of the yield stress lower than the one given by experimental data. However, it succeeds in predicting the critical grain size at which the Hall Petch law breaks down. The model takes into account both Coble creep and Grain boundary sliding accommodated by diffusion phenomena. Thus at very small grain sizes a redundancy of the diffusion phenomenon might occur. Also, the equations used to describe the two phenomena are based on a steady state creep rate which clearly overestimates the effect of diffusion. Hence, the model is expected to underestimates the stresses. However, it clearly reveals that Coble creep and grain boundary sliding can be responsible for the Breakdown of the Hall Petch law.

## 4. Conclusion

A model based on a self consistent scheme is developed to account for the evolution of the plastic deformation of crystalline materials with grain size. The material is assimilated as a composite with two phases. The model takes into account three mechanisms, (1) the glide of dislocations, (2) Coble creep and (3) grain boundary sliding. The last two mechanisms are incorporated in the description of the matrix phase. The steady state equations of Coble creep and of grain boundary sliding are used to describe the two phenomena. The global behavior predicted by the model is in agreement with experimental results in the early stages of deformation. When the strain is larger than one per cent, the model underestimates the stresses. The model predicts with success the breakdown of the Hall Petch law. The low values of the stress predicted by the model are due to both the redundancy of the Coble creep mechanism and of the Grain boundary sliding accommodated by grain boundary diffusion and by the fact that the steady state equations are used in describing the mechanisms. The model clearly shows that diffusion phenomena and grain boundary sliding can be responsible for the breakdown of the Hall Petch law. The model will be subject to further refinement, notably a dynamic equation for the diffusion of vacancies by the grain boundaries and for grain boundary sliding might be more suited to describe the behavior of the grain boundary phase.

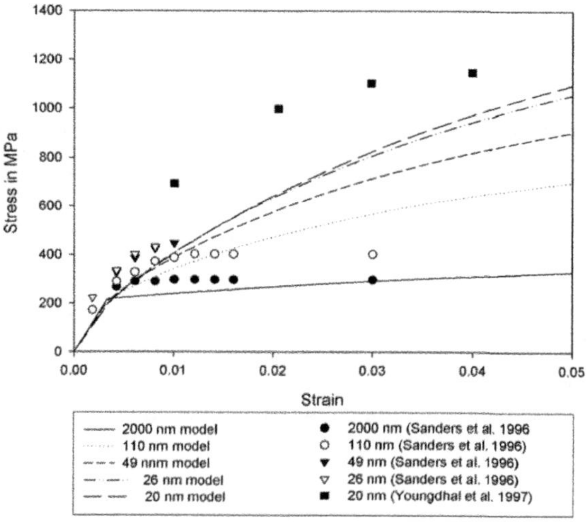


*Figure 1.* Stress Strain curves for different grain sizes.

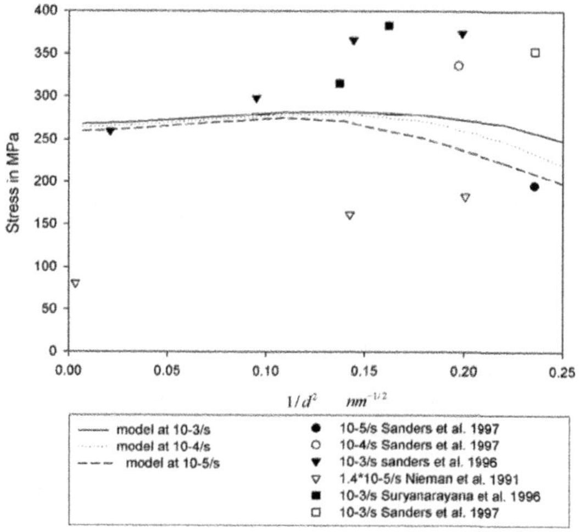


*Figure 2.* Evolution of the yield stress with grain size.

## 5. Acknowledgements

The research was supported in parts by NSF through the Packaging Research Center at Georgia Tech, by NSFC through under Granted No.10228204, and by Harbin Institute of Technology through a visiting professorship to JQ.

## 6. References


Ashby, M. F. (1970). "The deformation of plastically non homogeneous materials." Philosophical Magazine **21**: 399-424.

Birringer, R. (1989). "Nanocrystalline materials." Materials Science and Engineering A **117**: 33-43.

Carsley, J. E., J. Ning, W. W. Milligan, S. A. Hackney and E. C. Aifantis (1995). "A simple mixture based model for the grain size dependence of strength in nanophase metals." Nanostructured Materials **5**: 441-448.

Cherkaoui, M., Q. P. Sun and G. Q. Song (2000). "Micromechanics modeling of composite with ductile matrix and shape memory alloy reinforcement." International Journal of Solids and Structures **37**: 1577-1594.

Coble, R. L. (1963). "A model for boundary diffusion controlled creep in polycrystalline materials." Journal of Applied Physics **34**(6): 1679-1682.

Conrad, H. and J. Narayan (2000). "On the grain size softening in nanocrystalline materials." Scripta Materialia **42**: 1025-1030.

Eshelby, J. D. (1957). "The determination of an ellipsoidal inclusion and related problems." Proceedings of the Royal Society of London **A241**: 376-396.

Gutkin, M. Y., I. A. Ovid'ko and N. V. Skiba (2003). "Transofrmation of grain boundaries due to disclination motion and emission of dislocation pairs." Materials Science and Engineering **A339**: 73-80.

Hall, E. O. (1951). "The deformation and aging of mild steel." Proceedings of the Physical Society of London **B64**: 747.

Kim, H. S., Y. Estrin and M. B. Bush (2000). "Plastic deformation behaviour of fine grained materials." Acta Materialia **48**: 493-504.

Kocks, U. F. (1976). "Laws for work hardening and low temperature creep." Transactions of ASME: 76-85.

Konstantinidis, D. A. and E. C. Aifantis (1998). "On the anomalous hardness of nanocrystalline materials." Acra Metallurgica **10**: 1111-1118.

Kumar, K. S., S. Suresh, M. F. Chislom, J. A. Horton and P. Wang (2003). "Deformation of nanocrystalline nickel." Acta Materialia **51**: 387-405.

Kumar, K. S., H. Van Swygenhoven and S. Suresh (2003). "Mechanical behavior of nanocrystalline metals and alloys." Acta Materialia **51**.

Li, J. C. M. (1963). "Petch relation and grain boundary sources." Transactions of the Metallurigical Society of AIME **227**: 239.

Luthy, H., R. A. White and O. D. Sherby (1979). "Grain Boundary sliding and deformation mechanism maps." Materials Science and Engineering **39**: 211-216.

Murr, L. E. (1981). "Strain induced dislocation emission from grain boundaries in stainless steel." Materials Science and Engineering **51**: 71-79.

Petch, N. J. (1953). "The cleavage strength of polycrystals." Journal of Iron Steel Inst. **74**: 25.

Ranganathan, S., R. Divakar and V. S. Raghunathan (2000). "Interface structures in nanocrystalline materials." Scripta Materialia **44**: 1169-1174.

Sanders, P. G., J. A. Eastman and J. R. Weertman (1996). Processing and Properties of Nanocrystalline Materials, TMS.

Schuh, C. A., T. G. Nieh and T. Yamasaki (2002). "Hall Petch breakdown manifested in abrasive wera resistance of nanocrystalline nickel." Scripta Materialia **46**: 735-740.

Stokes, R. J. and A. H. Cottrel (1954). "Work softening in aluminium crystals." Acta Metallurgica **2**.

Taylor, G. I. (1934). "The mechanism of plastic deformation of crystals: Part I theoretical." Proceeedings of the Royal Society, A **CXLV**: 362-87.

Van Swygenhoven, H., M. Spaczer and A. Caro (1999). "Microscopic description of plasticity in computer generated metallic nanophase samples: a comparison between Cu and Ni." Acta Metallurgica **47**: 3117-3126.

Yamakov, V., D. Wolf, S. R. Phillpot and H. Gleiter (2002). "Deformation twinning in nanocrystalline Al by molecular dynamics simulation." Acta Materialia **50**: 5005-5020.

Yamakov, V., D. Wolf, M. Salazar, S. R. Phillpot and H. Gleiter (2001). "Length scale effects in the nucleation of extended dislocations in nanocrystalline Al by molecular dynamics simulation." Acta Materialia **49**(2713-2722).

Youngdahl, C. J., P. G. Sanders and J. R. Eastman (1997). "Compressive yield strengths of nanocrystalline Cu and Pd." Scripta Materialia **37**: 809-813.

# INSTABILITY AND HYSTERESIS IN THERMOELASTIC TENSION STRIPS

Y.J. HE and Q.P. SUN
*Department of Mechanical Engineering*
*The Hong Kong University of Science and Technology,*
*Hong Kong, China*

## Abstract

Macroscopic domain formations, usually accompanying elastic instability, appear in NiTi polycrystalline shape memory alloy which undergoes a typical thermoelastic phase transition [1-3]. Different from the plasticity approach (e.g. Shaw & Kyriakides 1998 [4]), this paper applies the Ginzburg-Landau model to a non-local elastic FEM simulation on the thermoelastic tension strips. Many key features of the stress-induced domain formation and evolution observed in the experiments have been reproduced. Based on the simulations, we discuss the effects of the domain wall energy on the domain formation; and the relation between the elastic instability and the hysteresis during the superelastic deformation (domain formation, evolution and annihilation) in the thermoelastic system.

## 1. Introduction

NiTi polycrystalline shape memory alloy (SMA) is one of the typical thermoelastic materials. It has the unique properties of superelasticity and shape memory effect due to the thermoelastic (temperature or stress induced) martensitic phase transition. For example in Figure 1, during the phase transformation of a NiTi polycrystalline strip under tensile loading, the specimen is divided into a high strain region (martensite with a strain of about 6%) and a low strain region (austenite with a strain of about 1%) [3]. The martensite domain grows via the propagation of the inclined macroscopic visible interface while the tensile stress of the specimen is (roughly) on a stress plateau. The appearance/disappearance of the localized deformation (domain formation/annihilation) during loading/unloading accompanies the macroscopic elastic instability—stress drop/jump. The strip can completely recover to the initial state without residual (permanent) deformation after unloading due to superelasticity [1-3].

A number of models have been used to describe the domain formation and evolution in polycrystal. For example, Shaw & Kyriakides 1998 [4] simulated these phenomena via a plasticity approach. Although the two-phase coexistence had been captured by their simulation, the interface properties (interface thickness and interface energy) were not described in their simulation. It is well known that interface energy plays an important role in the domain formation and evolution. To include the effect of interface energy, several Ginzburg-Landau type models have been proposed to simulate the phenomena from the solitons and twins at the lattice level [5,6] to the microstructure



*Q. P. Sun and P. Tong (eds.), IUTAM Symposium on Size Effects on Material and Structural Behavior at Micron- and Nano-Scales,* 227–238.

evolutions in polycrystals [7]. Focussing on the macroscopic behaviours of the polycrystal in this paper, we adopt the thermoelastic continuum Ginzburg-Landau framework, taking the NiTi polycrystal as a continuum since the grain size (40~100nm) is much smaller than the dimensions of the strips [3]. Then the macroscopic strain (the average strain of the representative volume containing thousands of grains) is taken as the state variable.

Based on our previously proposed Ginzburg-Landau type strain energy function for the NiTi polycrystal [8, 9], we simulated a series of equilibrium configurations of the domain formation in NiTi strips under displacement-controlled tensile loading. The simulation is implemented by the non-local elastic Finite Element codes which we developed, where the free energy of the strip is minimized to obtain the equilibrium state of the domain. Although the domain evolution of the dissipation system (e.g. NiTi polycrystal) may not strictly follow the principle of energy minimization, the series of the equilibrium domains can provide estimations of the "destination" states and the "direction" of the domain evolution.

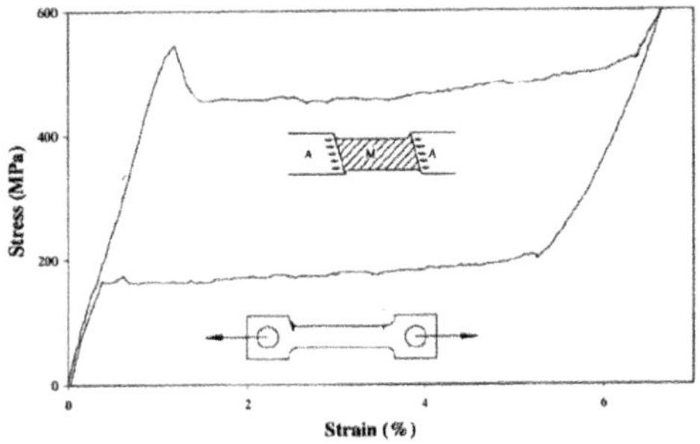


*Figure 1.* The nominal stress-nominal strain curve of the NiTi polycrystalline strip under displacement-controlled tensile loading (Tse 2000 [3]).

Although no dissipation is explicitly included in the energy minimization, there is still hysteresis in our elastic simulation because elastic instability happens. The fundamental relation between hysteresis (dissipation) and elastic instability has been studied by many researchers [10-16]. It is believed that the macroscopic hysteresis in many thermoelastic systems under quasi-static loadings originates from numerous microscopic elastic instabilities – fast transition processes from unstable and meta-stable to stable configuration states. For example in NiTi polycrystalline materials, the numerous microscopic instability processes inside grains during martensitic phase transition – the formation, switching and annihilation of microscopic martensite domains transfer a substantial part of the elastic energy into heat or waves which are then dissipated away. The purpose of our simulation on the tension strip in this paper is to provide a picture on this mechanism at the macroscopic scale level.

This paper is organized as follows. The Ginzburg-Landau type constitutive model and its implementation into FEM are described in Section 2. In Section 3, we report the results of the FEM simulations on the tension strips. The key features and factors of the domain evolution in the tension strip are also discussed. Conclusions are given in Section 4.

## 2. Model and FEM formulation

A Ginzburg-Landau type strain energy function for bulk polycrystal has been proposed by He & Sun 2005 [8] to describe the tension and shear deformations of NiTi polycrystals. The macroscopic strain of the polycrystalline aggregate is taken as a state variable and the strain energy function $W(\eta_1, \eta_3)$ was constructed where the material is incompressible ( $\eta_0 = \varepsilon_{xx} + \varepsilon_{yy} + \varepsilon_{zz} = 0$ ) and the difference of the two transversal contraction strains is ignored ($\eta_2 = \varepsilon_{zz} - \varepsilon_{xx} = 0$) during simple tension loading.

$$W(\eta_1, \eta_3) = (a\eta_1^2 + b\eta_1^3 + c\eta_1^4) + (s\eta_3^2 + d\eta_3^4 + q\eta_3^6) + \eta_3^2(\lambda\eta_1^2 + k\eta_1^3) \tag{1}$$

where x, y and z are the coordinates in the directions along the width, length and thickness of the strips, respectively. $\eta_1 = (2\varepsilon_{yy} - \varepsilon_{zz} - \varepsilon_{xx})/\sqrt{3}$ and $\eta_3 = \varepsilon_{xy}$ characterize the tension and shear deformation respectively. The material constants used in the simulations are: $a = 6.88 \times 10^3$, $b = -8.67 \times 10^4$, $c = 3.50 \times 10^5$, $s = 1.86 \times 10^4$, $d = -2.59 \times 10^7$, $q = 5.69 \times 10^{10}$, $\lambda = -1.40 \times 10^7$, $k = 1.98 \times 10^8$ (unit: MPa) which are selected based on the experimental data of the force-displacement curves under various loadings [9,17].

In order to account for domain wall energy, strain gradient energy (G) was included.

$$G = a_3\left[\left(\eta_{3,2} - \frac{1}{2\sqrt{3}}\eta_{1,1}\right)^2 + \left(\eta_{3,1} + \frac{1}{\sqrt{3}}\eta_{1,2}\right)^2\right] + a_4\left[2\eta_{3,1}^{\;2} + 2\eta_{3,2}^{\;2} + \frac{1}{2}\eta_{1,1}^{\;2} + \frac{1}{2}\eta_{1,2}^{\;2}\right] + a_5\left[\left(\eta_{3,2} - \frac{1}{2\sqrt{3}}\eta_{1,1}\right)^2 + \left(\eta_{3,1} + \frac{1}{\sqrt{3}}\eta_{1,2}\right)^2 + \sqrt{3}\left(\eta_{1,1}\eta_{3,2} - \eta_{1,2}\eta_{3,1}\right)\right] \tag{2}$$

Due to the lack of experimental data, we assume $a_3 = a_4 = a_5 = a(g)^2$ for simplicity, where a is the second order constant in Eq. (1) and g is a material intrinsic length. The effect of the magnitude of g on the pattern evolution will be discussed in section 3.

The tension strip is modeled as a two dimensional plane in the FEM simulation. The kinematic relationships are:

$$\varepsilon_{xx} = \frac{\partial u}{\partial x} \qquad \varepsilon_{yy} = \frac{\partial v}{\partial y} \qquad \varepsilon_{xy} = \frac{1}{2}\frac{\partial v}{\partial x} + \frac{1}{2}\frac{\partial u}{\partial y}$$
$$\varepsilon_{xz} = 0 \qquad \varepsilon_{yz} = 0 \qquad \varepsilon_{zz} = \varepsilon_{xx} \quad (\eta_2 = 0) \tag{3}$$

where u and v are the displacement components along the width (x) and length (y) directions, respectively. The energy functional is:

$$\Pi = \iint (W + G)dxdy = \iint F(u_{,x}, u_{,y}, u_{,xx}, u_{,yy}, u_{,xy}, v_{,x}, v_{,y}, v_{,xx}, v_{,yy}, v_{,xy})dxdy \tag{4}$$

where $F = W + G$. Euler (equilibrium) equations (minimizing this functional) are:

$$\frac{\partial}{\partial x}\left[F_{ux}-\frac{\partial}{\partial x}F_{uxx}-\frac{1}{2}\frac{\partial}{\partial y}F_{uxy}\right]+\frac{\partial}{\partial y}\left[F_{uy}-\frac{\partial}{\partial y}F_{uyy}-\frac{1}{2}\frac{\partial}{\partial x}F_{uxy}\right]=0$$

$$\frac{\partial}{\partial x}\left[F_{vx}-\frac{\partial}{\partial x}F_{vxx}-\frac{1}{2}\frac{\partial}{\partial y}F_{vxy}\right]+\frac{\partial}{\partial y}\left[F_{vy}-\frac{\partial}{\partial y}F_{vyy}-\frac{1}{2}\frac{\partial}{\partial x}F_{vxy}\right]=0 \quad (5)$$

where $F_{vxx}=\frac{\partial F}{\partial v_{,xx}}$ and other notations are defined similarly.

For a convergent solution to Eq. (5) in the finite element method, $C^1$ interpolation functions are needed to describe u and v fields in an element. However, due to mathematical complex calculations and large amount of computation required for $C^1$ elements, incompatible elements are widely used in practice and recommended [18,19]. This paper adopts the semi-$C^1$ "4-nodes rectangle" element, whose shape function is a polynomial up to the 4th order (with nodal freedoms u, v, $u_{,x}$, $v_{,x}$, $u_{,y}$ and $v_{,y}$). To mimic the displacement-controlled tensile loading at the ends of the strip, $u = 0$, $v = v_{specified}$ and $u_{,x} = 0$, $v_{,x} = 0$, other gradient components are not constrained in order not to apply high order forces at the boundaries [20-22]. Material constants and geometry parameters (width = 2mm, thickness t = 0.2mm, length L = 16~24mm) are selected from the experiments [3,9]. The Newton-Raphson method was used to find solutions for the non-linear problem.

## 3. Results of FEM Simulations

This section is divided into two parts. In the first part, the non-local FEM simulations with the Ginzburg-Landau model capture many key features (Load drop/jump, two phases coexistence with a diffusive interfacial zone, the angle of the inclined domain wall, etc.) of the domain formation in the non-linear softening-and-rehardening thermo-elastic material under tension. With the objective (mesh-independent) description of domain wall properties (e.g. domain wall thickness) in the non-local simulations, we can quantify the domain wall (interface) energy and find its role in the domain formation. In the second part of this section, we will discuss the relation between the hysteresis and the elastic instability such as domain formation, annihilation and switching.

### 3.1. GENERAL FEATURES OF DOMAIN FORMATION AND EVOLUTION

Three different meshes (10 x 80, 20 x 160 and 40 x 320 elements) have been adopted in the non-local FEM simulations. The results of the deformation patterns (distributions of the tensile strain) are mesh-independent as shown in Figure 2. Therefore, the non-local FEM codes with the Ginzburg-Landau type free energy function provides an efficient tool for an objective description of the phenomena of the domain formation (especially the domain wall properties).

*(1) Important features of domain formation*

As shown in Figures 2 and 3, the FEM simulations capture the important features : the two-phases (tensile strains 1.2% and 6.0%) coexistence with a propagating diffusive interface, the stress drop/jump (for the domain formation/annihilation) and the stress plateau (for the interface propagation) in the tensile loading curve, and the jumps of the bending moments during domain switching (domain shape: parallelogram ↔ trapezium). Consistent with the experimental observations, the simulated deformation patterns have the domain walls (interfaces) inclined to the axis. The inclined angle of the domain wall depends on the interface energy while the shape of the martensite domain (parallelogram or trapezium) can change (switch to each other) during loading and unloading (see Sec 3.2).

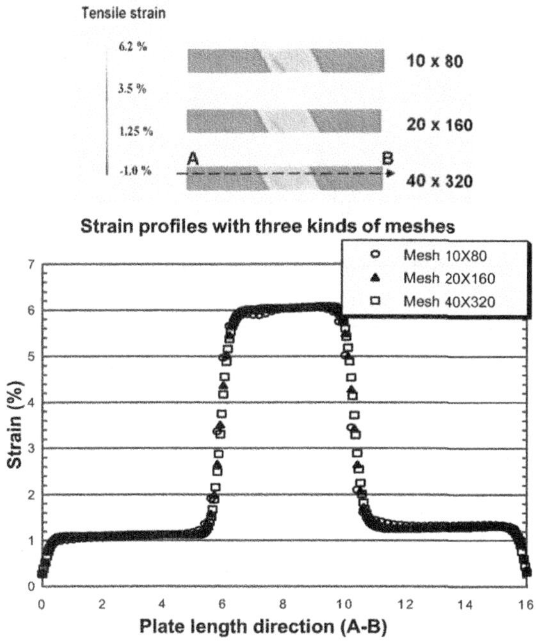


*Figure 2.* Mesh independent patterns and strain profiles (along A-B) in the non-local simulations (g = 70um and nominal strain 2.5%).

*(2) Interface energy and its role in domain formation*

As shown in Figure 2, there is a diffusive zone (interface) between the two coexisting phases, whose thickness ($l$) can be estimated by Eq. (6a) as suggested in Cahn-Hilliard's one-dimensional model [23]. Based on functional analysis, the interface energy (see Eq. (6b)) is two times the total strain gradient energy in the equilibrium diffusive interfacial zone (minimizing the system energy).

$$l = \frac{\varepsilon_M - \varepsilon_A}{\left(\dfrac{d\varepsilon}{dn}\right)_{\varepsilon = \frac{\varepsilon_M + \varepsilon_A}{2}}} = \frac{\varepsilon_M - \varepsilon_A}{\left(\sqrt{\dfrac{\Delta W}{\kappa}}\right)_{\varepsilon = \frac{\varepsilon_M + \varepsilon_A}{2}}} \tag{6a}$$

$$E_{interface} = 2\int_{\varepsilon_A}^{\varepsilon_M} \sqrt{\kappa \Delta W}\, d\varepsilon \tag{6b}$$

where n is the coordinate normal to the flat interface; $\Delta W$ characterizes the non-convex energy; $\varepsilon_A$ and $\varepsilon_M$ are the tensile strains of the two coexisting phases (austenite and martensite), satisfying the well-known Maxwell rule. If the gradient energy G in Eq. (2) is reduced to one-dimensional form (ignoring the shear strain gradient components) and applied to Eq. (6), Eq. (7) can be obtained, which reveals the linear relation between the interface energy (thickness) and the material intrinsic length g.

$$l = \frac{g(\varepsilon_M - \varepsilon_A)}{\left(\sqrt{\frac{\Delta W}{3a}}\right)_{\varepsilon=\frac{\varepsilon_M+\varepsilon_A}{2}}} \tag{7a}$$

$$E_{interface} = 2g\int_{\varepsilon_A}^{\varepsilon_M}\sqrt{3a\cdot\Delta W}\,d\varepsilon \tag{7b}$$

Substituting the strain energy function W of Eq. (1) (ignoring shear strain component) to Eq. (7), the interface thickness and energy for the domain wall can be calculated and compared with the results of the simulation on tension strips in Figure 4. Basically in agreement with the 1D model, the results of the simulation indicate that both the interface thickness and energy are proportional to g. Therefore, consistent with Cahn-Hilliard's 1D model, a material length (g) used in the present non-local simulations on the tension strips characterizes the interface properties – interface thickness and interface energy.

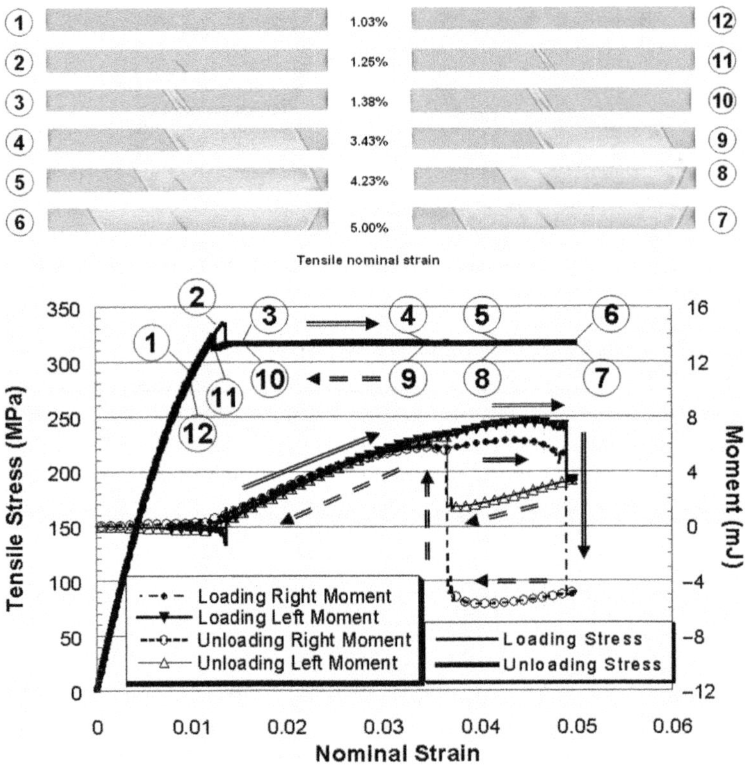


*Figure 3.* Simulated deformation patterns and the curves of the tensile nominal stress and the bending moments at clamping ends (g = 40um).

It is well known in material science that equilibrium microstructure evolution is the result of the competition between bulk (volume-related) energy and interface energy. For tension strips, when the interface energy approaches zero (g → 0), the angle of the inclined domain wall should be 35.26° (invariant plane with no jump of tangent normal strain across the interface) to minimize the elastic mismatch between the two phases. With nonzero interface energy, the simulated angle of the inclined domain wall depends on the magnitude of g as shown in Figure 5, where the angle changes from 35.26° to 0°. The physical meaning behind this relation is that: when the interface energy (density) is dominant (large g), the martensite domain prefers to reduce the interface area (reduce the angle) to minimize the total system energy, even at the expense of some increase in the elastic mismatch energy. In fact, (structural) boundary constraints also significantly affect the domain formation, which will be discussed in Section 3.2. By comparing the simulated patterns (where domain is far away from the clamping bounadries) with that in experiments, the material length g can be determined.

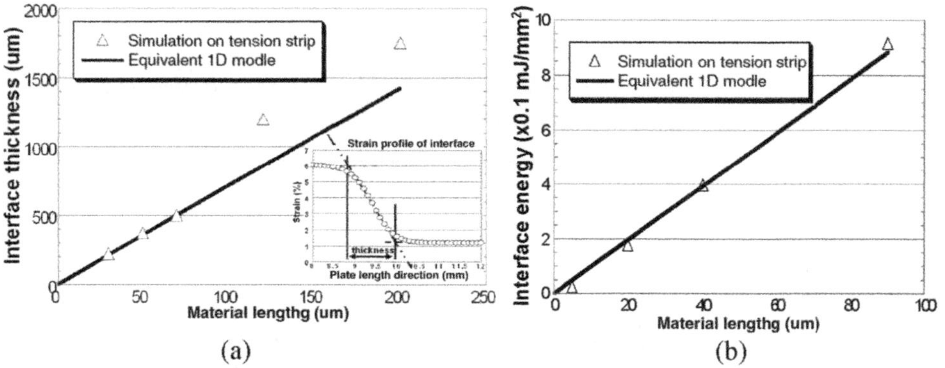


*Figure 4.* Comparison between the 1D model and the simulation on (a) interface thickness, (b) interface energy.

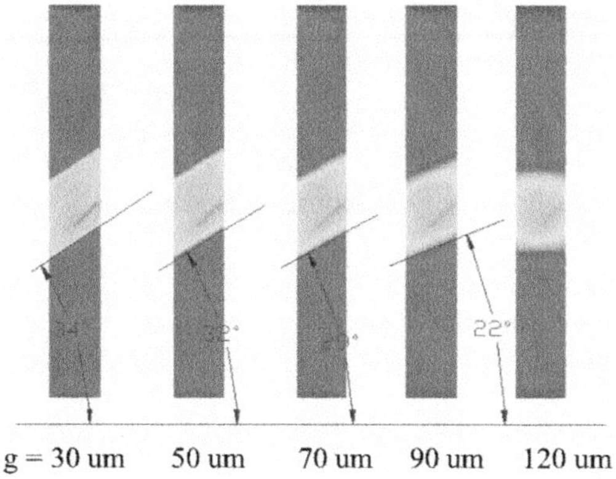


*Figure 5.* Effect of the magnitude of g on the inclined angle (0° ~35.27°) of the domain.

## 3.2 HYSTERESIS AND ELASTIC INSTABILITY

*(1) Domain formation and annihilation*

Although no explicit dissipation scheme is included in this elastic modeling and simulation, there is still a hysteretic loop in the tensile nominal stress-nominal strain curve with the stress drop/jump corresponding to the domain formation and annihilation as shown in Figure 3. As pointed out in [8-11] energy dissipation can be caused by elastic instability during transitions from meta-stable states to stable states. For example in Figure 3, two equilibrium states (No. "2" and "11") appear under the same tensile nominal strain 1.25% during loading and unloading, respectively. By calculating the stored elastic energies for these two states, it is seen that the stored energy of (stable) State "2" (with homogeneous deformation) is less than that of (meta-stable) State "11" (with localized deformation). Therefore, the domain tends to disappear so that the system transits into the stable state (homogenous deformation) dynamically in the next small unloading step.

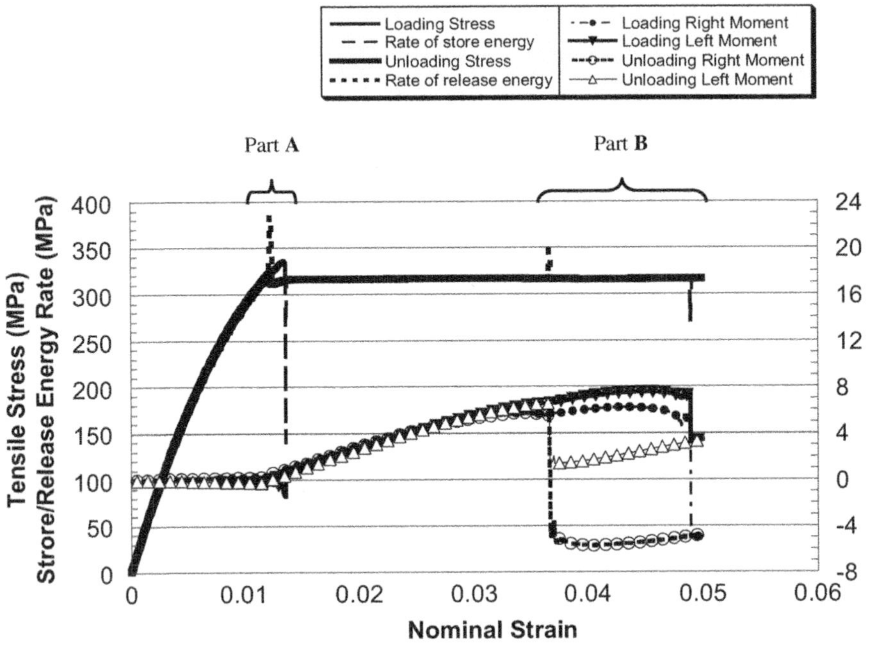


*Figure 6(a).* The tensile stress, the rate of stored/released energy and the bending moments during the loading and unloading.

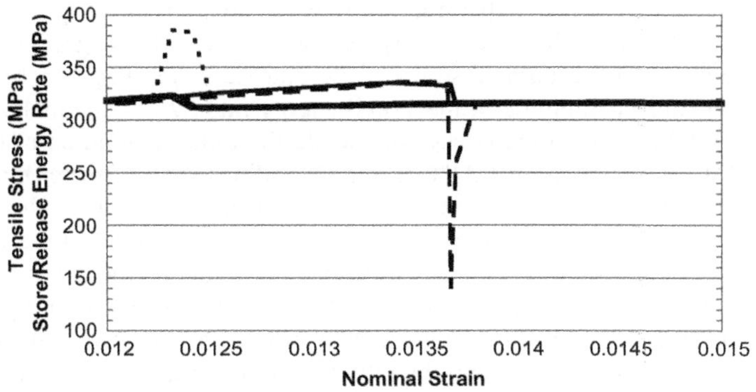


*Figure 6(b).* Magnified view of Part A in (a) on the tensile stress and the rate of stored/released energy in domain formation and annihilation.

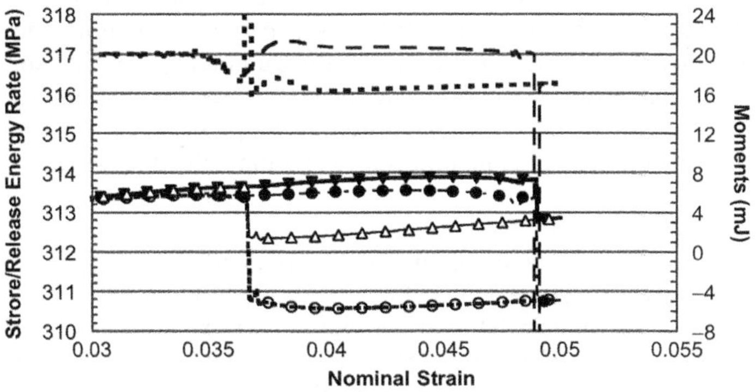


*Figure 6(c).* Magnified view of Part B in (a) on the bending moments and the rate of stored/released energy in domain switching.

In order to reveal the instability-induced dissipation more explicitly, we calculate the rate of the stored/released elastic energy of the tension strip during the loading/unloading based on the strain fields of the equilibrium states. Figure 6(a) shows the tensile stress, the rate of the stored/released energy and the bending moments at the clamping ends during the loading and unloading, where the rate of stored/released energy per unit volume per unit nominal strain has the same unit (MPa) as the tensile stress. The rates of stored/released energy curves exactly follow the loading/unloading tensile stress curves except on four positions: the stress drop (around the nominal strain of 1.37%), the stress jump (1.23%) and the two jumps of the bending moments (3.68% and 4.90%).

Under the uniform displacement-controlled tensile loading at the clamping ends, the strip exchanges energy with the loading machine only through the tensile stress (force). In other words, the machine does work on the strip only through the tensile force during loading while the strip returns energy back to the machine during unloading also through the tensile force. The hysteretic loop in loading and unloading tensile stress curves in Figures 3 and 6 means that not all the work done on the strip by the machine can be returned. There must be some energy dissipated. From the magnified view on the stress drop and jump (at the nominal strain 1.37% and 1.23%) in Figure 6(b) it is clearly indicated that the drop magnitude of the stored energy rate (at the nominal strain 1.37% during loading) is larger than the stress drop while the jump magnitude of the released energy rate (at the nominal strain 1.23% during unloading) is also greater than the stress jump. That means at the domain formation (stress drop), only a part of the work given by the loading machine in this loading step is stored into the strip and the other part is dissipated away in the form of elastic waves during the dynamic transition process from the meta-stable (unstable) state to the stable state. The amount of the dissipated energy (denoted as $E_f$) for the domain formation can be estimated by the area formed (enclosed) by the curves of the loading stress and the rate of stored energy. Likewise at the domain annihilation (stress jump), only a part of the released energy from the strip in the unloading step is returned to the loading machine and the other part is dissipated for the elastic instability process. The amount of the dissipated energy ($E_a$) for the domain annihilation can also be estimated by the area formed by the curves of the unloading stress and the rate of energy release. At each quasi-static loading/unloading step where there is no drastic change (no instability) in the pattern evolution, the rate of the stored/released energy coincides with the loading/unloading tensile stress (no dissipation). In fact, the total dissipation energy ($E_f$ + $E_a$) for domain formation and annihilation can be calculated from the hysteretic loop (between the stress drop and jump) formed by the curves of the loading stress and the unloading stress.

*(2) Domain switching*

Besides domain formation and annihilation, there is another kind of drastic changes in the pattern evolution – domain switching – which is the very fast change of the domain shape (trapezium to parallelogram or vice versa). As shown in Figure 3, domain switching happens at the loading states “5” ~ “6” with the domain shape change from a parallelogram to a trapezium and at the unloading states “8” ~ “9” with the shape change in the reverse direction. Accompanying the domain switching, the bending moments at the clamping ends suffer a jump where the tension strip takes a transition process from a meta-stable or unstable state to a stable state. When the tensile nominal strain is within the loop (3.68% ~ 4.90%) formed by the loading and unloading bending moments, we can always find a meta-stable state. For example, both the equilibrium states “5” and “8” are at 4.23% tensile nominal strain, but they have different domain shapes (parallelogram and trapezium, respectively), one of which must be meta-stable.

Like domain formation and annihilation, the domain switching also brings about energy dissipation as shown in Figures 6(a) and 6(c), where the drop/jump of the stored/released energy rates accompany the jumps of the bending moments. Moreover, the curves of the stored and released energy rates form a loop in the range between the two bending moment jumps. That means the energy given to the strip by the loading machine will not be fully returned back when domain switching (instability) happens.

In fact, the domain switchings may appear much more complex in real systems. For example in SMA polycrystals with small grains, where there are complex multi-scale microstructures and strong geometric incompatibility/constraints during the numerous nucleation and growth of martensite domains, the elastic strain energy significantly affects the polycrystal's macroscopic mechanical response, espeically in the hysteresis loop [10, 24].

In order to demonstrate the effect of boundary constaints on the domain evolution, we simulate a short strip (2 x 8 mm) under tensile loading. Due to the strong constraints on the short strip by the clamping ends, another martensite domain nucelates when the first-nucelated parallelogramic domain evolves to a certain volume (see State "3" in Figure 7). These domains would switch to the V-shaped domains (finger-like) to match the geometric accommodation (States "4" and "5") in further evolution. Both the new nucleation and domain switching lead to the jumps of the bending moments, dissipating energy into heat or waves. It is natural to think that the number of events (domain nucleation and switching) would increase with the contraints such as the grain boundary in polycrystals.

In summary, the domain formation, annihilation and switching are the elastic instability processes, transforming the system from the meta-stable (unstable) states to the stable states. During these processes, dissipations are generated leading to the hysteresis loops in the tensile stress (force) and the bending moments. Although the dissipation is relatively small for each of these events, the accumulated effects of a large number of these events are considerable. For example, if the tension strip consists of many grains within which the domain evolutions proceed (i.e. grains geometrically constrain each other), numerous instability processes of the domain formation, annihilation and switching in the numerous grains had to happen and dissipate considerable amounts of energy, which are macroscopically reflected by the large hysteresis loop as shown by the upper and lower force plateaus observed in the experiments.

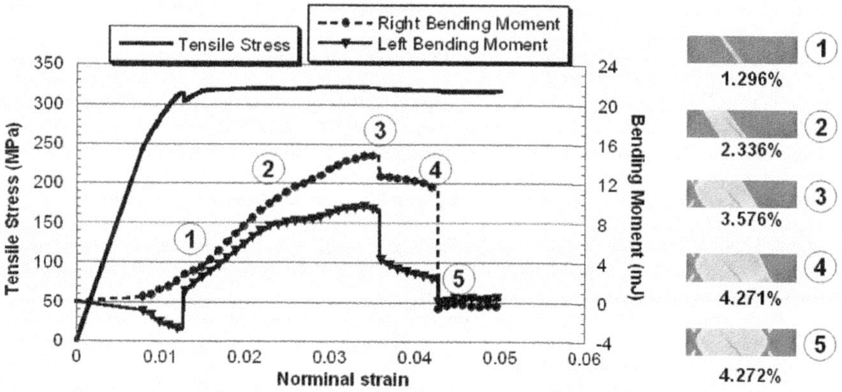


*Figure 7.* Domain switchings in a short strip under tensile loading (g = 15um).

## 4. Concluding Remarks

Within the continuum Ginzburg-Landau framework, a non-local elastic FEM simulation on thermoelastic strips successfully captured many key features in the domain formation and evolution. With the effect of the non-locality, the properties of a macroscopic domain wall (interface) can be objectively described and the results are consistent with that of Cahn-Hilliard's 1D model. Our simulation also demonstrates the mechanism relating the elastic instability process to hysteresis, which may serve as an example of the widely accepted paradigm of the macroscopic rate-independent hysteresis in SMA [10-15]. Analogous to the effect of grain size in SMA polycrystals, the effect of the specimen size on the domain evolution has been shown in the simulation on the tension strips.

## References


1. Shaw J. A. and Kyriakides S., 1995. Thermomechanical aspects of NiTi. *J. Mech. Phys. Solids* **43** (8), 1243-1281.
2. Shaw J. A. and Kyriakides S., 1997. On the nucleation and propagation of phase transformation fronts in a NiTi alloy. *Acta mater.* **45** (2), 683-700.
3. Tse K. K. K., 2000. Experimental study on superelastic deformation of NiTi shape memory alloy thin strips and wires under tension. MPhil thesis, The Hong Kong University of Science and Technology.
4. Shaw, J.A. and Kyriakides S., 1998. Initiation and propagation of Localized Deformation in Elasto-Plastic Strips under Uniaxial Tension. *International Journal of Plasticity* **13**(10), 837-871.
5. Falk F., 1983. Ginzbur-Landau theory of static domain walls in shape-memory alloys, *Z. Physik B – Condensed Matter* **51**, 177-185.
6. Barsch G. R. and Krumhansl, J. A., 1984. Twin boundaries in ferroelastic media without interface dislocations, *Phys. Rev. Lett.* **53** (11), 1069-1072.
7. Ahluwalia R., Lookamn T. and Saxena A. 2003. Elastic dformation of polycrystals, *Physical Review Letters* **91** (5), 055501.
8. He Y. J. and Sun Q. P., 2005. Modeling and simulation of deformation pattern evolution during stress-induced martensite phase transition in NiTi microtubing. *Proc. of Int. Conf. on Solid-Solid Phase Transformations in Inoganic Mater.*, May 29-June 3, Phoenix, Arizona, USA.
9. He Y. J., 2005. Non-local model for deformation patterns in NiTi microtubing. PhD thesis, Hong Kong University of Science and Technology.
10. Ball J. M., Chu C. and James R. D., 1995. Hysteresis during stress-induced variant rearrangement. *Journal De Physique* IV, C8, 245-251.
11. Truskinovsky L. and Vainchtein A., 2003. Peierls-Nabarro landscape for martensitic phase transitions. *Physical Review B* **67**, 172103.
12. Kresse O. and Truskinovsky L., 2004. Lattice friction for crystalline defects: from dislocations to cracks. *Journal of the Mechanics and Physics of Solids* **52**, 2521-2543.
13. Puglisi G. and Truskinovsky L. 2005. Thermodynamics of rate-independent plasticity. *Journal of the Mechanics and Physics of Solids* **53**, 655-679.
14. Ortin J and Delaey L, 2002. Hysteresis in shape-memory alloys. *International Journal of Non-linear Mechanics* **37**, 1275-1281.
15. Muller I. and Xu H., 1991. On the pseudo-elastic hysteresis. *Acta metall. Mater.* **39** (3) 263-271.
16. Wilmanski K. 1993. Note on the model of pseudoelastic hysteresis. *Models of hysteresis*, Visintin A (Ed), Longman Scientific & Technical, 207-221.
17. Sun Q. P. and Li Z. Q., 2002. Phase transformation in superelastic NiTi polycrystalline micro-tubes under tension and torsion–from localization to homogeneous deformation. *International Journal of Solids and Structures* **39**, 3797-3809.
18. Fung Y. C. and Tong P., 2001. *Classic and Computational Solid Mechanics* (World Scientific), 796-802.
19. Zienkiewicz O. C. and Taylor R. L., 2000. *The Finite Element Method* 5$^{th}$ Ed, vol. 2 (Butterworth-Heinemann), 111-127.
20. Mindlin R. D. and Eshel N. N., 1968. On first strain-gradient theories in linear elasticity, *Int. J. Solids Structures* **4**, 109-124.
21. Mindlin R. D., 1965. Second gradient of strain and surface-tension in linear elasticity. *Int. J. Solids Structures* **1**, 417-438.
22. M.R. Begley and Hutchinson, J.W. 1998. The mechanics of size-dependent indentation. *J. Mech. Phys. Solids*, **46**(10), 2049-2068.
23. Cahn, J. W. and Hilliard J. E., 1958. Free Energy of a nonuniform system I. Interfacial free energy. *The Journal of Chemical Physics* **28**(2), 258-267.
24. Salzbrenner R. J. and Cohen M., 1979. On the thermodynamics of thermoelastic martensitic transformations. *Acta Metallurgica.* **27**, 739-748.

# EFFECTIVE ELASTIC MODULUS OF NANO-PARTICLES

R. DINGREVILLE[a], J. QU[a], and M. CHERKAOUI[b]
[a]*George W. Woodruff School of Mechanical Engineering, Atlanta, Georgia 30332-0405, USA*
[b]*LPMM-CNRS University of Metz, Ile du Saulcy, 57045, Metz, Cedex1 France*

## Abstract

*Atoms* at a free surface experience a different local environment than do atoms in the bulk of a material. As a result, the equilibrium position and energy of these atoms will, in general, be different from bulk positions and bulk energies. In traditional continuum mechanics, such interfacial free energy is typically neglected because it is associated with only a couple of layers of atoms near the surface, and the ratio of the volume occupied by the surface atoms and the total volume is extremely small. However, for nano-size particles, the surface to volume ratio becomes significant, so does the effects of surface free energy. In this paper, a framework is developed to incorporate the surface free energy into the continuum mechanics theory. Based on this approach, it is shown that the effective modulus of a particle does depend on the particle size. Although such size dependency is negligible for larger size particles, it becomes significant when the particle shrinks to nanometer size.

## 1. Surface Energy and Surface Stress

Atoms at a free surface or interface experience a different local environment than do atoms in the bulk of a material. As a result, the equilibrium position and energy of these atoms will, in general, be different from bulk positions and bulk energies. Properties of the solid which are sensitive to the atomic positions or energies will necessarily be affected at or near a surface or interface. For conventional materials where the number of atoms near the free surface or interface is small compared to the total number of atoms, such surface or interface effects are insignificant, and can be rightfully ignored. For nano-structured materials, however, a substantial amount of atoms are located near the interfaces. Therefore, interface properties become significant in nano-structured materials. Consequently, the overall behavior of nano-structured materials is strongly influenced by the properties of interfaces, such as grain boundary (including grain boundary junctions) in polycrystalline materials and particle/matrix interfaces in composites.

There are different ways in which the properties of interfaces can be defined and introduced.



*Q. P. Sun and P. Tong (eds.), IUTAM Symposium on Size Effects on Material and Structural Behavior at Micron- and Nano-Scales,* 239–248.

For example, if we consider an interface separating two otherwise homogeneous phases, the surface property may be defined either in terms of a *surface phase*, or by introducing the concept of a *dividing surface*. In the first method, the system is considered to be one in which there are three phases present – the two bulk phases and a surface phase; the boundaries of the surface phase are somewhat arbitrary and are usually chosen to be at locations at which the properties are no longer varying significantly with position. The surface phase then has a finite volume and may be assigned thermodynamic properties in the normal way. In the second method where a single dividing surface is used to separate the two homogeneous phases, the interface contribution to the thermodynamic properties is defined as the excesses over the values that would obtain if the bulk phases retained their properties constant up to an imaginary surface (of zero thickness) separating the two phases. In this paper, we adopt this second approach.

The concept of dividing surface was first introduced by Gibbs [3] through the use of Gibbs surface free energy (also called surface tension in some literature). The Gibbs density of surface free energy is defined as the reversible work involved in creating a unit area of *new* surface at constant temperature, volume, and total number of moles. Because the surface atoms are subjected to different inter-atomic forces, their inter-atomic spacing is different from that of the atoms deep in the interior of the crystal. In other words, the creation of a free surface has resulted in an absolute surface strain $\tilde{\varepsilon}^{s}_{\alpha\beta}$ for the surface lattice relative to the perfect crystal lattice. This state of the free surface will be called the self-equilibrium state of the surface and the corresponding absolute surface strain, $\tilde{\varepsilon}^{s}_{\alpha\beta}$, is called the self-equilibrium surface strain. In the context of Gibbs surface free energy, $\tilde{\varepsilon}^{s}_{\alpha\beta}$ can be considered as the surface *free* (*excess*) strain. Further deformation of the surface from its self-equilibrium state is measured by the relative surface strain $\varepsilon^{s}_{\alpha\beta}$. Clearly, relative surface strain must be imposed by external loading which will also induce bulk strain in the crystal. It is conceivable that the deformation induced by external loading is continuous near the surface. Now, let the bulk crystal be subjected to, without loss of generality, a uniform deformation given by the Lagrangian strain tensor $\varepsilon_{ij}$, measured from the undeformed perfect crystal lattice of an infinite crystal. In the rest of this paper, this strain will be called the absolute strain, referring to the strain relative to the undeformed bulk crystal lattice. The surface deformation is thus described by a two dimensional surface strain tensor $\varepsilon^{s}_{\alpha\beta}$, which is related to $\varepsilon_{ij}$ through a coordinate transformation involving the surface normal direction cosines. In the above and the rest of this paper, Roman indices range from 1 to 3 and Greek indices range from 1 to 2, unless other wise indicated.

The surface free (excess) energy, $w_n$, of a surface atom (atoms on or near the surface) is defined by the difference between its energy and that of an atom deep in the interior of a large crystal. Note that $w_n$ is a function of the intrinsic crystal surface properties.

If there are $N$ atoms within an area $A$, then the total surface free energy of the area $A$ is given by $\sum_{n=1}^{N} w_n$ . Thus, the Gibbs surface free energy density is defined by,

$$\gamma = \frac{1}{A}\sum_{n=1}^{N} w_n \tag{1}$$

Note that the above definition is in the deformed configuration. It can be viewed as the Eulerian description of the surface free energy density. For solid crystal surfaces, the Lagrange description of the surface free energy density can be defined by,

$$\Gamma = \frac{1}{A_0}\sum_{n=1}^{N} w_n \tag{2}$$

where $A_0$ is the area initially (before the application of $\varepsilon_{ij}$ ) occupied by the $N$ atoms within $A$ in the current (deformed) configuration. It can be easily shown that the two areas are related by,

$$A = A_0(1 + \varepsilon^s_{\eta\eta} + \tilde{\varepsilon}^s_{\eta\eta}) \tag{3}$$

where, again, $\tilde{\varepsilon}^s_{\alpha\beta}$ is the self-equilibrium strain and $\varepsilon^s_{\alpha\beta}$ is the relative strain from the self-equilibrium state of the surface. Together, they deform the area $A_0$ into $A$.

It is clear from these two definitions that,

$$\gamma A = \Gamma A_0 \tag{4}$$

Therefore, one has,

$$\Gamma = \gamma(1 + \varepsilon^s_{\eta\eta} + \tilde{\varepsilon}^s_{\eta\eta}) \tag{5}$$

Unfortunately, it is not always clear in the literature which of the two definitions of the surface free energy is being used. This may have contributed to the wide range of values reported.

Having defined the surface free energy density, we can now introduce the concept of surface stress. Surface free energy corresponds to the work to create a unit area of surface, whereas surface stress is involved in computing the work in deforming a surface. Specifically, the change in surface free energy should be equal to the work done by the surface stress as it deforms the surface area from $A_0$ to $A$, i.e.,

$$d(\Gamma A_0) = A_0 \Sigma^s_{\alpha\beta} d\varepsilon^s_{\alpha\beta} \tag{6}$$

where $\Sigma^s_{\alpha\beta}$ is the surface stress tensor. Note that,

$$d(\Gamma A_0) = \Gamma dA_0 + A_0 d\Gamma = A_0 d\Gamma \tag{7}$$

Therefore, one has,

$$\Sigma^s_{\alpha\beta} = \frac{d\Gamma}{d\varepsilon^s_{\alpha\beta}} \tag{8}$$

This defines the surface stress tensor in terms of the Lagrange description of the surface free energy density. Substituting (5) into (8) yields the surface stress tensor in terms of the Eulerian surface free energy density,

$$\Sigma^s_{\alpha\beta} = \gamma\delta_{\alpha\beta} + (1+\varepsilon^s_{\eta\eta} + \tilde{\varepsilon}^s_{\eta\eta})\frac{d\gamma}{d\varepsilon^s_{\alpha\beta}} \tag{9}$$

where $\delta_{\alpha\beta}$ is the two-dimensional Kronecker delta. The above is often referred as the Shuttleworth relationship [4,5].

In a liquid, where the atomic mobility is sufficiently high, atoms from the bulk will come to the surface when the surface is stretched. This way the microscopic configuration of the surface is preserved following deformation, i.e., the surface free energy density remains invariant to the surface strain [5]. Consequently, the second term in (9) vanishes and the surface free energy is numerically equal to surface stress. This is perhaps why historically surface free energy is also called surface tension. Adding to the confusion is the fact that both surface free energy density and surface stress have the same physical dimension of force per unit length.

For a solid, due to the long range correlation in atomic positions and low atomic mobility, it might not be possible, in any reasonable experimental time, to keep constant the local configuration around any particular atom in the surface region where the deformation of the surface area is performed. In other words, when a solid crystal deforms, its surface area may change. Such change of surface area is not accomplished by adding mass to (or subtracting from) the surface. Instead, the change of surface area is accompanied by the change of surface free energy density. Consequently, the surface free energy density becomes a function of the surface strain in the solid. In this case, surface stress is deferent from surface free energy density.

Just as in the case of surface free energy density, there is an alternative way of introducing the surface stress,

$$d(\gamma A) = A\sigma^s_{\alpha\beta}d\varepsilon^s_{\alpha\beta} = A_0(1+\varepsilon^s_{\eta\eta} + \tilde{\varepsilon}^s_{\eta\eta})\sigma^s_{\alpha\beta}d\varepsilon^s_{\alpha\beta} \tag{10}$$

where $\sigma^s_{\alpha\beta}$ may be called the Eulerian surface stress tensor. Note from (3) that $dA = A_0\delta_{\alpha\beta}d\varepsilon^s_{\alpha\beta}$. Therefore,

$$A\sigma^s_{\alpha\beta}d\varepsilon^s_{\alpha\beta} = d(\gamma A) = \gamma A_0\delta_{\alpha\beta}d\varepsilon^s_{\alpha\beta} + Ad\gamma \tag{11}$$

or,

$$\sigma^s_{\alpha\beta} = \frac{\gamma}{1+\varepsilon^s_{\eta\eta} + \tilde{\varepsilon}^s_{\eta\eta}}\delta_{\alpha\beta} + \frac{d\gamma}{d\varepsilon^s_{\alpha\beta}} \tag{12}$$

Although this expression has never appeared in the literature, it nevertheless is the direct consequence of (10). Clearly,

$$\Sigma^s_{\alpha\beta} = (1+\varepsilon^s_{\eta\eta} + \tilde{\varepsilon}^s_{\eta\eta})\sigma^s_{\alpha\beta} \tag{13}$$

The total strain energy stored in the deformed surface follows from (6),

$$U_{surface} = \iint_{S_0} \left[ \int_0^{\varepsilon^s_{\alpha\beta}} \Sigma^s_{\alpha\beta}(e_{\kappa\lambda}) de_{\alpha\beta} \right] dS_0 \tag{14}$$

where $\varepsilon^s_{\alpha\beta}$ is the surface strain in the final deformed configuration, $e_{\alpha\beta}$ is the integration variable representing the surface strain, and the fact that the surface stress is a function of the surface strain is explicitly indicated.

In this paper, we will use the Lagrange surface free energy density and surface stress tensor as defined by (2) and (8), respectively.

When the deformation is small, one may expand the surface free energy density into power series of the surface strain,

$$\begin{aligned}\Gamma &= \Gamma_0 + \frac{\partial \Gamma}{\partial \varepsilon^S_{\alpha\beta}} \varepsilon^s_{\alpha\beta} + \frac{1}{2} \frac{\partial^2 \Gamma}{\partial \varepsilon^S_{\alpha\beta} \partial \varepsilon^S_{\kappa\lambda}} \varepsilon^s_{\alpha\beta} \varepsilon^s_{\kappa\lambda} + \frac{1}{6} \frac{\partial^3 \Gamma}{\partial \varepsilon^S_{\alpha\beta} \partial \varepsilon^S_{\kappa\lambda} \partial \varepsilon^S_{\gamma\eta}} \varepsilon^s_{\alpha\beta} \varepsilon^s_{\kappa\lambda} \varepsilon^s_{\gamma\eta} \cdots \\ &= \Gamma_0 + \Gamma^{(1)}_{\alpha\beta} \varepsilon^s_{\alpha\beta} + \frac{1}{2} \Gamma^{(2)}_{\alpha\beta\kappa\lambda} \varepsilon^s_{\alpha\beta} \varepsilon^s_{\kappa\lambda} + \frac{1}{6} \Gamma^{(3)}_{\alpha\beta\kappa\lambda\gamma\eta} \varepsilon^s_{\alpha\beta} \varepsilon^s_{\kappa\lambda} \varepsilon^s_{\gamma\eta} \cdots\end{aligned} \tag{15}$$

where $\Gamma_0$, $\Gamma^{(1)}_{\alpha\beta}$, $\cdots$ are material and surface dependent. For a given material surface, they can be either measured experimentally or computed using molecular dynamic or atomistic simulations. Their values computed using molecular dynamic simulation will be reported in a separate publication [6]. Because of symmetries, one has $\Gamma^{(1)}_{\alpha\beta} = \Gamma^{(1)}_{\beta\alpha}$, $\Gamma^{(2)}_{\alpha\beta\kappa\lambda} = \Gamma^{(2)}_{\kappa\lambda\alpha\beta} = \Gamma^{(2)}_{\beta\alpha\kappa\lambda}$ and $\Gamma^{(3)}_{\alpha\beta\kappa\lambda\gamma\eta} = \Gamma^{(3)}_{\beta\alpha\kappa\lambda\gamma\eta} = \Gamma^{(3)}_{\kappa\lambda\alpha\beta\gamma\eta} = \Gamma^{(3)}_{\kappa\lambda\gamma\eta\alpha\beta}$. These conditions imply that there are at most three independent parameters in $\Gamma^{(1)}_{\alpha\beta}$, six in $\Gamma^{(2)}_{\alpha\beta\kappa\lambda}$ and 18 in $\Gamma^{(3)}_{\alpha\beta\kappa\lambda\gamma\eta}$.

Substitution of (15) into (8) yields, for small surface strain,

$$\Sigma^s_{\alpha\beta} = \Gamma^{(1)}_{\alpha\beta} + \Gamma^{(2)}_{\alpha\beta\kappa\lambda} \varepsilon^s_{\kappa\lambda} + \frac{1}{2} \Gamma^{(3)}_{\alpha\beta\kappa\lambda\gamma\eta} \varepsilon^s_{\kappa\lambda} \varepsilon^s_{\gamma\eta} \tag{16}$$

Clearly, $\Gamma^{(1)}_{\alpha\beta} = \Gamma^{(1)}_{\beta\alpha}$ is the internal stress of the surface. It represents the part of surface stress that exists even when the absolute surface strain is absent. The two dimensional fourth order tensor $\Gamma^{(2)}_{\alpha\beta\kappa\lambda} = \Gamma^{(2)}_{\kappa\lambda\alpha\beta} = \Gamma^{(2)}_{\beta\alpha\kappa\lambda} = \Gamma^{(2)}_{\alpha\beta\lambda\kappa}$ represents the surface stiffness, while the two dimensional sixth order tensor $\Gamma^{(2)}_{\alpha\beta\kappa\lambda\gamma\eta}$ can be viewed as the tensor of the third order elastic constants of the surface. In general, $\Gamma^{(1)}_{\alpha\beta}$ and $\Gamma^{(2)}_{\alpha\beta\kappa\lambda}$ are typically anisotropic in their surface (where they are defined). It can be shown [1] that $\Gamma^{(1)}_{\alpha\beta}$ is isotropic if $\Gamma^{(1)}_{12} = \Gamma^{(1)}_{21} = 0$ and $\Gamma^{(1)}_{11} = \Gamma^{(1)}_{22}$, and $\Gamma^{(2)}_{\alpha\beta\kappa\lambda}$ is isotropic if $\Gamma^{(2)}_{1112} = \Gamma^{(2)}_{1222} = 0$, $\Gamma^{(2)}_{1212} = \mu^s$, $\Gamma^{(2)}_{1122} = K^s - \mu^s$, and $\Gamma^{(2)}_{1111} = \Gamma^{(2)}_{2222} = \Gamma^{(2)}_{1122} + 2\Gamma^{(2)}_{1212} = K^s + \mu^s$.

Another important comment must be made here is that both surface free energy density and surface stress are macroscopic thermodynamic quantities. The basic idea of Gibbs surface energy is based on the concept of a dividing surface that separates the two adjacent phases. Under this assumption, the surface contributions to the thermodynamic quantities (e.g., surface free energy and surface stress) are defined as the excesses over the values that would be obtained if the bulk phases retained their properties constants up to dividing surface. In other words, the interface (not interphase) is a mathematical surface of zero thickness over which the thermodynamic properties changes discontinuously from one bulk phase to the other. The excess amount is associated only with the dividing surface. However, this is only an idealization of the realistic situation. In the case of a free surface, for example, the surface contributions to the surface free energy come from several layers of atoms near the surface [2]. So, theoretically speaking, the surface free energy is defined not just on the surface, but on a layer of mass near the surface. The idealization of the dividing surface is thus valid if and only if the bulk crystal is much larger than several atomic sizes. If the bulk crystal contains only a small number of atoms, the usefulness of macroscopic thermodynamic quantities such as surface free energy is questionable.

From the view point of continuum mechanics, the dividing surface idealization means that there is an "excess" amount of deformation on the dividing surface due to surface stress. In other words, the matters in the dividing surface deform differently from its neighboring matters in the bulk. This idealization justifies the requirement that the surface free energy $U_{surface}$ must be positive definite.

## 2. Effective Stiffness of a Single Particle

Conventionally, the elastic modulus of a material is an intensive property. It is defined as a point-wise quantity that relates the stresses and strains at each point in the material. When a material is not homogeneous, its elastic modulus may vary from point to point. In this case, the concept of effective modulus can be introduced. For example, effective modulus is used to characterize the overall stiffness of a fiber reinforced composite, where the fiber and matrix have different elastic modulus.

Now consider a particle made from a homogeneous material. On or near the particle surface, atomic structure is somewhat different from that of the bulk. Therefore, a particle, strictly speaking, is not a homogeneous body. The overall stiffness of the particle needs to be characterized by its effective modulus. However, when the particle size is large enough, the surface region is negligible in comparison to the particle size. In this case, the particle can be considered as a homogenous body. Therefore, its elastic modulus is uniform and is the same as that of the material from which the particle is made. This is no longer the case when the particle size shrinks to the nanometer range, where the surface region becomes significant in comparison to the particle size.

Consequently, the particle must be viewed as an inhomogeneous body and the effective modulus of the particle needs to be used to characterize the stiffness of the particle. In this section, a formulation is developed to compute the effective modulus of a particle that includes the effects of surface energy.

To this end, consider an ellipsoidal particle, $\Omega$. It is assumed that the particle, having initial volume $V_0$, is made from a homogeneous elastic solid with elastic potential $\Phi$. Let $\Omega$ be subjected to a given homogeneous displacement field $u_i$, measured from the undeformed perfect crystal lattice. Its corresponding uniform strain field is denoted by $\varepsilon_{ij}$, the absolute strain as defined in the previous section. After the applied deformation $\varepsilon_{ij}$, the corresponding total strain energy of the particle can be written as,

$$U = U_{bulk} + U_{surface} \tag{17}$$

where $U_{bulk}$ is the total strain energy in the bulk of the particle,

$$U_{bulk} = V_0 \left[ \frac{1}{2} C_{ijkl} \varepsilon_{ij} \varepsilon_{kl} + \frac{1}{6} C^{(3)}_{ijklmn} \varepsilon_{ij} \varepsilon_{kl} \varepsilon_{mn} \right] \tag{18}$$

where $C_{ijkl}$ and $C^{(3)}_{ijklmn}$ are, respectively, the tensor of second and third order elastic constants of the perfect crystal lattice.

The total surface free energy on the entire particle surface follows from (14),

$$U_{surface} = \int_{S_0} \left[ \int_0^{\varepsilon^s_{\alpha\beta}} \Sigma^s_{\alpha\beta}(e_{\kappa\lambda}) de_{\alpha\beta} \right] dS_0 \tag{19}$$

After some simplifications [6], the total strain energy can be written as,

$$U = \frac{V_0}{a} \tau_{ij} \varepsilon_{ij} + \frac{V_0}{2} \left( C_{ijkl} + \frac{1}{a} Q_{ijkl} \right) \varepsilon_{ij} \varepsilon_{kl} + \frac{V_0}{6} \left( C^{(3)}_{ijklmn} + \frac{1}{a} P_{ijklmn} \right) \varepsilon_{ij} \varepsilon_{kl} \varepsilon_{mn} \tag{20}$$

where $a$ is the smallest of the three semi-axes of the ellipsoid and,

$$\begin{aligned}
\tau_{ij} &= \frac{a}{V_0} \int_{S_0} \Gamma^{(1)}_{\alpha\beta} t_{\alpha i} t_{\beta j} dS_0 \\
Q_{ijkl} &= \frac{a}{V_0} \int_{S_0} \Gamma^{(2)}_{\alpha\beta\kappa\lambda} t_{\alpha i} t_{\beta j} t_{\kappa k} t_{\lambda l} dS_0 \\
P_{ijklmn} &= \frac{a}{V_0} \int_{S_0} \Gamma^{(3)}_{\alpha\beta\kappa\lambda\gamma\eta} t_{\alpha i} t_{\beta j} t_{\kappa k} t_{\lambda l} t_{\gamma m} t_{\eta n} dS_0
\end{aligned} \tag{21}$$

where $t_{ij}$ is the rotation tensor that relates the surface strain $\tilde{\varepsilon}^s_{\alpha\beta}$ to the bulk strain $\varepsilon_{ij}$.

The fourth order tensor $Q_{ijkl}$ can be viewed as the *surface rigidity tensor*. It represents a combined effect of the surface stiffness, $\Gamma^{(2)}_{\alpha\beta\kappa\lambda}$, and the surface geometry. Note that the surface rigidity tensor has the dimension of force per unit length. This gives the total strain energy of the particle when it deforms relative to the undeformed perfect crystal lattice of an infinite extent.

Because of the surface stress, the self-equilibrium state of the particle is different from the perfect crystal lattice of an infinite extent. The strain tensor that describes the deformation from the perfect crystal lattice to the equilibrium state of the particle can be found by minimizing the total strain energy,

$$\frac{\partial U}{V_0 \partial \varepsilon_{ij}} = \frac{1}{a}\tau_{ij} + \left(C_{ijkl} + \frac{1}{a}Q_{ijkl}\right)\varepsilon_{ij} + \frac{1}{2}\left(C^{(3)}_{ijklmn} + \frac{1}{a}P_{ijklmn}\right)\varepsilon_{kl}\varepsilon_{mn} = 0 \tag{22}$$

For small strain deformation, this yields the equilibrium strain, retaining the leading order of $1/a$,

$$\hat{\varepsilon}_{ij} \approx -\frac{1}{a}\left(C_{ijkl} + \frac{1}{a}Q_{ijkl}\right)^{-1}\tau_{kl} \approx -\frac{1}{a}C^{-1}_{ijkl}\tau_{kl} = -\frac{1}{a}M_{ijkl}\tau_{kl} \tag{23}$$

where $M_{ijkl} = C^{-1}_{ijkl}$ is the compliance tensor of the bulk crystal.

Now, the effective stiffness tensor of the particle at the state of self-equilibrium can be defined as,

$$\bar{C}_{ijkl} = \frac{\partial^2}{\partial \varepsilon_{ij}\partial \varepsilon_{kl}}\left(\frac{U}{V_0}\right)\Bigg|_{\varepsilon_{kl}=\hat{\varepsilon}_{kl}} \tag{24}$$

Substitution of (24) into (25) gives,

$$\bar{C}_{ijkl} = C_{ijkl} + \frac{1}{a}Q_{ijkl} + \left(C^{(3)}_{ijklmn} + \frac{1}{a}P_{ijklmn}\right)\hat{\varepsilon}_{mn} \tag{25}$$

Substituting (23) into (25) and retaining the terms up to $1/a$, we have,

$$\bar{C}_{ijkl} = C_{ijkl} + \frac{1}{a}\left(Q_{ijkl} - C^{(3)}_{ijklmn}M_{mnpq}\tau_{pq}\right) \tag{26}$$

This is the effective elastic modulus of the ellipsoidal particle in consideration. The contribution of the surface energy to the effective modulus of the particle is inversely proportional to the particle size.

## 2.1. EXAMPLES

When the geometry of the particle consists of an isotropic spherical particle with isotropic surface, one can easily find that the effective stiffness tensor is still isotropic. For such a particle of radius *a*, the effective bulk and shear moduli are,

$$\bar{K} = K + \frac{4}{3a}\left[ K^s - \frac{2\Gamma_1}{3K}(9L + 18M + 8N) \right] \tag{27}$$

$$\bar{\mu} = \mu + \frac{1}{a}\left[ \frac{1}{5}\left(K^s + 6\mu^s\right) - \frac{2\Gamma_1}{3K}(3M + 4N) \right] \tag{28}$$

where K and $\mu$ are, respectively, the bulk and shear moduli of the bulk material, L, M and N are the third order elastic constants, and $K^s$, $\mu^s$ are related to the surface energy.

Unlike the spherical particles, a thin isotropic fiber of radius $a$ with isotropic surface shows transverse anisotropy with its axis of symmetry in the fiber direction. One of the interesting quantities is the Young's modulus in the fiber direction (direction 3),

$$\bar{E}_L = \frac{1}{V_0}\frac{\partial^2 U}{\partial \varepsilon_1^2} = \bar{C}_{3333} - \frac{\bar{C}_{1133}^2}{\bar{C}_{1111}} - \frac{\left(\bar{C}_{1111}\bar{C}_{2233} - \bar{C}_{1122}\bar{C}_{1133}\right)^2}{\bar{C}_{1111}\left(\bar{C}_{1111}\bar{C}_{2222} - \bar{C}_{1122}^2\right)}$$

$$\bar{E}_L = E + \frac{1}{a}\frac{3\pi}{4(3K+\mu)^2}\left\{ \frac{1}{4}(3K+4\mu)^2\left(K^s + \mu^s\right) - 9\mu^s(5K+2\mu)(3K-2\mu) \right. \tag{29}$$

$$\left. - \frac{\Gamma_1}{3K\mu}\left(27K^3N + 2\mu^3(9L + 18M + 8N) + 27K^2\mu(3M + 4N)\right) \right\}$$

where $E$ is the bulk Young's modulus and $\nu$ is the Poisson's ratio of the fiber.

Similarly, the stiffness tensor of an isotropic film of thickness $t$ with isotropic surfaces is also transversely isotropic with the isotropic plane being the film plane. For a film in the $x_1 - x_2$ plane, the Young's modulus (Fig. 1) is given by,

$$\bar{Y} = \frac{1}{V_0}\frac{\partial^2 U}{\partial \varepsilon_{11}^2} = \bar{C}_{1111} - \frac{\bar{C}_{1133}^2}{\bar{C}_{3333}} - \frac{\left(\bar{C}_{1122}\bar{C}_{3333} - \bar{C}_{1133}\bar{C}_{2233}\right)^2}{\bar{C}_{3333}\left(\bar{C}_{2222}\bar{C}_{3333} - \bar{C}_{2233}^2\right)}$$

$$\bar{Y} = + \frac{1}{a}\frac{3}{(3K+\mu)^2}\left\{ \frac{81}{8}K^2\left(K^s + \mu^s\right) - K^s(3K+\mu)(3K-2\mu) \right. \tag{30}$$

$$\left. - \frac{\Gamma_1}{6K\mu}\left(27K^3N + 2\mu^3(9L + 18M + 8N) + 27K^2\mu(3M + 4N)\right) \right\}$$

where K and μ are the bulk and shear moduli of the bulk material.

As shown by Fig. 1 that this model clearly demonstrates that the effective elastic moduli of nano-size particle is affected by the surface energy and therefore depends on the particle size and shape. This dependency becomes significant only when the particle size reduces below 20 nanometers.

By introducing the surface energy, effective elastic modulus of nano-particles was studied. It was found that the effective modulus of a nano-particle does depend on its size. Explicit expressions of such size dependency were given for ellipsoidal particles of arbitrary anisotropy. Furthermore, self-equilibrium strain of the particle due to surface tension was also derived explicitly. By taking proper limits, the solution derived here can be applied to nano-wires and films [6].

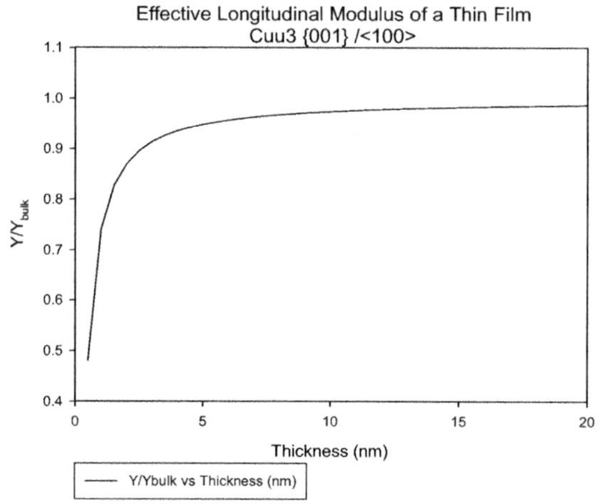


*Figure 1.* Effective Young's Modulus of a Thin Film as a function of Film Thickness for Copper.

## Acknowledgements

The research was supported in parts by NSF through the Packaging Research Center at Georgia Tech, by NSFC through under Granted No. 10228204, and by Harbin Institute of Technology through a visiting professorship to JQ.

## References

[1] Aris, R, Vectors, Tensors, and Basic Equations of Fluid Mechanics, Prentice-Hall, N.J., 1962.

[2] Blackely, J.M, Introduction to the Properties of Crystal Surfaces, Pergamon Press, New York, 1973.

[3] Gibbs, J.W, The Collected Works of J. Willard Gibbs, Longmans, New York. 1928.

[4] Sanfeld, A. and Steinchen, A, "Surface Energy, Stress, Capillary-Elastic Pressure and Chemical Equilibrium Constant in Nanoparticles," *Surface Science*, Vol. 463, 157–173, 2000.

[5] Shuttleworth, R, "The Surface Tension of Solids," *Proc. R. Soc. London*, Ser. A., Vol. 63, 444–457, 1950.

[6] Dingreville, R, Qu, J and Cherkaoui, M., "Mechanics of Nano-Structured and Nano-Composite Materials", in preparation, 2004.

## NANOINDENTATIONS AND THEORY OF STRAIN GRADIENT VISCOELASTICITY

L. H. KEUNG, D. C. C. LAM* and P. TONG[1]

[1]*Department of Mechanical Engineering, the Hong Kong University of Science and Technology*
*Clear Water Bay, Hong Kong, People's Republic of China*
[1]*Bioengineering Department, University of California at San Diego*
*La Jolla, California, USA*

### Abstract

Nanoindentations on polymeric viscoelastic materials indicated plastic deformation is size dependent when the indentation is small. The deformation behaviors were successfully modeled using strain gradient plasticity. In this investigation, dynamic nanoindentations on polymeric viscoelastic materials indicated that the storage and loss moduli are size dependent when indent size is small. This suggests that higher order strain gradient contributions to deformation need to be included in deformation of small scaled structures. In this paper, we extended the strain gradient theory to treat visco-elastic materials. A strain gradient viscoelastic bending theory for plane-strain beams is then developed based on this strain gradient viscoelastic theory. In classical bending theory, the normalized bending rigidity is independent of the length and the thickness of the beam. The theory predicts that the normalized creep compliance and the normalized bending relaxation modulus (related to the regular bending rigidity) depends on a higher-order bending parameter, $b_h$, and the thickness of the beam.

### 1. Introduction

Micrometer-scaled films are widely used in engineering applications for their electrical, optical, hardness and corrosion-resistance properties for non-structural functions. On the other hand, microelectromechanical systems (MEMS) and nanoelectromechanical systems (NEMS) are created to probe surfaces, study cells and neurons, move micro-liters of fluids and divert photons in optoelectronics (Bashir *et al.*, 2000; Carr and Craighead, 1997. These structures are microns and nanometers in size (called small-scaled structures in this paper). The beams and plates in these devices carry load and deform. The focuses were on metals and silicon and this attracted the attentions of the



*Q. P. Sun and P. Tong (eds.), IUTAM Symposium on Size Effects on Material and Structural Behavior at Micron- and Nano-Scales,* 249–261.

materials and mechanics community. Because of the small size of the structures, higher-order theories of plasticity and elasticity, which include contributions from strain gradients, have been developed by many authors to examine the elastic and plastic deformations of the micro-scaled structures, *e.g.*, Fleck and Hutchinson (1997, 2001), Yang *et al.* (2002), Lam *et al.* (2003), in which many additional references can be found.

From a performance perspective, metal, ceramic or semiconductor based systems are suited for system applications with high engineering demands, but the system costs are unsuited for implementation in consumer health and related products. For consumer-oriented systems where temperature requirements are less demanding, polymeric based systems maybe considered as cost-effective options. In prior works, we investigated the behavior of polymeric micron-scaled beams and found that the deformation behavior is dependent on strain gradient. A strain gradient elasticity theory was developed to describe the behavior in the elastic regime. In this paper, the viscoelastic behavior of polymer is investigated using nanoindentation. It will be shown that the polymer behaves in the conventional fashion when the indent size is large and has negligible strain gradients. When the indent size is small, i.e., when the strain gradient contributions are non-negligible, the viscoelastic behaviors becomes size dependent, indicating strain gradient dependence. Following the nanoindentation experiments, the higher order viscoelastic strain gradient theory and the higher order viscoelastic solution for bending of a small scale viscoelastic beam are presented.

## 2. Nanoindentation Experiments on Polymer

A commercial liquid epoxy resin (Shell Epikote 828 epoxy resin (bisphenol–A epichlorohydrin)) consisting of 2,2-bis (4%-hydroxyphenyl) propane (bisphenol A) and 1-chloro-2,3-epoxy-propane, known as epichlorohydrin with an average molecular weight less than 700 g $mol^{-1}$ is used. The molecular species is a linear polyether with terminal glycidyl ether groups and secondary hydroxyl groups occurring at regular intervals along the length of the macromolecule. The epoxy resin was cured by an amine curing agent (hardener), Shell's Epikure T (diethylene triamine (DETA). Diethylenetriamine is a reactive primary aliphatic amine with five active hydrogen atoms available for cross-linking.

Bisphenol-A epichlorohydrin resin was cured with DETA. Both liquid epoxy and hardener were mixed at room temperature and cast into a circular Teflon crucible. The specimens were then placed in a vacuum chamber to remove trapped air, cured in an oven at 100°C for 1 h and slowly cooled to room temperature followed by annealed at 80°C for 0.5 h. The nanoindentations were conducted in a Hysitron Triboindenter with a Berkovich head. Loads in the range of 10 μM – 1000 μM with amplitudes between 1 μN to 10 μN and frequencies between 1– 300 Hz, were used.

The variation of the storage and loss moduli calculated using Hysitron's software are plotted for an epoxy made with 1:6 ratio of curing agent to resin ratio. At low frequencies of 10 and 100 Hz, the loss moduli are near zero with no significant size dependence observed; but the storage modulus exhibited size dependence at indent size less than 100 μm. At 300 Hz, the loss moduli have risen above zero, which indicated that the loss is becoming significant at higher frequencies, which is in line with conventional viscoelastic behaviors. At small indents, the loss moduli are observed to be size dependence when strain gradients are non-negligible. In the following section, the development of higher order viscoelasticity theory is detailed.

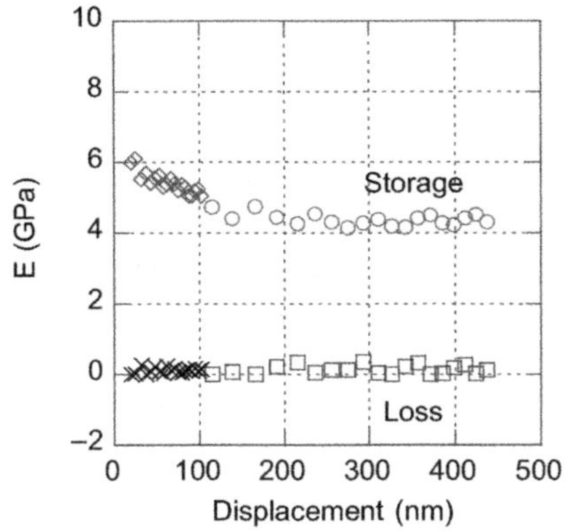


*Figure 1.* Storage and loss modulus of epoxy indented at 10Hz

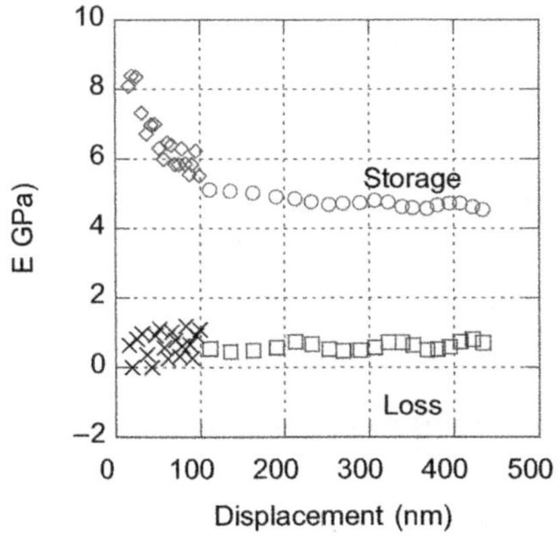


*Figure 2.* Storage and loss modulus of epoxy indented at 100Hz

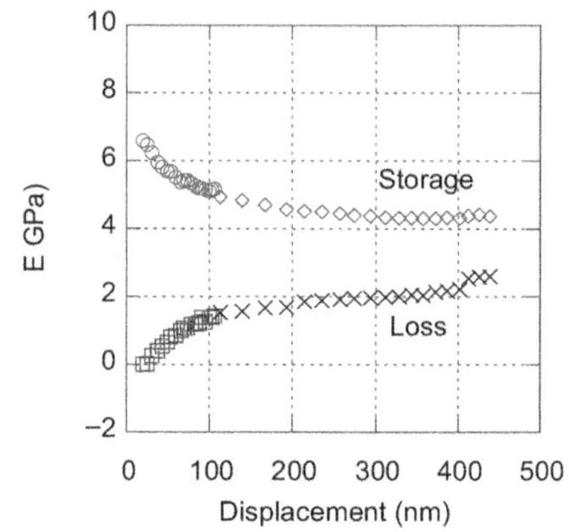


*Figure 3.* Storage and loss modulus of epoxy indented at 100Hz

## 3. Strain Gradient Viscoelasticity

The description of the deformation behavior of solids is not limited to relations between stresses and strains. Higher-order theories of plasticity and elasticity, which include contributions from strain gradients, have been developed by many authors. Lam *et al.* (2003) further developed a linear elastic bending theory for plane-strain beams. In the following, higher order viscoelasticity theory will be developed on basis of the framework of strain gradient elasticity.

### A Summary of Strain Gradient Theory for Linear Elasticity

In classical linear elasticity for isotropic materials, the stress $\sigma_{ij}$ and the strain $\varepsilon_{ij}$ are related via the elastic shear modulus, $\mu$, giving

$$\sigma_{ij} = 2\mu\left[\varepsilon_{ij} + \frac{\nu}{1-2\nu}\varepsilon_{kk}\delta_{ij}\right], \tag{1}$$

where $\nu$ is Poisson's ratio. (In the equation above and in subsequent equations, the index notation will be used with repeated indices denoting summation from 1 to 3.) $\mu$ and v are material constants and are independent of the geometry of the test specimen.

In the strain gradient theory, the deformation measures include the strain tensor, $\varepsilon_{ij}$, and the second-order deformation gradient tensors: the dilatation gradient vector, $\gamma_i$, the deviatoric stretch gradient tensor, $\eta_{ijk}$, and symmetric rotational gradient (curvature) tensor, $\chi_{ij}$. These strain measures are defined as

$$\varepsilon_{ij} = \frac{1}{2}(\partial_i u_j + \partial_j u_i), \quad \gamma_i = \partial_i \varepsilon_{mm},$$
$$\eta_{ijk} = \frac{1}{3}(\partial_i \varepsilon_{jk} + \partial_j \varepsilon_{ki} + \partial_k \varepsilon_{ij}) \tag{2}$$
$$- \frac{1}{15}[\delta_{ij}(\partial_k \varepsilon_{mm} + 2\partial_m \varepsilon_{mk}) + \delta_{jk}(\partial_i \varepsilon_{mm} + 2\partial_m \varepsilon_{mi}) + \delta_{ki}(\partial_j \varepsilon_{mm} + 2\partial_m \varepsilon_{mj})],$$
$$\chi_{ij} = \frac{1}{2}(e_{ipq}\partial_p \varepsilon_{qj} + e_{jpq}\partial_p \varepsilon_{qi}).$$

where $\partial_i$ is the forward gradient operator, $u_i$ the displacement vector, $\delta_{ij}$ the Kronecker delta and $e_{ijk}$ the alternating tensor. The strain tensor has six independent symmetric components. The number of independent components in $\gamma_i$ is three, in $\eta_{ijk}$ is seven, and in $\chi_{ij}$ is five. Thus the second-order deformation gradient tensors have a total of 15 independent components. The stress measures, the work-conjugates to the strain measures, are the classical stress tensor, $\sigma_{ij}$, and the higher-order stresses, $p_i$, $\tau_{ijk}$ and $m_{ij}$, respectively.

The application of the principle of virtual work gives the equilibrium equation in the volume $V$ as

$$\partial_i \sigma_{ik} - \frac{1}{2} e_{jlk} \partial_{il} m_{ij} - \partial_{ik} p_i - \partial_{ij} \tau_{ijk}^{(1)} + f_k = 0 \tag{3}$$

and the appropriate boundary conditions on $\partial V$ with $f_k$ being the body force per unit volume.

For linear elastic center-symmetric isotropic materials, the constitutive relations are

$$\sigma_{ij} = \lambda \delta_{ij} \varepsilon_{mm} + 2\mu \varepsilon_{ij}$$
$$p_i = 2\mu l_0^2 \varepsilon_{mm,i}, \quad \tau_{ijk} = 2\mu l_1^2 \eta_{ijk}, \quad m_{ij} = 2\mu l_2^2 \chi_{ij}. \tag{4}$$

where $\lambda$ and $\mu$ are Lame's constant and the shear modulus, respectively, and $l_n$ ($n$ = 0,1,2) are the three additional length scale material parameters associated with the dilatation, deviatoric stretch and rotation gradients, respectively.

**Viscoelastic constitutive relations**

For linear isotropic viscoelastic materials, we can define the constitutive relations in convolution integral form as

$$\sigma_{ij}(\mathbf{x},t) = \int_{-\infty}^{t} \lambda(\mathbf{x},t-\tau)\delta_{ij}\frac{\partial \varepsilon_{mm}(\mathbf{x},\tau)}{\partial \tau}d\tau + 2\int_{-\infty}^{t} \mu(\mathbf{x},t-\tau)\frac{\partial \varepsilon_{ij}(\mathbf{x},\tau)}{\partial \tau}d\tau \tag{5}$$

$$p_i(\mathbf{x},\tau) = 2l_0^2 \int_{-\infty}^{t} \mu_0(\mathbf{x},t-\tau)\frac{\partial \varepsilon_{mm,i}(\mathbf{x},\tau)}{\partial \tau}d\tau \tag{6}$$

$$\tau_{ijk} = 2l_1^2 \int_{-\infty}^{t} \mu_1(\mathbf{x},t-\tau)\frac{\partial \eta_{ijk}(\mathbf{x},\tau)}{\partial \tau}d\tau \tag{7}$$

$$m_{ij} = 2l_2^2 \int_{-\infty}^{t} \mu_2(\mathbf{x},t-\tau)\frac{\partial \chi_{ij}(\mathbf{x},\tau)}{\partial \tau}d\tau \tag{8}$$

where $\lambda$ and $\mu$'s are relaxation functions. A simple viscoelastic model is

$$\lambda(\mathbf{x},\tau) = \frac{2v}{1-2v}\mu(\mathbf{x},\tau) \tag{9}$$

$$\mu_0(\mathbf{x},\tau) = \mu_1(\mathbf{x},\tau) = \mu_2(\mathbf{x},\tau) \tag{10}$$

in which Poisson's ratio, $v$, is independent of time.

The constitutive relations can be put in differential equation forms as

$$P_3(\frac{\partial}{\partial t})\sigma_{kk} = Q_3(\frac{\partial}{\partial t})\varepsilon_{kk} \tag{11}$$

$$P_4(\frac{\partial}{\partial t})\sigma'_{ij} = Q_4(\frac{\partial}{\partial t})\varepsilon'_{ij} \tag{12}$$

$$P_0(\frac{\partial}{\partial t})p_i = 2l_0^2 Q_0(\frac{\partial}{\partial t})\varepsilon_{mm,i} \tag{13}$$

$$P_1(\frac{\partial}{\partial t})\tau_{ijk} = 2l_1^2 Q_1(\frac{\partial}{\partial t})\eta_{ijk} \tag{14}$$

$$P_2(\frac{\partial}{\partial t})m_{ij}^s = 2l_2^2 Q_2(\frac{\partial}{\partial t})\chi_{ij}^s \tag{15}$$

where $\sigma'_{ij}$, $\varepsilon'_{ij}$ are the stress and strain deviations, and $P$'s and $Q$'s are polynomials of the time-derivative operator. Again a simple viscoelastic model is

$$P_3 = P_4 \tag{16}$$

$$P_0 = P_1 = P_2 \tag{17}$$

$$Q_4 = \frac{2(1+\nu)}{3(1-2\nu)} Q_3 \tag{18}$$

$$Q_0 = Q_1 = Q_2 \tag{19}$$

As an example, the Kelvin model gives

$$\sigma_{kk} = \frac{2(1+\nu)\mu}{3(1-2\nu)}(\varepsilon_{kk} + \eta \frac{\partial \varepsilon_{kk}}{\partial t})$$

$$\sigma'_{ij} = 2\mu(\varepsilon'_{ij} + \eta \frac{\partial \varepsilon'_{ij}}{\partial t})$$

where η is the viscosity coefficient with dimension of time.

## 4. Bending Theory for Plane-strain Beams

The derivation follows that given by Lam *et al.* (2003). The bending stress resultants are first defined in terms of the stress measures in strain gradient theory. Then, the bending equilibrium equations and force-prescribed boundary conditions are derived. The bending deformation measures and constitutive relations are established using Taylor expansions of displacements with respect to the beam thickness.

With the presence of higher-order stress measures, the stress resultants on a beam are defined as

$$\begin{aligned}
N &= \int_{-h/2}^{h/2} \sigma_{11} dz, \\
Q &= \int_{-h/2}^{h/2} (\sigma_{13} + \frac{1}{2}\partial_1 m_{12} - \partial_1 \tau_{113}) dz, \\
M &= \int_{-h/2}^{h/2} (z\sigma_{11} + m_{12} + p_3 + \tau_{113}) dz, \\
N^h &= \int_{-h/2}^{h/2} (p_1 + \tau_{111}) dz, \\
M^h &= \int_{-h/2}^{h/2} z(p_1 + \tau_{111}) dz
\end{aligned} \tag{20}$$

where $N$ is the axial stress resultant per unit width, $Q$ is the shear stress resultant per unit width, and $M$ is the moment per unit width in the beam. The higher-order stress resultant per unit width, $N^h$, and the higher-order moment resultant per unit width, $M^h$, are introduced to account for the higher-order effect. The shear stress resultant, $Q$, and the moment resultant, $M$, now contain higher-order stress terms. Combining the equilibrium equations, the boundary conditions and the stress resultant definitions in Eq. (20), we find that

$$\frac{\partial N}{\partial x}-\frac{\partial^2 N^k}{\partial x^2}=m\frac{\partial^2 u_0}{\partial t^2}, \tag{21}$$

$$\frac{\partial M}{\partial x}-\frac{\partial^2 M^k}{\partial x^2}-Q=0 \tag{22}$$

$$\frac{\partial Q}{\partial x}+\bar{q}=m\frac{\partial^2 w_0}{\partial t^2} \tag{23}$$

where $\bar{q} = \bar{q}_{\frac{h}{2}} + \bar{q}_{-\frac{h}{2}}$ with $\bar{q}_{\pm\frac{h}{2}}$ being the prescribed normal force-traction on the top or bottom surfaces in units of $N/m^2$, $u_0$ and $w_0$ are the mid-plane beam axial and normal displacements and $m$ is the mass per unit length of the beam. Equations (22) and (23) can be combined into one equation as

$$\frac{\partial^2 M}{\partial x^2}-\frac{\partial^3 M}{\partial x^3}=m\frac{\partial^2 w_0}{\partial t^2}-\bar{q} \tag{24}$$

The stress resultants at the boundary where the tractions are

$$N-\frac{\partial N^k}{\partial x}=\bar{N} \tag{25}$$

$$N^h=\bar{N}_0^h \tag{26}$$

$$Q=\bar{Q} \tag{27}$$

$$M-\frac{\partial M^k}{\partial x}=\bar{M} \tag{28}$$

$$M^h=\bar{M}^h \tag{29}$$

where $\bar{N}$ is the prescribed total axial force in units of N/m; $\bar{Q}$ is the prescribed shear force in units of N/m; $\bar{M}$ is the prescribed moment per unit width in units of N-m/m; $\bar{N}^h$ is the prescribed higher-order axial force in units of N-m/m; and $\bar{M}^h$ is the prescribed higher-order moment per unit width in units of N-$m^2$/m. In bending, the prescribed axial stress resultant, $\bar{N}$, and the higher-order axial stress resultant, $\bar{N}^h$, which correspond to pure tension, are both zero.

**Deformation and constitutive relations**

The expressions of deformation and constitutive relations for thin beams can be simplified by expanding the displacements as a power series of the beam thickness, $h$. Utilizing the axial equilibrium equation and neglecting the higher order terms of $h$ (Lam *et al.* 2003), one can show that the beam deformation measures are the curvature, $\kappa$, and the curvature gradient, $\kappa^h$, respectively, defined as

$$\kappa = -w_0''(x,t), \qquad \kappa^h = -w_0'''(x,t), \tag{30}$$

where $w_0$ is the normal deflection of the beam at mid-surface and $(\cdot)''$ and $(\cdot)'''$ denote the second and third partial derivatives with respect to $x$.

Similar to strain gradient elastic bending (Lam *et al.* 2003), we can then derive the constitutive equations for the beam bending in terms of the curvature and the curvature gradients. For the simple viscoelastic model as defined in Eqs. (5)-(8), the Laplace transform of the constitutive relations are

$$\hat{\sigma}_{ij}(\mathbf{x},s) = s\hat{\lambda}(\mathbf{x},s)\hat{\varepsilon}_{mm}(\mathbf{x},s)\delta_{ij} + 2s\hat{\mu}(\mathbf{x},s)\hat{\varepsilon}_{ij}(\mathbf{x},s) \tag{31}$$

$$\hat{p}_i(\mathbf{x},\tau) = 2l_0^2 s\hat{\mu}_0(\mathbf{x},s)\hat{\varepsilon}_{mm,i}(\mathbf{x},s) \tag{32}$$

$$\hat{\tau}_{ijk} = 2l_1^2 s\hat{\mu}_1(\mathbf{x},s)\hat{\eta}_{ijk}(\mathbf{x},s) \tag{33}$$

$$\hat{m}_{ij} = 2l_2^2 s\hat{\mu}_2(\mathbf{x},s)\hat{\chi}_{ij}(\mathbf{x},s) \tag{34}$$

where $(\hat{\ })$ denotes the Laplace transform of the associated quantity. For beam bending, we can show that

$$\hat{M}(x,s) = -\hat{D}(x,s)\hat{w}_0''(x,s)\,;\ \bar{M}^h(x,s) = -\bar{D}^h(x,s)\bar{w}_0'''(x,s), \tag{35}$$

where the bending rigidities are

$$\hat{D}(x,s) = \hat{D}_3(x,s) + \frac{b_h^2}{h^2}\hat{D}_0(x,s) = \frac{h^3 s\hat{\mu}(x,s)}{6(1-\nu)}\left[1 + \frac{b_h^2}{h^2}\frac{\hat{\mu}_0(x,s)}{\hat{\mu}(x,s)}\right], \tag{36}$$

$$\hat{D}^h(x,s) = \delta^2\hat{D}_0(x,s) \tag{37}$$

where

$$\hat{D}_i(x,s) = \frac{h^3}{6(1-\nu)}s\hat{\mu}_i(x,s), \qquad (i = 0 \text{ and } 3), \tag{38}$$

$$b_h^2 = 6(1-2\nu)l_0^2 + \frac{2}{5}(4-\nu)l_1^2\frac{\hat{\mu}_1}{\hat{\mu}_0} + 3(1-\nu)l_2^2\frac{\hat{\mu}_2}{\hat{\mu}_0} \tag{39}$$

$$\delta^2 = (1-2\nu)\frac{l_0^2}{2} + \frac{l_1^2}{10}\frac{\hat{\mu}_1}{\hat{\mu}_0}\left(\frac{8l_1^2\hat{\mu}_1}{15} + \frac{l_2^2\hat{\mu}_2}{4}\right)^{-1}\left[\frac{2l_1^2\hat{\mu}_1}{3} + \frac{l_2^2\hat{\mu}_1}{2} - \left(\frac{2l_1^2\hat{\mu}_1}{3} - \frac{l_2^2\hat{\mu}_2}{4}\right)\nu\right] \tag{40}$$

in which $\bar{\mu}_3 = \mu$. In the above equations, $b_h$ and $\delta$ are the higher-order bending parameters, which characterize the thickness dependence of beam bending and $D_3$ becomes the conventional bending rigidity when $Q_3$ is a constant equal to the shear modulus $\mu$. The boundary conditions are

$$\begin{aligned} Q &= \bar{Q} & &\text{or} & w_0 &= \bar{w}_0, \\ M - \frac{\partial M^k}{\partial x} &= \bar{M} & &\text{or} & w_0' &= \bar{w}_0', \\ M^h &= \bar{M}^h & &\text{or} & w_0'' &= \bar{w}_0''. \end{aligned} \tag{41}$$

The displacement boundary conditions at the fixed end are

$$w_0 = \bar{w}_0, \quad w_0' = \bar{w}_0', \quad w_0'' = \bar{w}_0'', \tag{42}$$

at a simply supported end are

$$w_0 = \bar{w}_0, \quad M - \frac{\partial M^h}{\partial x} = \bar{M}, \quad M^h = 0, \tag{43}$$

and at a free end are

$$Q = \frac{\partial M}{\partial x} - \frac{\partial^2 M^h}{\partial x^2} = \bar{Q}, \quad M - \frac{\partial M^h}{\partial x} = \bar{M}, \quad M^h = \bar{M}^h \tag{44}$$

After solving the bending problems, the stresses and higher-order stresses can be obtained from constitutive relations, except for $\sigma_{13}$ and $\sigma_{33}$, which are determined by integrating equilibrium equation.

**Cantilever Beam with Prescribed Deflection or Force**

Consider a uniform viscoelastic cantilever beam clamped at $x = 0$ and subjected to forced displacements at $x = l$. Neglecting the inertia effect, from Eq. (24) we have

$$\frac{\partial^2 \hat{M}}{\partial x^2} - \frac{\partial^3 \hat{M}}{\partial x^3} = -\hat{D}(s)\frac{\partial^2 \hat{w}_0(x,s)}{\partial x^2} + \hat{D}^h(s)\frac{\partial^6 \hat{w}_0(x,s)}{\partial x^6} = 0. \tag{45}$$

The solution for $\hat{w}_0(x,s)$ is quite complicated. We will the case that all the length scaled parameters are of order micron-scale or smaller and that the term associated with $\hat{D}^h(x,s)$ in Eq. (45) can be neglected. In this case, based on the correspondence principle, the force and deflection relation in the thickness direction at x = $l$ is simply

$$\hat{P}(s) = \frac{6\hat{D}(s)}{l^3}\hat{w}_0(l,s). \tag{46}$$

**Bending Relaxation Modulus**

For the relaxation test with

$$w_0(l,t) = \overline{w}H(t) \tag{47}$$

where $H$ is a unit step function defined as

$$\begin{aligned} H(t) &= 0 \quad \text{for} \quad t < 0 \\ &= 1 \quad \text{for} \quad t \geq 0 \end{aligned} \tag{48}$$

we have

$$\hat{w}_0(l,s) = \frac{\overline{w}}{s} \tag{49}$$

Then

$$P(t) = \frac{h^3\overline{w}}{l^3(1-\nu)}Y(t) \tag{50}$$

where $Y(t)$ is called the normalized bending relaxation modulus that

$$Y(t) = \mathcal{L}^{-1}\left(\hat{\mu}(s)(1+\frac{b_h^2}{h^2}\frac{\hat{\mu}_0(s)}{\hat{\mu}(s)})\right) \tag{51}$$

which exhibits an inverse squared dependence on the beam's thickness. The symbol $\mathcal{L}^{-1}$ denotes the inverse Laplace transform of the referred function.

**Creep Compliance**

Consider a creep test by an applied load $P(t) = \overline{P}H(t)$ . Then

$$\hat{w}_0(l,s) = \frac{(1-\nu)l^3\overline{P}}{6h^3}[1+\frac{b_h^2}{h^2}\frac{\hat{\mu}_0(s)}{\hat{\mu}(s)}]^{-1}\frac{\hat{P}(s)}{\hat{\mu}(s)} \tag{52}$$

or

$$w_0(l,t) = \frac{(1-\nu)l^3\overline{P}}{6h^3} J(t) \tag{53}$$

where

$$J(t) = \mathcal{L}^{-1}\left( [1 + \frac{b_h^2}{h^2}\frac{\hat{\mu}_0(s)}{\hat{\mu}(s)}]^{-1} \frac{1}{s^2\hat{\mu}(s)} \right) \tag{54}$$

is called the normalized creep compliance.

## 5. Discussion and Conclusions

In this paper, we showed that dynamic nanoindentation results indicated that the storage and loss moduli become size dependent at small indent, indicating strain gradient dependence. With this observation, we extended the strain gradient elasticity theory to visco-elasticity. From the strain gradient viscoelasticity theory, a viscoelastic bending theory for thin plane-strain beams was developed based on a simple viscoelasticity model. In the constitutive relations for beams, similar to that of elastic case, both the normalized creep compliance and the normalized regular bending relaxation modulus and the normalized higher-order complex bending rigidity $\hat{D}(x,s) / \hat{D}_3(x,s)$, and $\hat{D}^h(x,s) / \hat{D}_0(x,s)$ depend on the material length scale parameters. Only the normalized creep compliance and the normalized higher-order complex bending rigidity depends on the beam thickness that $D/D_3$ is a function of the inverse square of the thickness. On the basis of the higher order separability, the normalized higher-order bending complex bending rigidity is expected to be on the micron-scale.

## Acknowledgements

This work was supported by the Research Grants Council of the Hong Kong Special Administrative Region, the People's Republic of China.

## References

Bashir, R., Gupta, A., Neudeck, G.W., McElfresh, M., Gomez, R., (2000) On the design of piezoresistive silicon cantilevers with stress concentration regions for scanning probe microscopy applications, *Journal of Micromechanics & Microengineering* **10**, 483-491.

Carr, D.W., Craighead, H.G., (1997) Fabrication of nanoelectromechanical systems in single crystal silicon using silicon on insulator substrates and electron beam lithography *Journal of Vacuum Science & Technology* **B 15**, 2760-2763.

Fleck, N.A., Hutchinson, J.W., (1997) Strain gradient plasticity, Hutchinson, J.W., Wu, T.Y. (Eds.) *Advances in Applied Mechanics*. **33** Academic Press, New York, 295-361.

Fleck, N.A., Hutchinson, J.W., (2001) A reformulation of strain gradient plasticity, *J. Mech. Phys. Solids.* **49**, 2245-2271.

Lam, D.C.C., Chong, A.C.M., Yang, F., Wang, J., and Tong, P. (2003). Experiments and theory in strain gradient elasticity. *J. Mech. Phys. Solids.,* **51**, 1477-1503.

Yang, F., Chong, A.C.M., Lam, D.C.C., Tong, P., (2000) Couple stress Based Strain gradient theory for elasticity, *Int. J. Solids & Struct.,* **39**, 2731-2743.

# Author Index

# Mechanics

## *SOLID* MECHANICS AND ITS APPLICATIONS

*Series Editor*: G.M.L. Gladwell

*Aims and Scope of the Series*

The fundamental questions arising in mechanics are: *Why?*, *How?*, and *How much?* The aim of this series is to provide lucid accounts written by authoritative researchers giving vision and insight in answering these questions on the subject of mechanics as it relates to solids. The scope of the series covers the entire spectrum of solid mechanics. Thus it includes the foundation of mechanics; variational formulations; computational mechanics; statics, kinematics and dynamics of rigid and elastic bodies; vibrations of solids and structures; dynamical systems and chaos; the theories of elasticity, plasticity and viscoelasticity; composite materials; rods, beams, shells and membranes; structural control and stability; soils, rocks and geomechanics; fracture; tribology; experimental mechanics; biomechanics and machine design.

1. R.T. Haftka, Z. Gürdal and M.P. Kamat: *Elements of Structural Optimization*. 2nd rev.ed., 1990 ISBN 0-7923-0608-2
2. J.J. Kalker: *Three-Dimensional Elastic Bodies in Rolling Contact*. 1990 ISBN 0-7923-0712-7
3. P. Karasudhi: *Foundations of Solid Mechanics*. 1991 ISBN 0-7923-0772-0
4. *Not published*
5. *Not published.*
6. J.F. Doyle: *Static and Dynamic Analysis of Structures*. With an Emphasis on Mechanics and Computer Matrix Methods. 1991 ISBN 0-7923-1124-8; Pb 0-7923-1208-2
7. O.O. Ochoa and J.N. Reddy: *Finite Element Analysis of Composite Laminates*. ISBN 0-7923-1125-6
8. M.H. Aliabadi and D.P. Rooke: *Numerical Fracture Mechanics*. ISBN 0-7923-1175-2
9. J. Angeles and C.S. López-Cajún: *Optimization of Cam Mechanisms*. 1991 ISBN 0-7923-1355-0
10. D.E. Grierson, A. Franchi and P. Riva (eds.): *Progress in Structural Engineering*. 1991 ISBN 0-7923-1396-8
11. R.T. Haftka and Z. Gürdal: *Elements of Structural Optimization*. 3rd rev. and exp. ed. 1992 ISBN 0-7923-1504-9; Pb 0-7923-1505-7
12. J.R. Barber: *Elasticity*. 1992 ISBN 0-7923-1609-6; Pb 0-7923-1610-X
13. H.S. Tzou and G.L. Anderson (eds.): *Intelligent Structural Systems*. 1992 ISBN 0-7923-1920-6
14. E.E. Gdoutos: *Fracture Mechanics*. An Introduction. 1993 ISBN 0-7923-1932-X
15. J.P. Ward: *Solid Mechanics*. An Introduction. 1992 ISBN 0-7923-1949-4
16. M. Farshad: *Design and Analysis of Shell Structures*. 1992 ISBN 0-7923-1950-8
17. H.S. Tzou and T. Fukuda (eds.): *Precision Sensors, Actuators and Systems*. 1992 ISBN 0-7923-2015-8
18. J.R. Vinson: *The Behavior of Shells Composed of Isotropic and Composite Materials*. 1993 ISBN 0-7923-2113-8
19. H.S. Tzou: *Piezoelectric Shells*. Distributed Sensing and Control of Continua. 1993 ISBN 0-7923-2186-3
20. W. Schiehlen (ed.): *Advanced Multibody System Dynamics*. Simulation and Software Tools. 1993 ISBN 0-7923-2192-8
21. C.-W. Lee: *Vibration Analysis of Rotors*. 1993 ISBN 0-7923-2300-9
22. D.R. Smith: *An Introduction to Continuum Mechanics*. 1993 ISBN 0-7923-2454-4
23. G.M.L. Gladwell: *Inverse Problems in Scattering*. An Introduction. 1993 ISBN 0-7923-2478-1

# Mechanics

## *SOLID* MECHANICS AND ITS APPLICATIONS

*Series Editor*: G.M.L. Gladwell

24. G. Prathap: *The Finite Element Method in Structural Mechanics.* 1993 ISBN 0-7923-2492-7
25. J. Herskovits (ed.): *Advances in Structural Optimization.* 1995 ISBN 0-7923-2510-9
26. M.A. González-Palacios and J. Angeles: *Cam Synthesis.* 1993 ISBN 0-7923-2536-2
27. W.S. Hall: *The Boundary Element Method.* 1993 ISBN 0-7923-2580-X
28. J. Angeles, G. Hommel and P. Kovács (eds.): *Computational Kinematics.* 1993 ISBN 0-7923-2585-0
29. A. Curnier: *Computational Methods in Solid Mechanics.* 1994 ISBN 0-7923-2761-6
30. D.A. Hills and D. Nowell: *Mechanics of Fretting Fatigue.* 1994 ISBN 0-7923-2866-3
31. B. Tabarrok and F.P.J. Rimrott: *Variational Methods and Complementary Formulations in Dynamics.* 1994 ISBN 0-7923-2923-6
32. E.H. Dowell (ed.), E.F. Crawley, H.C. Curtiss Jr., D.A. Peters, R. H. Scanlan and F. Sisto: *A Modern Course in Aeroelasticity.* Third Revised and Enlarged Edition. 1995 ISBN 0-7923-2788-8; Pb: 0-7923-2789-6
33. A. Preumont: *Random Vibration and Spectral Analysis.* 1994 ISBN 0-7923-3036-6
34. J.N. Reddy (ed.): *Mechanics of Composite Materials.* Selected works of Nicholas J. Pagano. 1994 ISBN 0-7923-3041-2
35. A.P.S. Selvadurai (ed.): *Mechanics of Poroelastic Media.* 1996 ISBN 0-7923-3329-2
36. Z. Mróz, D. Weichert, S. Dorosz (eds.): *Inelastic Behaviour of Structures under Variable Loads.* 1995 ISBN 0-7923-3397-7
37. R. Pyrz (ed.): *IUTAM Symposium on Microstructure-Property Interactions in Composite Materials.* Proceedings of the IUTAM Symposium held in Aalborg, Denmark. 1995 ISBN 0-7923-3427-2
38. M.I. Friswell and J.E. Mottershead: *Finite Element Model Updating in Structural Dynamics.* 1995 ISBN 0-7923-3431-0
39. D.F. Parker and A.H. England (eds.): *IUTAM Symposium on Anisotropy, Inhomogeneity and Nonlinearity in Solid Mechanics.* Proceedings of the IUTAM Symposium held in Nottingham, U.K. 1995 ISBN 0-7923-3594-5
40. J.-P. Merlet and B. Ravani (eds.): *Computational Kinematics '95.* 1995 ISBN 0-7923-3673-9
41. L.P. Lebedev, I.I. Vorovich and G.M.L. Gladwell: *Functional Analysis.* Applications in Mechanics and Inverse Problems. 1996 ISBN 0-7923-3849-9
42. J. Menčik: *Mechanics of Components with Treated or Coated Surfaces.* 1996 ISBN 0-7923-3700-X
43. D. Bestle and W. Schiehlen (eds.): *IUTAM Symposium on Optimization of Mechanical Systems.* Proceedings of the IUTAM Symposium held in Stuttgart, Germany. 1996 ISBN 0-7923-3830-8
44. D.A. Hills, P.A. Kelly, D.N. Dai and A.M. Korsunsky: *Solution of Crack Problems.* The Distributed Dislocation Technique. 1996 ISBN 0-7923-3848-0
45. V.A. Squire, R.J. Hosking, A.D. Kerr and P.J. Langhorne: *Moving Loads on Ice Plates.* 1996 ISBN 0-7923-3953-3
46. A. Pineau and A. Zaoui (eds.): *IUTAM Symposium on Micromechanics of Plasticity and Damage of Multiphase Materials.* Proceedings of the IUTAM Symposium held in Sèvres, Paris, France. 1996 ISBN 0-7923-4188-0
47. A. Naess and S. Krenk (eds.): *IUTAM Symposium on Advances in Nonlinear Stochastic Mechanics.* Proceedings of the IUTAM Symposium held in Trondheim, Norway. 1996 ISBN 0-7923-4193-7
48. D. Ieşan and A. Scalia: *Thermoelastic Deformations.* 1996 ISBN 0-7923-4230-5

# Mechanics

## *SOLID* MECHANICS AND ITS APPLICATIONS

*Series Editor*: G.M.L. Gladwell

49. J.R. Willis (ed.): *IUTAM Symposium on Nonlinear Analysis of Fracture*. Proceedings of the IUTAM Symposium held in Cambridge, U.K. 1997 ISBN 0-7923-4378-6
50. A. Preumont: *Vibration Control of Active Structures*. An Introduction. 1997 ISBN 0-7923-4392-1
51. G.P. Cherepanov: *Methods of Fracture Mechanics: Solid Matter Physics*. 1997 ISBN 0-7923-4408-1
52. D.H. van Campen (ed.): *IUTAM Symposium on Interaction between Dynamics and Control in Advanced Mechanical Systems*. Proceedings of the IUTAM Symposium held in Eindhoven, The Netherlands. 1997 ISBN 0-7923-4429-4
53. N.A. Fleck and A.C.F. Cocks (eds.): *IUTAM Symposium on Mechanics of Granular and Porous Materials*. Proceedings of the IUTAM Symposium held in Cambridge, U.K. 1997 ISBN 0-7923-4553-3
54. J. Roorda and N.K. Srivastava (eds.): *Trends in Structural Mechanics*. Theory, Practice, Education. 1997 ISBN 0-7923-4603-3
55. Yu.A. Mitropolskii and N. Van Dao: *Applied Asymptotic Methods in Nonlinear Oscillations*. 1997 ISBN 0-7923-4605-X
56. C. Guedes Soares (ed.): *Probabilistic Methods for Structural Design*. 1997 ISBN 0-7923-4670-X
57. D. François, A. Pineau and A. Zaoui: *Mechanical Behaviour of Materials*. Volume I: Elasticity and Plasticity. 1998 ISBN 0-7923-4894-X
58. D. François, A. Pineau and A. Zaoui: *Mechanical Behaviour of Materials*. Volume II: Viscoplasticity, Damage, Fracture and Contact Mechanics. 1998 ISBN 0-7923-4895-8
59. L.T. Tenek and J. Argyris: *Finite Element Analysis for Composite Structures*. 1998 ISBN 0-7923-4899-0
60. Y.A. Bahei-El-Din and G.J. Dvorak (eds.): *IUTAM Symposium on Transformation Problems in Composite and Active Materials*. Proceedings of the IUTAM Symposium held in Cairo, Egypt. 1998 ISBN 0-7923-5122-3
61. I.G. Goryacheva: *Contact Mechanics in Tribology*. 1998 ISBN 0-7923-5257-2
62. O.T. Bruhns and E. Stein (eds.): *IUTAM Symposium on Micro- and Macrostructural Aspects of Thermoplasticity*. Proceedings of the IUTAM Symposium held in Bochum, Germany. 1999 ISBN 0-7923-5265-3
63. F.C. Moon: *IUTAM Symposium on New Applications of Nonlinear and Chaotic Dynamics in Mechanics*. Proceedings of the IUTAM Symposium held in Ithaca, NY, USA. 1998 ISBN 0-7923-5276-9
64. R. Wang: *IUTAM Symposium on Rheology of Bodies with Defects*. Proceedings of the IUTAM Symposium held in Beijing, China. 1999 ISBN 0-7923-5297-1
65. Yu.I. Dimitrienko: *Thermomechanics of Composites under High Temperatures*. 1999 ISBN 0-7923-4899-0
66. P. Argoul, M. Frémond and Q.S. Nguyen (eds.): *IUTAM Symposium on Variations of Domains and Free-Boundary Problems in Solid Mechanics*. Proceedings of the IUTAM Symposium held in Paris, France. 1999 ISBN 0-7923-5450-8
67. F.J. Fahy and W.G. Price (eds.): *IUTAM Symposium on Statistical Energy Analysis*. Proceedings of the IUTAM Symposium held in Southampton, U.K. 1999 ISBN 0-7923-5457-5
68. H.A. Mang and F.G. Rammerstorfer (eds.): *IUTAM Symposium on Discretization Methods in Structural Mechanics*. Proceedings of the IUTAM Symposium held in Vienna, Austria. 1999 ISBN 0-7923-5591-1

# Mechanics

## *SOLID* MECHANICS AND ITS APPLICATIONS

*Series Editor*: G.M.L. Gladwell

69. P. Pedersen and M.P. Bendsøe (eds.): *IUTAM Symposium on Synthesis in Bio Solid Mechanics.* Proceedings of the IUTAM Symposium held in Copenhagen, Denmark. 1999 ISBN 0-7923-5615-2
70. S.K. Agrawal and B.C. Fabien: *Optimization of Dynamic Systems.* 1999 ISBN 0-7923-5681-0
71. A. Carpinteri: *Nonlinear Crack Models for Nonmetallic Materials.* 1999 ISBN 0-7923-5750-7
72. F. Pfeifer (ed.): *IUTAM Symposium on Unilateral Multibody Contacts.* Proceedings of the IUTAM Symposium held in Munich, Germany. 1999 ISBN 0-7923-6030-3
73. E. Lavendelis and M. Zakrzhevsky (eds.): *IUTAM/IFToMM Symposium on Synthesis of Nonlinear Dynamical Systems.* Proceedings of the IUTAM/IFToMM Symposium held in Riga, Latvia. 2000 ISBN 0-7923-6106-7
74. J.-P. Merlet: *Parallel Robots.* 2000 ISBN 0-7923-6308-6
75. J.T. Pindera: *Techniques of Tomographic Isodyne Stress Analysis.* 2000 ISBN 0-7923-6388-4
76. G.A. Maugin, R. Drouot and F. Sidoroff (eds.): *Continuum Thermomechanics.* The Art and Science of Modelling Material Behaviour. 2000 ISBN 0-7923-6407-4
77. N. Van Dao and E.J. Kreuzer (eds.): *IUTAM Symposium on Recent Developments in Non-linear Oscillations of Mechanical Systems.* 2000 ISBN 0-7923-6470-8
78. S.D. Akbarov and A.N. Guz: *Mechanics of Curved Composites.* 2000 ISBN 0-7923-6477-5
79. M.B. Rubin: *Cosserat Theories: Shells, Rods and Points.* 2000 ISBN 0-7923-6489-9
80. S. Pellegrino and S.D. Guest (eds.): *IUTAM-IASS Symposium on Deployable Structures: Theory and Applications.* Proceedings of the IUTAM-IASS Symposium held in Cambridge, U.K., 6–9 September 1998. 2000 ISBN 0-7923-6516-X
81. A.D. Rosato and D.L. Blackmore (eds.): *IUTAM Symposium on Segregation in Granular Flows.* Proceedings of the IUTAM Symposium held in Cape May, NJ, U.S.A., June 5–10, 1999. 2000 ISBN 0-7923-6547-X
82. A. Lagarde (ed.): *IUTAM Symposium on Advanced Optical Methods and Applications in Solid Mechanics.* Proceedings of the IUTAM Symposium held in Futuroscope, Poitiers, France, August 31–September 4, 1998. 2000 ISBN 0-7923-6604-2
83. D. Weichert and G. Maier (eds.): *Inelastic Analysis of Structures under Variable Loads.* Theory and Engineering Applications. 2000 ISBN 0-7923-6645-X
84. T.-J. Chuang and J.W. Rudnicki (eds.): *Multiscale Deformation and Fracture in Materials and Structures.* The James R. Rice 60th Anniversary Volume. 2001 ISBN 0-7923-6718-9
85. S. Narayanan and R.N. Iyengar (eds.): *IUTAM Symposium on Nonlinearity and Stochastic Structural Dynamics.* Proceedings of the IUTAM Symposium held in Madras, Chennai, India, 4–8 January 1999 ISBN 0-7923-6733-2
86. S. Murakami and N. Ohno (eds.): *IUTAM Symposium on Creep in Structures.* Proceedings of the IUTAM Symposium held in Nagoya, Japan, 3-7 April 2000. 2001 ISBN 0-7923-6737-5
87. W. Ehlers (ed.): *IUTAM Symposium on Theoretical and Numerical Methods in Continuum Mechanics of Porous Materials.* Proceedings of the IUTAM Symposium held at the University of Stuttgart, Germany, September 5-10, 1999. 2001 ISBN 0-7923-6766-9
88. D. Durban, D. Givoli and J.G. Simmonds (eds.): *Advances in the Mechanis of Plates and Shells The Avinoam Libai Anniversary Volume.* 2001 ISBN 0-7923-6785-5
89. U. Gabbert and H.-S. Tzou (eds.): *IUTAM Symposium on Smart Structures and Structonic Systems.* Proceedings of the IUTAM Symposium held in Magdeburg, Germany, 26–29 September 2000. 2001 ISBN 0-7923-6968-8

# Mechanics

## *SOLID* MECHANICS AND ITS APPLICATIONS

*Series Editor*: G.M.L. Gladwell

90. Y. Ivanov, V. Cheshkov and M. Natova: *Polymer Composite Materials – Interface Phenomena & Processes*. 2001 ISBN 0-7923-7008-2
91. R.C. McPhedran, L.C. Botten and N.A. Nicorovici (eds.): *IUTAM Symposium on Mechanical and Electromagnetic Waves in Structured Media*. Proceedings of the IUTAM Symposium held in Sydney, NSW, Australia, 18-22 Januari 1999. 2001 ISBN 0-7923-7038-4
92. D.A. Sotiropoulos (ed.): *IUTAM Symposium on Mechanical Waves for Composite Structures Characterization*. Proceedings of the IUTAM Symposium held in Chania, Crete, Greece, June 14-17, 2000. 2001 ISBN 0-7923-7164-X
93. V.M. Alexandrov and D.A. Pozharskii: *Three-Dimensional Contact Problems*. 2001 ISBN 0-7923-7165-8
94. J.P. Dempsey and H.H. Shen (eds.): *IUTAM Symposium on Scaling Laws in Ice Mechanics and Ice Dynamics*. Proceedings of the IUTAM Symposium held in Fairbanks, Alaska, U.S.A., 13-16 June 2000. 2001 ISBN 1-4020-0171-1
95. U. Kirsch: *Design-Oriented Analysis of Structures*. A Unified Approach. 2002 ISBN 1-4020-0443-5
96. A. Preumont: *Vibration Control of Active Structures*. An Introduction ($2^{nd}$ Edition). 2002 ISBN 1-4020-0496-6
97. B.L. Karihaloo (ed.): *IUTAM Symposium on Analytical and Computational Fracture Mechanics of Non-Homogeneous Materials*. Proceedings of the IUTAM Symposium held in Cardiff, U.K., 18-22 June 2001. 2002 ISBN 1-4020-0510-5
98. S.M. Han and H. Benaroya: *Nonlinear and Stochastic Dynamics of Compliant Offshore Structures*. 2002 ISBN 1-4020-0573-3
99. A.M. Linkov: *Boundary Integral Equations in Elasticity Theory*. 2002 ISBN 1-4020-0574-1
100. L.P. Lebedev, I.I. Vorovich and G.M.L. Gladwell: *Functional Analysis*. Applications in Mechanics and Inverse Problems ($2^{nd}$ Edition). 2002 ISBN 1-4020-0667-5; Pb: 1-4020-0756-6
101. Q.P. Sun (ed.): *IUTAM Symposium on Mechanics of Martensitic Phase Transformation in Solids*. Proceedings of the IUTAM Symposium held in Hong Kong, China, 11-15 June 2001. 2002 ISBN 1-4020-0741-8
102. M.L. Munjal (ed.): *IUTAM Symposium on Designing for Quietness*. Proceedings of the IUTAM Symposium held in Bangkok, India, 12-14 December 2000. 2002 ISBN 1-4020-0765-5
103. J.A.C. Martins and M.D.P. Monteiro Marques (eds.): *Contact Mechanics*. Proceedings of the $3^{rd}$ Contact Mechanics International Symposium, Praia da Consolação, Peniche, Portugal, 17-21 June 2001. 2002 ISBN 1-4020-0811-2
104. H.R. Drew and S. Pellegrino (eds.): *New Approaches to Structural Mechanics, Shells and Biological Structures*. 2002 ISBN 1-4020-0862-7
105. J.R. Vinson and R.L. Sierakowski: *The Behavior of Structures Composed of Composite Materials*. Second Edition. 2002 ISBN 1-4020-0904-6
106. Not yet published.
107. J.R. Barber: *Elasticity*. Second Edition. 2002 ISBN Hb 1-4020-0964-X; Pb 1-4020-0966-6
108. C. Miehe (ed.): *IUTAM Symposium on Computational Mechanics of Solid Materials at Large Strains*. Proceedings of the IUTAM Symposium held in Stuttgart, Germany, 20-24 August 2001. 2003 ISBN 1-4020-1170-9

# Mechanics

## *SOLID* MECHANICS AND ITS APPLICATIONS

*Series Editor*: G.M.L. Gladwell

109. P. Ståhle and K.G. Sundin (eds.): *IUTAM Symposium on Field Analyses for Determination of Material Parameters – Experimental and Numerical Aspects.* Proceedings of the IUTAM Symposium held in Abisko National Park, Kiruna, Sweden, July 31 – August 4, 2000. 2003 ISBN 1-4020-1283-7
110. N. Sri Namachchivaya and Y.K. Lin (eds.): *IUTAM Symposium on Nonlinear Stochastic Dynamics.* Proceedings of the IUTAM Symposium held in Monticello, IL, USA, 26 – 30 August, 2000. 2003 ISBN 1-4020-1471-6
111. H. Sobieckzky (ed.): *IUTAM Symposium Transsonicum IV.* Proceedings of the IUTAM Symposium held in Göttingen, Germany, 2–6 September 2002, 2003 ISBN 1-4020-1608-5
112. J.-C. Samin and P. Fisette: *Symbolic Modeling of Multibody Systems.* 2003 ISBN 1-4020-1629-8
113. A.B. Movchan (ed.): *IUTAM Symposium on Asymptotics, Singularities and Homogenisation in Problems of Mechanics.* Proceedings of the IUTAM Symposium held in Liverpool, United Kingdom, 8-11 July 2002. 2003 ISBN 1-4020-1780-4
114. S. Ahzi, M. Cherkaoui, M.A. Khaleel, H.M. Zbib, M.A. Zikry and B. LaMatina (eds.): *IUTAM Symposium on Multiscale Modeling and Characterization of Elastic-Inelastic Behavior of Engineering Materials.* Proceedings of the IUTAM Symposium held in Marrakech, Morocco, 20-25 October 2002. 2004 ISBN 1-4020-1861-4
115. H. Kitagawa and Y. Shibutani (eds.): *IUTAM Symposium on Mesoscopic Dynamics of Fracture Process and Materials Strength.* Proceedings of the IUTAM Symposium held in Osaka, Japan, 6-11 July 2003. Volume in celebration of Professor Kitagawa's retirement. 2004 ISBN 1-4020-2037-6
116. E.H. Dowell, R.L. Clark, D. Cox, H.C. Curtiss, Jr., K.C. Hall, D.A. Peters, R.H. Scanlan, E. Simiu, F. Sisto and D. Tang: *A Modern Course in Aeroelasticity.* 4th Edition, 2004 ISBN 1-4020-2039-2
117. T. Burczyński and A. Osyczka (eds.): *IUTAM Symposium on Evolutionary Methods in Mechanics.* Proceedings of the IUTAM Symposium held in Cracow, Poland, 24-27 September 2002. 2004 ISBN 1-4020-2266-2
118. D. Ieşan: *Thermoelastic Models of Continua.* 2004 ISBN 1-4020-2309-X
119. G.M.L. Gladwell: *Inverse Problems in Vibration.* Second Edition. 2004 ISBN 1-4020-2670-6
120. J.R. Vinson: *Plate and Panel Structures of Isotropic, Composite and Piezoelectric Materials, Including Sandwich Construction.* 2005 ISBN 1-4020-3110-6
121. *Forthcoming*
122. G. Rega and F. Vestroni (eds.): *IUTAM Symposium on Chaotic Dynamics and Control of Systems and Processes in Mechanics.* Proceedings of the IUTAM Symposium held in Rome, Italy, 8–13 June 2003. 2005 ISBN 1-4020-3267-6
123. E.E. Gdoutos: *Fracture Mechanics. An Introduction. 2nd edition.* 2005 ISBN 1-4020-3267-6
124. M.D. Gilchrist (ed.): *IUTAM Symposium on Impact Biomechanics from Fundamental Insights to Applications.* 2005 ISBN 1-4020-3795-3
125. J.M. Huyghe, P.A.C. Raats and S. C. Cowin (eds.): *IUTAM Symposium on Physicochemical and Electromechanical Interactions in Porous Media.* 2005 ISBN 1-4020-3864-X
126. H. Ding, W. Chen and L. Zhang: *Elasticity of Transversely Isotropic Materials.* 2005 ISBN 1-4020-4033-4
127. W. Yang (ed): *IUTAM Symposium on Mechanics and Reliability of Actuating Materials.* Proceedings of the IUTAM Symposium held in Beijing, China, 1–3 September 2004. 2005 ISBN 1-4020-4131-6

# Mechanics

## *SOLID* MECHANICS AND ITS APPLICATIONS

*Series Editor*: G.M.L. Gladwell

128. J.-P. Merlet: *Parallel Robots.* 2006 ISBN 1-4020-4132-2
129. G.E.A. Meier and K.R. Sreenivasan (eds.): *IUTAM Symposium on One Hundred Years of Boundary Layer Research.* Proceedings of the IUTAM Symposium held at DLR-Göttingen, Germany, August 12–14, 2004. 2006 ISBN 1-4020-4149-7
130. H. Ulbrich and W. Günthner (eds.): *IUTAM Symposium on Vibration Control of Nonlinear Mechanisms and Structures.* 2006 ISBN 1-4020-4160-8
131. L. Librescu and O. Song: *Thin-Walled Composite Beams.* Theory and Application. 2006 ISBN 1-4020-3457-1
132. G. Ben-Dor, A. Dubinsky and T. Elperin: *Applied High-Speed Plate Penetration Dynamics.* 2006 ISBN 1-4020-3452-0
133. X. Markenscoff and A. Gupta (eds.): *Collected Works of J. D. Eshelby.* Mechanics and Defects and Heterogeneities. 2006 ISBN 1-4020-4416-X
134. R.W. Snidle and H.P. Evans (eds.): *IUTAM Symposium on Elastohydrodynamics and Microelastohydrodynamics.* Proceedings of the IUTAM Symposium held in Cardiff, UK, 1–3 September, 2004. 2006 ISBN 1-4020-4532-8
135. T. Sadowski (ed.): *IUTAM Symposium on Multiscale Modelling of Damage and Fracture Processes in Composite Materials.* Proceedings of the IUTAM Symposium held in Kazimierz Dolny, Poland, 23–27 May 2005. 2006 ISBN 1-4020-4565-4
136. A. Preumont: *Mechatronics.* Dynamics of Electromechanical and Piezoelectric Systems. 2006 ISBN 1-4020-4695-2
137. M.P. Bendsøe, N. Olhoff and O. Sigmund (eds.): *IUTAM Symposium on Topological Design Optimization of Structures, Machines and Materials.* Status and Perspectives. 2006 ISBN 1-4020-4729-0
138. A. Klarbring: *Models of Mechanics.* 2006 ISBN 1-4020-4834-3
139. H.D. Bui: *Fracture Mechanics.* Inverse Problems and Solutions. 2006 ISBN 1-4020-4836-X
140. M. Pandey, W.-C. Xie and L. Xu (eds.): *Advances in Engineering Structures, Mechanics and Construction.* Proceedings of an International Conference on Advances in Engineering Structures, Mechanics & Construction, held in Waterloo, Ontario, Canada, May 14–17, 2006. 2006 ISBN 1-4020-4890-4
141. G.Q. Zhang, W.D. van Driel and X. J. Fan: *Mechanics of Microelectronics.* 2006 ISBN 1-4020-4934-X
142. Q.P. Sun and P. Tong (eds.): *IUTAM Symposium on Size Effects on Material and Structural Behavior at Micron- and Nano-Scales.* Proceedings of the IUTAM Symposium held in Hong Kong, China, 31 May–4 June, 2004. 2006 ISBN 1-4020-4945-5

GPSR Compliance

*The European Union's (EU) General Product Safety Regulation (GPSR) is a set of rules that requires consumer products to be safe and our obligations to ensure this.*

*If you have any concerns about our products, you can contact us on ProductSafety@springernature.com*

In case Publisher is established outside the EU, the EU authorized representative is:

Springer Nature Customer Service Center GmbH
Europaplatz 3
69115 Heidelberg, Germany

**Batch number: 10755292**

Printed by Printforce, the Netherlands